Principles of Modern Communication Systems

Written specifically for a one-semester course, this highly readable textbook explains the physical and engineering principles of communication systems using an accessible, yet mathematically rigorous, approach. Beginning with valuable background material on signals and systems and random processes, the text then guides students through the core topics including amplitude modulation, frequency modulation, pulse modulation, and noise.

Key features include:

- A range of worked examples, practice problems, and review questions to reinforce concepts and enable students to develop confidence in solving problems on their own
- Key terms and formulas highlighted throughout to help students easily identify essential points
- MATLAB-based exercises and examples provided throughout, supported by an introductory appendix for those who are new to Matlab
- End-of-chapter practical applications, showing students how concepts are applied to real-life communication scenarios and devices

Samuel Agbo is a Professor at California Polytechnic State University. His teaching and research are in Communications, Fiber Optic Communications, and Electronics. He is a member of the IEEE, the IEEE Communications Society, and the IEEE Photonics Society.

Matthew N. O. Sadiku is a Professor at Prairie View A&M University. His awards include the McGraw-Hill/Jacob Millman Award for outstanding contributions in the field of electrical engineering (2000), and the Regents Professor Award from Texas A&M University (2012-2013). He is the author of numerous papers as well as over 70 books, and is a registered Professional Engineer and a Fellow of the IEEE.

Principles of Modern Communication Systems

SAMUEL O. AGBO
California Polytechnic State University

MATTHEW N. O. SADIKU
Prairie View A&M University

CAMBRIDGE
UNIVERSITY PRESS

CAMBRIDGE
UNIVERSITY PRESS

University Printing House, Cambridge CB2 8BS, United Kingdom

One Liberty Plaza, 20th Floor, New York, NY 10006, USA

477 Williamstown Road, Port Melbourne, VIC 3207, Australia

4843/24, 2nd Floor, Ansari Road, Daryaganj, Delhi – 110002, India

79 Anson Road, #06–04/06, Singapore 079906

Cambridge University Press is part of the University of Cambridge.

It furthers the University's mission by disseminating knowledge in the pursuit of
education, learning and research at the highest international levels of excellence.

www.cambridge.org
Information on this title: www.cambridge.org/9781107107922

First published 2017

Printed in the United States of America by Sheridan Books, Inc.

A catalog record for this publication is available from the British Library

Library of Congress Cataloging in Publication Data
Names: Agbo, Samuel O., author. | Sadiku, Matthew N. O., author.
Title: Principles of modern communication systems / Samuel O. Agbo, California
Polytechnic State University, Matthew N. O. Sadiku, Prairie View A&M University.
Description: Cambridge, United Kingdom : Cambridge University Press, 2017. |
Includes bibliographical references and index.
Identifiers: LCCN 2015045740 | ISBN 9781107107922 (Hardback : alk. paper)
Subjects: LCSH: Telecommunication.
Classification: LCC TK5101 .A37 2016 | DDC 621.382–dc23 LC record available at
http://lccn.loc.gov/2015045740

ISBN 978-1-107-10792-2 Hardback

Additional resources for this publication at www.cambridge.org/Agbo

Cambridge University Press has no responsibility for the persistence or accuracy
of URLs for external or third-party internet websites referred to in this publication,
and does not guarantee that any content on such websites is, or will remain,
accurate or appropriate.

Dedicated to our wives:

Ifeoma Esther and Kikelomo Esther

CONTENTS

PREFACE

We live in the information age – news, weather, sports, shopping, financial, business inventory, and other sources make information available to us almost instantly via communication systems. The field of communication is perhaps the fastest growing area in electrical engineering. This is why a course on communication systems is an important part of most engineering curricula.

Most books on communication systems are designed for a two-semester course sequence. Unfortunately, electrical engineering has grown considerably, and its curriculum is so crowded that there is no room for a two-semester course on communication systems. This book is designed for a three-hour semester course on communication systems. It is intended as a textbook for senior-level students in electrical and computer engineering. The prerequisites for a course based on this book are standard engineering mathematics (including calculus and differential equations), electronics, and electric circuit analysis.

This book is intended to present communication systems to electrical and computer engineering students in a manner that is clearer, more interesting, and easier to understand than other texts. This objective is achieved in the following ways:

- We have included several features to help students feel at home with the subject. Each chapter opens with a historical profile or technical note. This is followed by an introduction that links the chapter with the previous chapters and states the chapter objectives. The chapter ends with a summary of key points and formulas.

- All principles are presented in a lucid, logical, step-by-step manner. As much as possible, we avoid wordiness and giving too much detail that could hide concepts and impede overall understanding of the material.

- Important formulas are boxed as a means of helping students sort out what is essential from what is not. Also, to ensure that students clearly get the gist of the matter, key terms are defined and highlighted.

- Thoroughly worked examples are liberally given at the end of every section. The examples are regarded as a part of the text and are clearly explained without asking the reader to fill in missing steps. Thoroughly worked examples give students a good understanding of the material and the confidence to solve problems themselves.

- To give students practice opportunity, each illustrative example is immediately followed by a practice problem with the answer. The students can follow the example step by step to solve the practice problem without flipping pages or looking at the end of the book for answers.

The practice problem is also intended to test whether students understand the preceding example. It will reinforce their grasp of the material before they move on to the next section.

- The last section in each chapter is devoted to application aspects of the concepts covered in the chapter. The material covered in the chapter is applied to at least one or two practical problems or devices. This helps the students see how the concepts are applied to real-life situations.

- Ten review questions in the form of multiple-choice objective items are provided at the end of each chapter with answers. The review questions are intended to cover the little "tricks" which the examples and end-of-chapter problems may not cover. They serve as a self-test device and help students determine how well they have mastered the chapter.

In recognition of the requirements by ABET (Accreditation Board for Engineering and Technology) on integrating computer tools, the use of MATLAB is encouraged in a student-friendly manner. We have introduced MATLAB in Appendix B and applied it gradually throughout this book. MATLAB has become a standard software package in electrical engineering curricula.

We would like to thank Dr. Siew Koay for going over the entire manuscript and spotting errors. Special thanks are due to Dr. Pamela H. Obiomon (head) and Dr. Kendall T. Harris (dean) at Prairie View A&M University for their support.

Samuel Agbo and Matthew N. O. Sadiku

SI PREFIXES AND CONVERSION FACTORS

Power	Prefix	Symbol
10^{18}	Exa	E
10^{15}	Peta	P
10^{12}	Tera	T
10^{9}	Giga	G
10^{6}	Mega	M
10^{3}	kilo	k
10^{2}	hecto	h
10^{1}	deka	da
10^{-1}	deci	d
10^{-2}	centi	c
10^{-3}	milli	m
10^{-6}	micro	μ
10^{-9}	nano	n
10^{-12}	pico	p
10^{-15}	femto	f
10^{-18}	atto	a

Unit	Symbol	Conversion
Micron	μm	10^{-6} m
Mil	mil	10^{-3} in = 25.4 μm

1 Introduction

Nothing in the world can take the place of perseverance.
Talent will not; nothing is more common than unsuccessful men with talent.
Genius will not; unrewarded genius is almost a proverb.
Education will not; the world is full of educated derelicts.
Persistence and determination alone are omnipotent.

CALVIN COOLIDGE

HISTORICAL PROFILES

Samuel F. B. Morse (1791–1872), an American painter, invented the telegraph, the first practical, commercialized application of a communication system.

Morse was born in Charlestown, Massachusetts, and studied at Yale College (now Yale University) and the Royal Academy of Arts in London to become an artist. In 1832, he conceived the basic idea of an electromagnetic telegraph while on a ship returning from Europe. He had a working model by 1836 and applied for a patent in 1838. In 1843 US Congress voted $30,000 for Morse to construct a telegraph experimental line between Baltimore and Washington DC. On May 24, 1844, he sent the famous first message: "What hath God wrought!" The telegraph spread across the USA more quickly than had the railroads, whose routes the wires often followed. By 1854, there were 23,000 miles of telegraph wire in operation. Morse also developed a code of dots and dashes for letters and numbers, for sending messages on the telegraph. Though it went through some changes, the Morse code became a standard throughout the world. The development of the telegraph led to the invention of the telephone.

Claude Elwood Shannon (1916–2001), American mathematician and engineer, is the founding father of information theory. The term "bit", today used to describe individual units of information processed by a computer, was coined from Shannon's research in the 1940s.

Born in Petoskey, Michigan, he showed an affinity for both engineering and mathematics from an early age. He graduated from the Massachusetts

Institute of Technology in 1940 with a doctorate degree in mathematics. He joined the staff of Bell Telephone Laboratories, where he spent 15 years. Soon he discovered the similarity between Boolean algebra and telephone switching circuits. He thought that by reducing information to a series of ones and zeros, information could be processed by using on-off switches. The mathematical theory of communication was the climax of Shannon's mathematical and engineering investigations.

Shannon was renowned for his eclectic interests and capabilities. He designed and built chess-playing, maze-solving, juggling, and mind-reading machines. These activities bear out Shannon's claim that he was more motivated by curiosity than by usefulness. Throughout his life, Shannon received many honors, including the Alfred Noble Prize (1939), the National Medal of Science (1966), the IEEE Medal of Honor (1966), the Golden Plate Award (1967), and the Kyoto Prize (1985), along with numerous other prizes and over a dozen honorary degrees.

1.1 INTRODUCTION

From early times, man has felt the need to communicate. To meet this need, several devices such as drums, ram's horns, church bells, books, newspapers, signal lamps, optical signals, and the telegraph have been invented. (Churches still use bells today.) In modern times, we use radios, televisions, phones, computer networks to communicate. In fact, we live in the information age, where we are bombarded with information from all angles. News, weather, sports, shopping, financial, business inventory, and other sources make information available to us almost instantly. This is all made possible by the various communications systems we have in place.

> A **communication system** is device that conveys information from a source (the transmitter) to a destination (the receiver) via a channel (the propagation medium).

Some obvious examples of communications systems are the regular telephone network, mobile cellular telephones, radio, cable TV, satellite TV, fax, and radar. Mobile radio, used by the police and fire departments, aircraft, and various businesses, is another example.

    The field of communications is a very exciting branch of electrical engineering. The merging of the communications field with computer technology in recent years has made the field more exciting and led to digital data communications networks such as local area networks, metropolitan area networks, and broadband integrated services digital networks. For example, the Internet (the "information superhighway") allows students, educators, business people, and others to send electronic mails from their computers worldwide, log onto remote databases, and transfer files. The Internet is hitting the entire world like a tidal wave and is drastically changing the way people do business, communicate, and get information. This trend will continue. More and more government agencies, academic departments, and businesses are demanding faster and more accurate transmission of information. To meet these needs, communications engineers are highly in demand.

But how is this information created, processed, stored, transmitted, and used? This is the question we will try to answer in this book. We begin by introducing some basic concepts in this chapter. These concepts include components of a communication system, noise, frequency allocation and designation, and channel capacity. First, let us take a quick look at the historical development of communication systems.

1.2 COMMUNICATION MILESTONES

The key events in the evolution of communication systems are summarized in Table 1.1. Such a historical overview should have two effects on the reader. First, it should give us a sense of appreciation of what it took to get us to where we are today. Second, it should provide a sense of motivation to learn what we have in this text.

Note that the discovery of the electrical telegraph by Samuel F. B. Morse in 1838 ushered in the era of electrical communication, where our main interest lies. Until the telephone was invented by Alexander Graham Bell in 1874 and became generally available in 1880, the telegraph was the standard means of communication. Since the 1980s, the communication links of the telephone network have been gradually replaced by high-capacity optical fiber. This has led to the establishment of a standard signal format known as *Synchronous Optical Network* (SONET) in North American and *Synchronous Digital Hierarchy* (SDH) in other parts of the world. Perhaps the most exciting era of communication is the integration of voice, data, and video through the introduction of Integrated Services Digital Networks (ISDN).

1.3 COMMUNICATION PROCESS

The fundamental objective of a communication system is to exchange information between two parties. In other words, a communication system is designed to transfer information from one point to another. The key components of such a system are source of information, transmitter, channel, receiver, and user of information, as shown in Figure 1.1 and explained below.

1.3.1 Source of information

The source generates the information to be transmitted. There are four major sources of information: speech, television, fax, and computer. The information generated by these sources can be classified into three categories: voice, data, and video.

- Audio/voice: this is information in acoustic form corresponding to normally audible sound waves as we hear in radio. Voice is the most common type of information

Table 1.1. **Key dates in the development of communications**

Year	Event
1838	William F. Cooke and Charles Wheatstone build the electric telegraph
1844	Samuel F. B. Morse successfully transmits on Baltimore–Washington DC telegraph
1858	First transatlantic cable is laid but fails after 26 days
1864	James C. Maxwell predicts electromagnetic (EM) waves
1874	Alexander G. Bell discovers the principle of the telephone
1887	Thomas Edison patents motion picture camera
1888	Heinrich Hertz produces the first EM wave, verifying Maxwell's prediction
1897	Ferdinand Braun develops the cathode ray tube
1900	Guglielmo Marconi transmits the first transatlantic wireless signal
1905	Reginald Fessenden transmits music by radio
1906	Lee de Forest invents the vacuum tube triode
1918	Edwin H. Armstrong invents the superheterodyne receiver
1920	Radio station KDKA in Pittsburgh, PA, provides the first broadcast
1923	Vladimir K. Zworkykin invents the "iconoscope" TV tube
1928	Philo T. Farnsworth demonstrates the first all-electronic TV
1933	Edwin H. Armstrong invents frequency modulation (FM)
1934	The Federal Communication Commission (FCC) is created in the USA
1947	Transistor is invented at Bell Lab by Walter H. Brattain, John Bardeen, and William Shockley
1953	First transatlantic telephone cable is laid
1957	Sputnik I, the first Earth satellite, is launched by USSR
1962	Telstar I, the first active satellite, relays signals between Europe and the USA
1968	Cable TV systems are developed
1972	Cellular telephone is demonstrated to FCC by Motorola
1976	Personal computers are developed
1980s	Fiber optic technology is developed

Table 1.1. (cont.)

Year	Event
1984	Integrated Services Digital Network (ISDN) is approved by CCITT
1995	The Internet and World Wide Web become popular
2001	Satellite radio begins broadcasting
2003	Skype Internet telephony
2004	SpaceShipOne reaches an altitude of more than 100 km
2012	Launching of the fourth generation (4G) of cellular communication

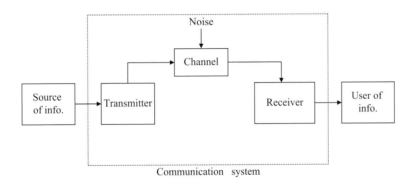

Figure 1.1. Components of a communication system.

worldwide. With the recent introduction of the mobile apps, data communication has overtaken voice communication.

- Data: this computer-generated type of information is in digital form, nearly always represented in binary form, 0s and 1s. This kind of information is characterized by burstiness, meaning that information is transferred from one terminal to another in bursts with periods of silence between them. Data on its own has no meaning, until it is interpreted. For example, 1234.56 is data, while "Joe owes me $1234.56" is information.
- Video: this is the electronic representation of stationary or moving images or pictures. The visual portion of a televized broadcast is video, while the audible portion is audio.

Information sources may also be classified as either analog or digital. An analog source produces a continuous electronic waveform as its output. A typical example of an analog source is a human being using the microphone. The bulk of the telephone network is analog. A digital source produces a sequence of data symbols as its output. We become a digital source of information when we dial a telephone number, send a fax, use the MAC machine, or use the credit card. The type of information source dictates the type of communication system we have.

An **analog (or digital) communication system** is one that transfers information from an analog (or digital) source to the intended receiver.

1.3.2 Transmitter

Most often the information generated by the source is not suitable for direct transmission because it is not electrical in nature. It requires a transmitter to encode and possibly modulate it to produce an electromagnetic signal. Depending on the requirements, the modulation may be amplitude modulation (AM), frequency modulation (FM), pulse modulation, or any combination of these.

1.3.3 Channel

A **channel** is the path through which the electrical signal flows.

A channel may also be regarded as the propagation medium. Channels may also be classified in terms of the medium involved: wire and wireless. Typical examples of wire channels are twisted pairs used as telephone lines, coaxial cables used in computer networks, waveguides, and optical fibers. Some examples of wireless channels are vacuum, atmosphere/air, and sea.

Another common way of characterizing channels[1] is by their transmission mode. There are three types of transmission modes as illustrated in Figure 1.2: simplex, half duplex, full duplex.

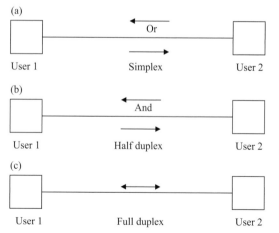

Figure 1.2. Modes of channel operation.

---

[1] The term "channel" is also used to mean the frequency range allocated to a particular transmission.

- A *simplex* system allows transmission of information in one direction only. One can transmit but not receive or receive but not transmit. Examples of simplex systems include radio broadcast, TV, and public address systems.
- A *half duplex* system is capable of sending information in either direction but not simultaneously. This requires that the users take turns. The sending end first transmits to the receiver, and then the receiver transmits to the sender, thereby reversing their roles. This is the most common mode of transmission used today in data communications. Another example is the citizen-band radio transmission.
- A *full duplex* system allows transmission in both directions simultaneously. Examples of full duplex systems include the telephone systems and many computer systems.

1.3.4 Noise

It is inevitable that the signal encounters noise as it flows from the information source to the information user. Noise plays a major role in communication systems.[2] Noise refers to unwanted energy, usually of random character, present in the transmission system. It may be due to any cause. Although it appears in Figure 1.1 that noise interferes with the channel, this is just to simplify the diagram. Noise may interfere with the signal at any point in the communication system. Noise usually has its greatest impact where the signal is weakest such as at the input of the receiver.

1.3.5 Receiver

The receiver accepts the transmitted signal, amplifies it, demodulates and decodes it if necessary, and converts it to a form suitable to the reception of the user. Demodulation and decoding are the reverse of the corresponding processes of modulation and encoding at the transmitter and will be discussed later in the book.

1.3.6 User of information

The user of information takes the signal from the receiver. The communication engineers design the transmitter and receiver and have much control over them; they have little control over the source, channel, or user.

1.4   FREQUENCY ALLOCATION

In order to maintain order and minimize interference in wireless communication, frequency allocation is done by regulatory bodies. At the international level, frequency assignment and

---

[2] In fact, the theory of communications systems can be divided into two parts: (1) how the systems work, and (2) how they perform in the presence of noise (B. P. Lathi).

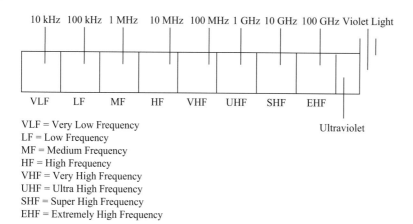

Figure 1.3. Frequency bands.

technical standards are set by the International Telecommunication Union, Telecommunication Standardization Sector (ITU-T), which is one of the three sectors of the International Telecommunication Union (ITU), headquartered in Geneva, Switzerland (see www.itu.int/ITU-T). ITU itself is an organ of the United Nations. Although each member nation abides by the jointly agreed frequency allocation and standards, each nation has sovereignty over the use of the frequency spectrum and technical regulations.

In the United States, the Federal Communications Commission (FCC) is responsible for frequency allocation and for issuing licenses to users for specific purposes (see www.fcc.gov). It decides the kind of information that one can transmit over specific frequency bands. Table 1.2 shows how the frequency spectrum is allocated for different applications in the United States. This is partly illustrated in Figure 1.3.

For historical reasons, some band assignments are referred to by their wavelength λ (in meters) instead of their frequency f (in hertz). The two are related by

$$\lambda = \frac{c}{f} \tag{1.1}$$

where $c = 3 \times 10^8$ m/s is the speed of light in vacuum.

Note the following popular frequency ranges:

Audio frequencies	– 15 Hz to 20 MHz
Radio frequency	– 3 kHz to 300 GHz
Microwave frequencies	– 3 to 30.0 GHz
Optical frequencies	– 100 THz range

To help understand the prefixes k, M, T, etc., the SI prefixes and common conversion factors are given after the Preface. The International System (SI) of units and prefixes is used throughout this book.

Table 1.2. **Frequency designation and bands**

Designation	Frequency band	Typical Applications
Extremely low frequency (ELF)	0 to 3 kHz	Power transmission
Very low frequency (VLF)	3 to 30 kHz	Submarine communications
Low frequency (LF)	30 to 300 kHz	Radio navigation, US govt.
Medium frequency (MF)	300 to 3000 kHz	AM radio broadcast
High frequency (HF)	3 to 30 MHz	Shortwave radio broadcast
Very high frequency (VHF)	30 to 300 MHz	VHF TV channels 2–13, FM radio
Ultra-high frequency (UHF)	300 to 3000 MHz	UHF TV channels 14–70, PCS
L	500 to 1500 MHz	GPS, police, fire dept.
Superhigh frequencies (SHF)	3 to 30.0 GHz	Microwaves, satellite comm.
C	3600 to 7025 MHz	
X	7.25 to 8.4 GHz	
Ku	10.7 to 14.5 GHz	
Ka	17.3 to 31.0 GHz	
R	26.5 to 40 GHz	
Q	33 to 56 GHz	
V	40 to 75 GHz	
W	75 to 110 GHz	
Extremely high frequencies (EHF)	30.0 o 300 GHz	Millimeter waves, radar
Infrared radiation	300 GHz to 810 THz	
Visible light	430 to 750 THz	
Ultraviolet radiation	1.62 to 30 PHz	
X-rays	30 PHz to 30 EHz	
Gamma rays	30 to 3000 EHz	

PCS – personal communication services, an alternative to cellular service.
GPS – global positioning system.

1.5   INFORMATION CAPACITY

The performance of a communication system is judged by the quality of information transmitted through it. The quality of information is in turn determined by two factors: channel bandwidth and signal-to-noise ratio.

Carrying information requires bandwidth, and carrying more information at higher rates requires more bandwidth. Every communication system has a limited amount of bandwidth B (in Hz) that controls the rate of signal variations.

> The **bandwidth** (in Hz) of a channel is the range of frequencies it can transmit with reasonable fidelity.

For example, music employs the frequency range of 0 to 20 kHz so that a communication system must have at least 20 kHz bandwidth to transmit the music signal with high fidelity.

Noise constitutes a major limitation to communication systems. As said earlier, noise is unavoidable in communication. Noise is measured relative to the information signal power in terms of signal-to-noise ratio S/N. The larger the S/N, the longer is the distance over which we can communicate.

In view of these two limitations, Claude Shannon stated in 1948 that for a continuous analog communications channel subject to additive Gaussian noise the rate of information transmission R (in bits/s) cannot exceed the channel capacity C (in bits/s), where[3]

$$C = B \log_2 \left(1 + \frac{S}{N} \right)$$

(1.2)

B is the channel bandwidth (in Hz), S is the signal power, and N is the noise power. This is known as the Shannon–Hartley theorem. If the information transmission rate R is less than the channel capacity C ($R < C$), then the probability of error would be small. If $R > C$, the probability of error would be close to unity and reliable transmission is not possible.

There are two important points to remember about this theorem. First, the theorem provides the maximum rate (a theoretical upper bound on the channel) at which information can be transmitted; we can strive to achieve this when designing practical communication systems. The upper limit on communication rate was first applied in telephone channels, then in optical fibers, and now in wireless. Second, it provides a trade-off between bandwidth and signal-to-noise ratio. For a given channel capacity C, we may reduce S/N by increasing the bandwidth B. Equation (1.2) indicates that a noiseless channel (with $N = 0$) has an infinite capacity, which is not practicable. Thus, for a finite capacity, a minimum amount of noise is a necessary evil.

The signal-to-noise ratio is sometimes expressed in decibel (dB), which is the base 10 logarithm measure of power ratio. In general, the ratio G of power P_2 to power P_1 is expressed in decibel as

---

[3] Note that $\log_2 x = \frac{\ln x}{\ln 2}$

$$G_{dB} = 10 \log_{10} G = 10 \log_{10} \frac{P_2}{P_1} \tag{1.3}$$

Alternatively, given G_{dB}, we can find G as

$$G = 10^{G_{dB}/10} \tag{1.4}$$

The logarithmic nature of the decibel allows us to express a large ratio in a simple manner.

The decibel is also used to express values of power relative to a reference. If 1 W is used as a reference, dBW is expressed as dB relative to 1 W. Thus,

$$P \text{ (in dBW)} = 10 \log_{10} P \text{ (in W)} \tag{1.5}$$

If 1 mW is used as reference, dBm is expressed as dB relative to 1 mW. Thus,

$$P \text{ (in dBm)} = 10 \log_{10} P \text{ (in mW)} \tag{1.5}$$

Therefore, we may write

$$0.1 \text{ W} = -10 \text{ dBW} = 20 \text{ dBm}$$
$$1 \text{ W} = 0 \text{ dBW} = 30 \text{ dBm}$$
$$10 \text{ W} = 10 \text{ dBW}$$

EXAMPLE 1.1

Express the optical output power 300 μW in dBm units.

Solution

$$P \text{ (dBm)} = 10 \log_{10} \left(\frac{P(W)}{1 \text{ mW}} \right) = 10 \log_{10} \left(\frac{300 \times 10^{-6} \text{ W})}{1 \times 10^{-3} \text{ W}} \right) = -5.23 \text{ dBm}$$

PRACTICE PROBLEM 1.1

Express 10 mW in dBW and dBm units.

Answer: −20 dBW, 10 dBm.

EXAMPLE 1.2

A telephone line transmits frequencies from 300 Hz to 3400 Hz with a signal-to-ratio of 45 dB. (a) Determine the capacity of the line. (b) What must the S/N ratio be if the channel capacity increases by 10% while the bandwidth remains the same?

Solution

(a) $B = 3400 - 300 = 3100$ Hz,

$$45 \text{ dB} = 10 \log_{10} S/N \rightarrow S/N = 10^{45/10} = 31{,}623$$

$$C = B \log_2(1 + S/N) = 3100 \log_2(1 + 31{,}623) = 46.341 \text{ kbps}$$

where 1 kbps = 1000 bits per second.

(b) $C = (1 + 10\%) \times 46{,}341 = 50{,}975$

$$\frac{C}{B} = \frac{50{,}975}{3{,}100} = 16.444 = \log_2(1 + S/N) \to 1 + \frac{S}{N} = 2^{16.444} = 89{,}127$$

or $S/N = 89{,}126$. In decibels, this is equivalent to

$$10 \log_{10} 89{,}126 = 49.5 \text{ dB}.$$

PRACTICE PROBLEM 1.2

What is the required bandwidth to support a channel with a capacity of 20 kbps when the signal-to-noise ratio is 20 dB?

Answer: 3003.8 Hz

Summary

1. A communication system consists of the transmitter, channel, and receiver.
2. Noise is unwanted signal that corrupts the desired information signal.
3. Bandwidth is the span of frequencies occupied by a signal.
4. The decibel (dB) is a relative power measurement. Power P_2 relative to power P_1 is defined as

$$dB = 10 \log_{10} \frac{P_2}{P_1}$$

5. The Shannon–Hartley theorem gives the maximum rate at which information can be reliably transmitted over a channel. The channel capacity C is related to the bandwidth as

$$C = B \log_2 \left(1 + \frac{S}{N} \right)$$

where S/N is the signal-to-noise ratio.

Review questions

1.1 Which of the following communication devices was invented first?
(a) Radio. (b) TV. (c) Telegraph. (d) Telephone.

1.2 Which type of information is produced by cable TV?
(a) Audio. (b) Video. (c) Data.

1.3 The telephone system can be regarded as:
(a) Simplex. (b) Half duplex. (c) Full duplex. (d) Echo-plex.

1.4 AM radio broadcast (540 to 1630 kHz) is in the following frequency designation:
(a) LF. (b) MF. (c) HF. (d) VHF.

1.5 Microwaves fall in the following frequency designation:
(a) L-band. (b) X-band. (c) UHF. (d) SHF.

1.6 Optical frequencies have frequencies in the 10^{14} Hz range and corresponding to wavelengths of the order of microns (10^{-6} m).
(a) True. (b) False.

1.7 1000 is equivalent to:
(a) 10 dB. (b) 30 dB. (c) 50 dB. (d) 100 dB.

1.8 The power ratio of 20 dB is the same as:
(a) 20,000. (b) 200. (c) 100. (d) 10.

1.9 A power of 100 W is equivalent to 50 dBm or 20 dBW.
(a) True. (b) False.

1.10 For a prescribed channel capacity, it is advantageous to use a broad bandwidth to transmit information.

(a) True.   (b) False.

Answers: 1.1c, 1.2a,b, 1.3c, 1.4b, 1.5d, 1.6a, 1.7b, 1.8c, 1.9a, 1.10a,

Problems

1.1 Describe simplex, half duplex, and full duplex modes of transmission. Give an example of each.

1.2 What is noise? Mention where it can mostly affect the signal.

1.3 Express the following numbers in dB:

(a) 0.036   (b) 42   (c) 508   (d) 3×10^5

1.4 Convert the following decibels to numbers:

(a) -140 dB   (b) -3 dB   (c) 20 dB   (d) 42 dB

1.5 Convert the following to watts:

(a) -3 dBm   (b) -12 dBW   (c) 65 dBm   (d) 35 dBW

1.6 Express the following powers in dB and dBm:

(a) 4 mW   (b) 0.36 W   (c) 2 W   (d) 110 W

1.7 When a 1 pW reference level is used, the power level is expressed in dBrn.

(a) Express the following in dBrn: 0 dBm, -1.5 dBm, -60 dBm.

(b) Show that in general dBrn = dBm + 90.

1.8 A channel transmits only frequencies between 4 MHz and 6 MHz with a signal-to-noise ratio of 25 dB. What is the channel capacity? How long does it take to transmit 50,000 ASCII characters of information using the channel? Assume each character has 8 bits.

1.9 (a) What is the channel capacity for signal power of 250 W, noise power of 20 W, and bandwidth of 3 MHz?

(b) What is the channel capacity when the S/N ratio is doubled but the bandwidth is reduced by a factor of 3?

1.10 Calculate the bandwidth required for a channel capacity of 25 kbps when the S/N is 500.

1.11 Evaluate the bandwidth of a channel with capacity 36,000 bps and a signal-to-noise ratio of 30 dB.

2 Signals and systems

Knowing is not enough; we must apply. Willing is not enough, we must do.

JOHANN WOLFGANG VON GOETHE

HISTORICAL PROFILES

Jean-Baptiste Joseph Fourier (1768–1830), a French mathematician, first presented the series and transform (covered in this chapter) that bears his name. Fourier's results were not enthusiastically received by the scientific world. He could not even get his work published as a paper.

Born in Auxerre, France, as the son of a tailor, Fourier was orphaned at age 8. He attended a local military college run by Benedictine monks, where he decided to train for the priesthood. He soon distinguished himself as a student and made rapid progress, especially in mathematics. By the age of 14, he had completed a study of the six volumes of Bézout's *Cours de mathématiques.* Up until this time there had been a conflict inside Fourier about whether he should follow a religious life or one of mathematical research. However, in 1793 a third element was added to this conflict when he became involved in politics and joined the local Revolutionary Committee.

Like most of his contemporaries, Fourier was swept into the politics of the French Revolution. Due to his political involvement, he narrowly escaped death twice. In 1798 Fourier joined Napoleon's army in its invasion of Egypt as scientific adviser. While in Cairo, Fourier helped found the Cairo Institute. Fourier returned to France in 1801 and resumed his post as Professor of Analysis at the École Polytechnique. It was while there that he did his important mathematical work on the propagation of heat. In 1822, he published his *Théorie analytique de la chaleur.* In 1826, Fourier became a member of the French Academy, and in 1827 succeeded Laplace as the president of the council of the École Polytechnique.

As a politician, Fourier achieved uncommon success, but his fame chiefly rests on his strikingly original contributions to science and mathematics. His life was not without problems, however, since his theory of heat still provoked controversy.

Heinrich Rudolf Hertz (1857–1894), a German experimental physicist, demonstrated that electromagnetic waves obey the same fundamental laws as light.

Hertz was born into a prosperous family in Hamburg, Germany. He attended the University of Berlin, where he received his doctoral degree magna cum laude in 1880 under the prominent physicist Hermann von Helmholtz. He became a professor of physics at Karlsruhe, where he began his quest for electromagnetic waves. Their existence had been predicted in 1873 by the mathematical equations of James Clerk Maxwell, a British scientist. Hertz successfully generated and detected electromagnetic waves; he was the first to establish the fact that light is a form of electromagnetic radiation. But Hertz saw no practical use for his discovery. Hertz also simplified Maxwell's theory to a mathematical formalism and that led to its widespread acceptance. In 1887, Hertz noted for the first time the photoelectric effect of electrons in a molecular structure. Although Hertz died of blood poisoning at the age of 37, his discovery of electromagnetic waves paved the way for the practical use of such waves in wireless telegraph, radio, television, radar, and other communication systems. The unit of frequency, the hertz, bears his name.

2.1   INTRODUCTION

Communication theory involves applying mathematical models and techniques to communication systems. The purpose of the communication system is to convey a message (information), which is usually in the form of signals. In other words, signals are means of representing information, while systems are means of processing the signals. Although a signal is not necessarily a function of time, in this book we deal exclusively with signals that are time-dependent. In general,

> A **signal** is a function of time that represents a physical quantity.

For our purposes, a signal is a waveform that describes or encodes information. In other words, a signal is a function that carries information. Examples of signals include sine or cosine signals, audio or radio signals, video or image signals, speech signals, seismic signals, and radar signals. For example, the voice of my wife is a signal that makes me react in a certain manner. A signal may be voltage waveform $v(t)$, current waveform $i(t)$, or any other variable.

Signals may be classified broadly as either deterministic or random. A deterministic (or nonrandom) signal is one whose value is known precisely at all instants of time. Examples of such signals include sine waves, square waves, and pulses. A random (or stochastic) signal is one whose values at any instant of time can only be described statistically using the mean value, variance, etc. Noise is a good example of a random signal. We will restrict ourselves to deterministic signals in this chapter; random signals will be treated in Chapter 6.

There are two ways of describing a deterministic signal: time-domain and frequency-domain techniques. The frequency-domain approach applies line spectral analysis based on the Fourier series expansion of periodic signals and continuous spectral analysis based on the Fourier transform of nonperiodic signals. Thus spectral analysis should be regarded as an

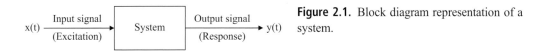

Figure 2.1. Block diagram representation of a system.

inestimable mathematical tool for studying communication systems. The spectral analysis can be done in the laboratory using the spectrum analyzer.

The term *system* is broad. It is used in politics, education, economics, engineering, etc. For our purposes, a system is a combination of several components to perform a given task. It may also be regarded as a device that can manipulate, change, or transmit signals. In fact, any complete set of mathematical relationships between input and output variables constitutes a system.

> A **system** is a mathematical model of a physical entity that processes signals (inputs) to produce new signals (outputs).

The block diagram of a system is shown in Figure 2.1. Examples include a filter, a camera, an automobile ignition system, an aircraft control system, an audio compact disk (CD) player, radar, sonar, a computer (hardware), and a computer program (software).

We begin this chapter by understanding the concepts of signals and systems. Then we review spectral analysis of the signals using Fourier series and Fourier transform, assuming that the reader has some familiarity with them. We apply the concepts learned in this chapter to filters, which are important components of communication systems. We finally use MATLAB to perform some of the signal analysis covered in this chapter.

2.2 CLASSIFICATION OF SIGNALS

As mentioned earlier, the term *signal* is used for an electric quantity such as a voltage or current (or even an electromagnetic wave) when it is used for conveying information. Engineers prefer to call such variables signals rather than mathematical functions of time because of their importance in communications and other disciplines. Table 2.1 provides a list of common signals we will encounter.

There are several ways of looking at the same thing. Signals are no exceptions. In addition to classifying signals as deterministic or random, signals may also be classified as follows:

- *Continuous or discrete*: if the independent variable t is continuous (defined for *every* value of t, i.e. over a continuum of values of t), the corresponding signal $x(t)$ is said to be a continuous-time signal. If the independent variable assumes *only* discrete values $t = nT$, where T is fixed and n is a set of integers (e.g. $n = 0, \pm 1, \pm 2, \pm 3, \ldots$), the corresponding signal $x(t)$ is a discrete-time signal. Examples of continuous and discrete signals are shown in Figure 2.2.

Table 2.1. **A list of common functions**

Signals	Description
Sinusoidal signal	$f(t) = A \cos t$
Unit step function	$u(t) = \begin{cases} 1, & t \geq 0 \\ 0, & \text{otherwise} \end{cases}$
Unit ramp function	$r(t) = \begin{cases} t, & t \geq 0 \\ 0, & \text{otherwise} \end{cases}$
Impulse (or delta) function	$\delta(t) = 0, \quad t \neq 0$
Exponential function	$f(t) = e^{-at}$
Rectangular pulse	$\Pi\left(\dfrac{t}{\tau}\right) = \begin{cases} 1, & -\tau/2 < t < \tau/2 \\ 0, & \text{otherwise} \end{cases}$
Triangular pulse	$\Delta\left(\dfrac{t}{\tau}\right) = \begin{cases} 1 - \dfrac{|t|}{\tau}, & -\tau < t < \tau \\ 0, & \text{otherwise} \end{cases}$
Sinc function	$\text{sinc}(t) = \dfrac{\sin t}{t}$
Signum function	$\text{sgn}(t) = \begin{cases} 1, & t > 0 \\ -1, & t < 0 \end{cases}$
Gaussian function	$g(t) = e^{-t^2/2\sigma^2}$

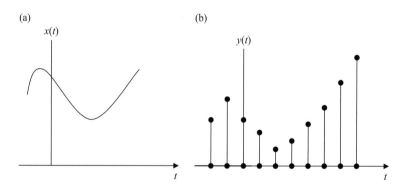

(a)   (b)

Figure 2.2. (a) Continuous-time signal; (b) discrete-time signal.

- *Analog or digital*: while terms *continuous* and *discrete* describe the nature of a signal along the time (horizontal axis), the terms *analog* and *digital* describe the signal amplitude (vertical axis). A digital signal can take a finite number of values, e.g. 0 and 1. An analog signal has an amplitude that assumes any value.
- *Periodic or aperiodic*: a signal $x(t)$ is called *periodic* if for a constant positive T,

$$x(t) = x(t + T), \quad \text{for all} \quad t \tag{2.1}$$

The small value of T that satisfies Eq. (2.1) is known as the *period* of $x(t)$. A signal which cannot satisfy Eq. (2.1) for any T is said to be *aperiodic* or *nonperiodic*. Examples of periodic signals are sinusoidal functions (sine and cosine functions), while examples of aperiodic functions are exponential and singularity functions (unit step, impulse function, etc.). Periodic functions are very useful in science and engineering, especially in communications. The average or mean value X_{ave} of a periodic signal $x(t)$ is given by

$$X_{ave} = \frac{1}{T} \int_0^T x(t)dt \tag{2.2}$$

- *Energy or power*: let $x(t)$ represent the voltage signal across a resistance R. The corresponding current produced is $i(t) = x(t)/R$ so that the instantaneous power dissipated is $Ri^2(t) = x^2(t)/R$. Since we may not know whether $x(t)$ is a voltage or current signal, it is customary to normalize power calculations by assuming $R = 1$ ohm. Thus, we may express the instantaneous power associated with signal $x(t)$ as $x^2(t)$. The total energy of the signal over a time interval of $2T$ is

$$E = \lim_{T \to \infty} \int_T^T |x(t)|^2 dt = \int_{-\infty}^{\infty} |x(t)|^2 dt \tag{2.3}$$

where the magnitude square has been used in case $x(t)$ is a complex-valued signal. Since power is the time average of the energy, the average power of the signal is

$$P = \lim_{T \to \infty} \frac{1}{2T} \int_{-T}^T |x(t)|^2 dt \tag{2.4}$$

A signal is said to be an *energy signal* when the total energy E of the signal satisfies the condition

$$0 < E < \infty \tag{2.5}$$

Similarly, a signal is called a *power signal* when its average power satisfies the condition

$$0 < P < \infty \tag{2.6}$$

Thus, an energy signal must have finite power, non-zero energy ($P = 0$, $0 < E < \infty$) and a power signal must have finite, non-zero power ($0 < P < \infty$, $E = 0$). Deterministic and aperiodic signals are energy signals, whereas periodic and random signals are power signals. From Eqs. (2.3) and (2.4), we note that an energy signal has zero average power, while a power signal has infinite energy. In other words, a signal with finite energy has zero average power, whereas a signal with finite power has infinite energy. Thus, energy signals and power signals are mutually exclusive. If a signal is a power signal, it cannot be an energy signal and vice versa. Of course, a signal may be neither an energy nor a power signal. Thus,

> $x(t)$ is an energy signal if it has finite energy, $0 < E < \infty$, so that $P = 0$
> $x(t)$ is a power signal if it has finite power, $0 < P < \infty$, so that $E = 0$
> $x(t)$ is neither an energy nor a power signal if neither property is satisfied.

EXAMPLE 2.1

Show that the signal

$$x(t) = \begin{cases} 10, & t \geq 0 \\ 0, & t < 0 \end{cases}$$

is a power signal.

Solution
From Eq. (2.4),

$$P = \lim_{T \to \infty} \frac{1}{2T} \int_{-T}^{T} x^2(t)dt = \lim_{T \to \infty} \frac{1}{2T} \int_{0}^{T} 100\, dt = \lim_{T \to \infty} \frac{1}{2T}(100T) = 50$$

Similarly, from Eq. (2.3), $E = \infty$. Since P is finite while the energy is infinite, we conclude that $x(t)$ is a power signal.

PRACTICE PROBLEM 2.1

Show that the signal $y(t) = 2 \cos(\pi t)$ is a power signal.

Answer: Proof.

2.3 OPERATIONS ON SIGNALS

Having examined different ways a signal can be classified, we will now consider some important basic operations on signals. The operations include time shifting, time reflecting, and time scaling.

- *Shifting operation*: the signal $x(t-a)$ denotes a time-shifted version of $x(t)$. If $a > 0$, the signal is delayed (or shifted right) by a seconds, as shown in Figure 2.3(a). If a is negative, the signal is advanced (or shifted left) by a seconds, as illustrated in Figure 2.3(b).

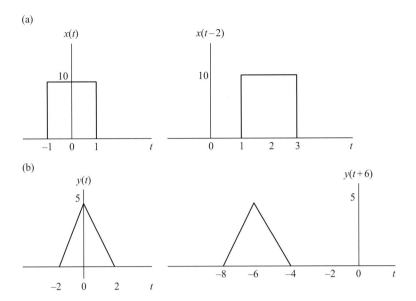

Figure 2.3. Time-shifting a signal: (a) $x(t)$ is delayed by 2 s; (b) $y(t)$ is advanced by 6 s.

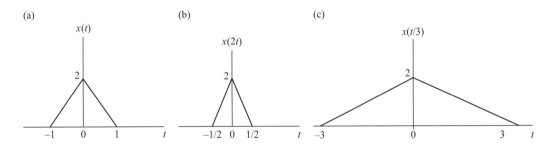

Figure 2.4. Time-scaling a signal.

- *Scaling operation*: the signal $x(at)$ is known as the time-scaled version of $x(t)$. If $|a| > 1$, the signal $x(at)$ is compressed because it exists in a smaller time interval than $x(t)$. If $|a| < 1$, the signal $x(at)$ is expanded since the signal exists in a larger time interval. For example, given the signal in Figure 2.4(a), its compressed version $x(2t)$ and its expanded version $x(t/3)$ are shown in Figure 2.4(b) and 2.4(c) respectively. If we regard $x(t)$ as the signal from a tape recorder, then $x(2t)$ is the signal obtained when the recorder plays twice as fast, and $x(t/3)$ is the signal we get when the recorder plays one-third the speed.
- *Inverting operation*: the signal $x(-t)$ is the inverted or reflected version of $x(t)$ about $t = 0$. In other words, $x(-t)$ is the mirror image of $x(t)$ about the vertical axis. For example, if $x(t)$ is the signal from a tape recorder when played forward, then $x(-t)$ is the signal when the recorder plays backward. Note that time inversion is a special case of time scale with the scaling factor $a = -1$. Thus, to time-invert $x(t)$, we merely replace t with $-t$, as illustrated in Figure 2.5.

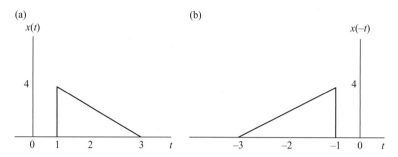

(a)

$x(t)$

(b)

$x(-t)$

Figure 2.5. Time-inverting a signal.

EXAMPLE 2.2
Given the signal

$$x(t) = \begin{cases} t, & 0 \le t \le 1 \\ 1, & 1 \le t \le 2 \\ -t + 3, & 2 \le t \le 3 \end{cases}$$

shown in Figure 2.6, obtain and sketch: (a) $x(-t + 4)$, (b) $x(2t - 1)$.

Solution
Each of the required signals can be obtained from $x(t)$ in two ways.

(a) Method 1: (Graphical approach) $x(-t + 4)$ combines both time inverting and time-shifting. Since $x(-t + 4) = x(-[t - 4])$, we first time-invert $x(t)$ to get $x(-t)$ as shown in Figure 2.7(a) and then shift $x(-t)$ to the right by 4 seconds to get $x(-t + 4)$ as shown in Figure 2.7(b). Method 2: (Analytic approach) To obtain $x(-t + 4)$ from the given $x(t)$, replace every t with $-t + 4$. So we get

$$x(-t + 4) = \begin{cases} -t + 4, & 0 \le -t + 4 \le 1 \\ 1, & 1 \le -t + 4 \le 2 \\ -(-t + 4) + 3, & 2 \le -t + 4 \le 3 \end{cases}$$

Note that the equality $0 \le -t + 4 \to t \le 4$ and $-t + 4 \le 1 \to 3 \le t$ so that $0 \le -t + 4 \le 1 \to 3 \le t \le 4$. By treating other inequalities the same way, we obtain

$$x(-t + 4) = \begin{cases} -t + 4, & 3 \le t \le 4 \\ 1, & 2 \le t \le 3 \\ t - 1, & 1 \le t \le 2 \end{cases} = \begin{cases} t - 1, & 1 \le t \le 2 \\ 1, & 2 \le t \le 3 \\ 4 - t & 3 \le t \le 4 \end{cases}$$

which is what we have in Figure 2.7(b).

(b) Method 1: (Graphical approach) $x(2t - 1)$ is both time-scaled (compressed) and time-shifted. Since $x(2t - 1) = x(2[t - 1/2])$, we first time-scale $x(t)$ to get $x(2t)$ as shown in Figure 2.8(a) and then shift $x(2t)$ to the right by 1/2 second to get $x(2t - 1)$ as shown in Figure 2.8(b).

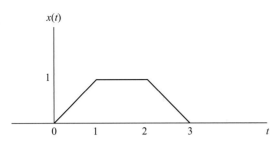

Figure 2.6. For Example 2.2.

(a)

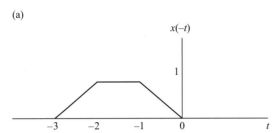

Figure 2.7. For Example 2.2.

(b)

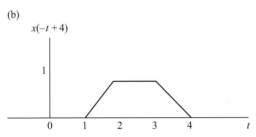

(a)

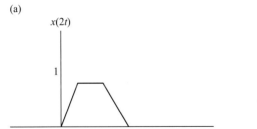

Figure 2.8. For Example 2.2.

(b)

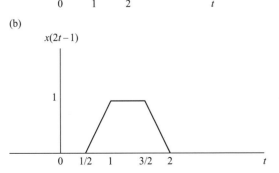

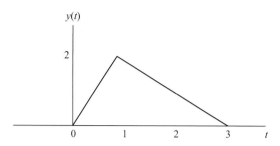

Figure 2.9. For Practice problem 2.2.

(a)

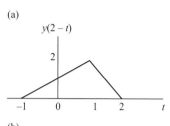

Figure 2.10. For Practice problem 2.2.

(b)

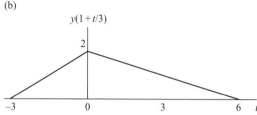

Method 2: (Analytic approach) Similarly, to obtain $x(2t - 1)$ from $x(t)$, replace every t with $2t - 1$.

$$x(2t - 1) = \begin{cases} 2t - 1, & 0 \leq 2t - 1 \leq 1 \\ 1, & 1 \leq 2t - 1 \leq 2 \\ -(2t - 1) + 3, & 2 \leq 2t - 1 \leq 3 \end{cases} = \begin{cases} 2t - 1, & 1/2 \leq t \leq 1 \\ 1, & 1 \leq t \leq 3/2 \\ 4 - 2t, & 3/2 \leq t \leq 2 \end{cases}$$

which is the same as the sketch in Figure 2.8(b).

PRACTICE PROBLEM 2.2
Let

$$y(t) = \begin{cases} 2t, & 0 \leq t \leq 1 \\ 3 - t, & 1 \leq t \leq 3 \end{cases}$$

which is sketched in Figure 2.9. Determine and sketch: (a) $y(2 - t)$, (b) $y(1 + t/3)$.

Answers: (a) $y(2 - t) = \begin{cases} 1 + t, & -1 \leq t \leq 1 \\ 4 - 2t, & 1 \leq t \leq 2 \end{cases}$

(b) $y(1 + t/3) = \begin{cases} 2 + 2t/3, & -3 \leq t \leq 0 \\ 2 - t/3, & 0 \leq t \leq 6 \end{cases}$

They are both shown in Figure 2.10.

2.4 CLASSIFICATION OF SYSTEMS

As said earlier, a system relates an input signal $x(t)$ (known as the excitation) to the output signal $y(t)$ (known as the response). We may write the input-output relationship as

$$y(t) = f[x(t)] \tag{2.7}$$

On the basis of this relationship, we may classify systems as follows:

- *Linear or nonlinear*: a system is linear when its output is linearly related to its input. It is nonlinear otherwise. A linear system has an advantage in that superposition principle applies. If $y_1(t)$ is the response of $x_1(t)$ and $y_2(t)$ is the response of $x_2(t)$, then a system is linear if it satisfies the following two conditions:
  (1) the response of $x_1(t) + x_2(t)$ is $y_1(t) + y_2(t)$;
  (2) the response of $ax_1(t)$ is $ay_1(t)$, where a is a constant.

  An electric network consisting of linear elements such as resistors, capacitors, and inductors is a linear system, where an electronics network consisting of nonlinear elements such as diodes and transistors is a nonlinear system.

- *Continuous or discrete*: a system is continuous-time when the input and output signals are time-continuous. It is a discrete-time system if input and output signals are discrete-time.
- *Time-varying or time-invariant:* a system is said to be time-invariant or fixed if its input-output relationship does not vary with time. Otherwise, it is said to be time-varying. In other words, if a time shift in the input signal produces a corresponding time shift in the output signal so that

$$y(t - \tau) = f[x(t - \tau)] \tag{2.8}$$

  the system is time-invariant.

  A system is time-varying if it does not satisfy Eq. (2.8). For example, suppose the input-output relationship of a system is described by

$$y(t) = Ax(t) + B \tag{2.9}$$

  The system is time-invariant if A and B are time-independent; it is time-varying if A and B vary with time.

- *Causal or noncausal*: a signal is causal if it has zero values when $t < 0$. A causal system is one whose output signal (response) does not start before the input signal (excitation) is applied. Causal systems are also referred to as physically realizable or nonanticipatory systems. A noncausal system is one in which the response depends on the future values of the input. Such a system is not physically realizable; it does not exist in real life but is mathematically modeled using a time delay. For example, a system described by $y(t) = f[x(t - 1)]$ is causal, whereas a system described by $y(t) = f[x(t + 1)]$ is noncausal.
- *Analog or digital*: the input signal to a system determines whether the system is analog or digital. An analog system is one whose input signal is analog. Examples of analog systems include analog switches, analog filters, and analog cellular telephone systems. A digital

system is one whose input is in the form of a sequence of digits. Typical examples of digital systems include the digital computer, digital filters, and digital audiotape (DAT) systems.

2.5 TRIGONOMETRIC FOURIER SERIES

The Fourier techniques, named after the French physicist Jean-Baptiste Fourier (1768–1830) who first investigated them, are important tools for scientists and engineers. Fourier representation (Fourier series and Fourier transform) of signals plays a major role in the analysis of communication systems for at least two main reasons. First, they help characterize signals in terms of frequency-domain parameters such as bandwidth which are of major concern to communication engineers. Second, they provide direct physical interpretation. We will consider Fourier series in this section and treat Fourier transform later in this chapter.

Fourier series allow us to represent a periodic function exactly in terms of sinusoids. Any practical periodic function $f(t)$ with period T can be expressed as an infinite sum of sine or cosine functions that are integral multiples of ω_0. Thus, $f(t)$ can be expressed as

$$f(t) = \underbrace{a_0}_{\text{dc}} + \underbrace{\sum_{n=1}^{\infty} (a_n \cos n\omega_0 + b_n \sin n\omega_0)}_{\text{ac}} \tag{2.10}$$

where

$$\omega_0 = \frac{2\pi}{T} \tag{2.11}$$

is called the *fundamental frequency* in radians per second. Equation (2.10) is known as *quadrature* Fourier series. The sinusoid $\sin n\omega_0 t$ or $\cos n\omega_0 t$ is called the nth harmonic of $f(t)$; it is an odd harmonic if n is odd or an even harmonic if n is even. The constants a_n and b_n are the *Fourier coefficients*. The coefficient a_0 is the dc component or the average value of $f(t)$. (Recall that sinusoids have zero average value.) The coefficients a_n and b_n (for $n \neq 0$) are the amplitudes of the sinusoids in the ac component. Thus,

> The **Fourier series** of a periodic function $f(t)$ is a representation that resolves $f(t)$ into a dc component and an ac component comprising an infinite series of harmonic sinusoids.

A function that can be represented by a Fourier series as in Eq. (2.10) must meet certain requirements because the infinite series in Eq. (2.10) may or may not converge. These conditions on $f(t)$ to yield a convergent Fourier series are as follows:

(1) $f(t)$ is single-valued everywhere;
(2) $f(t)$ has a finite number of finite discontinuities in any one period;

(3) $f(t)$ has a finite number of maxima and minima in any one period;

(4) The integral $\int_{t_o}^{t_o+T} |f(t)|dt$ is finite for any t_o.

These conditions are called *Dirichlet conditions*. Although they are not necessary conditions, they are sufficient conditions for a Fourier series to exist. Luckily, all practical periodic functions obey these conditions.

A major task in Fourier series is the determination of the Fourier coefficients a_0, a_n, and b_n. The process of determining the coefficients, known as *Fourier analysis*, involves evaluating the following integrals

$$a_0 = \frac{1}{T}\int_0^T f(t)dt \tag{2.12a}$$

$$a_n = \frac{2}{T}\int_0^T f(t)\cos n\omega_o t\, dt \tag{2.12b}$$

$$b_n = \frac{2}{T}\int_0^T f(t)\sin n\omega_o t\, dt \tag{2.12c}$$

Note that the dc component in Eq. (2.12a) is the average value of the signal $f(t)$. In Eq. (2.12), the interval $0 < t < T$ is chosen for convenience. The integrals would be the same if we chose instead the interval $-T/2 < t < T/2$ or $t_o - T/2 < t < t_o + T/2$, where t_o is a constant. Some trigonometric integrals, provided in Appendix A, are very helpful in Fourier analysis. It can be shown that the Fourier series of an even periodic function consists of the dc term and cosine terms only ($b_n = 0$) since cosine is an even function. Similarly, the Fourier series expansion of an odd function has only sine terms ($a_0 = 0$, $a_n = 0$). Table 2.2 provides the Fourier series of some common periodic signals.

An alternative form of Eq. (2.10) is the *amplitude-phase* form

$$f(t) = A_0 + \sum_{n=1}^{\infty} A_n \cos(n\omega_0 + \varphi_n) \tag{2.13}$$

where

$$A_0 = a_o, \quad A_n = \sqrt{a_n^2 + b_n^2}, \quad \varphi_n = -\tan^{-1}\frac{b_n}{a_n} \tag{2.14}$$

The plot of the coefficients A_n versus $n\omega_0$ is called the *amplitude spectrum* of $f(t)$; while the plot of the phase ϕ_n versus $n\omega_0$ is the *phase spectrum* of $f(t)$. Both the amplitude and phase spectra form the *frequency spectrum* or *line spectrum* of $f(t)$. Equation (2.13) is also known as *polar Fourier series*. Thus, there are two forms of trigonometric Fourier series: the quadrature form and the polar form.

Table 2.2. **The Fourier series of common signals**

Signal	Fourier series
1. Square wave	$f(t) = \dfrac{4A}{\pi} \displaystyle\sum_{n=1}^{\infty} \dfrac{1}{2n-1} \sin(2n-1)\omega_o t$
2. Rectangular pulse train	$f(t) = \dfrac{A\tau}{T} + \dfrac{2A}{T} \displaystyle\sum_{n=1}^{\infty} \dfrac{1}{n} \sin \dfrac{n\pi\tau}{T} \cos n\omega_o t$
3. Sawtooth wave	$f(t) = \dfrac{A}{2} - \dfrac{A}{\pi} \displaystyle\sum_{n=1}^{\infty} \dfrac{\sin n\omega_o t}{n}$
4. Triangular wave	$f(t) = \dfrac{A}{2} - \dfrac{4A}{\pi^2} \displaystyle\sum_{n=1}^{\infty} \dfrac{\cos(2n-1)\omega_o t}{(2n-1)^2}$
5. Half-wave rectified sine	$f(t) = \dfrac{A}{\pi} + \dfrac{A}{2} \sin \omega_o t - \dfrac{4A}{\pi} \displaystyle\sum_{n=1}^{\infty} \dfrac{\cos 2n\omega_o t}{4n^2 - 1}$
6. Full-wave rectified sine	$f(t) = \dfrac{2A}{\pi} - \dfrac{4A}{\pi} \displaystyle\sum_{n=1}^{\infty} \dfrac{\sin n\omega_o t}{4n^2 - 1}$

The average power of signal $f(t)$ is defined as

$$P = \frac{1}{T} \int_0^T f^2(t)\,dt \qquad (2.15)$$

If the Fourier series expansion of $f(t)$ in Eq. (2.10) is substituted into Eq. (2.15), we can readily show that

$$P = a_0^2 + \frac{1}{2}\sum_{n=1}^{\infty}\left(a_n^2 + b_n^2\right) \tag{2.16}$$

indicating that the average power of a periodic signal is the sum of the squares of its Fourier coefficients. In other words, the total average power is the sum of the powers in the dc component and the powers in the harmonic components. This confirms the fact that a periodic signal is a power signal and that every component of its Fourier series expansion is also a power signal contributing its individual power to the total power.

The same conclusion can be reached using the polar Fourier series. By substituting Eq. (2.13) into Eq. (2.15) or substituting Eq. (2.14) into Eq. (2.16), we obtain

$$P = A_0^2 + \frac{1}{2}\sum_{n=1}^{\infty} A_n^2 \tag{2.17}$$

Thus, the signal power can be found either in the time domain using Eq. (2.15) or in the frequency domain using Eq. (2.16) or (2.17). This is known as *Parseval's theorem*:

$$P = \frac{1}{T}\int_0^T f^2(t)\,dt = a_0^2 + \frac{1}{2}\sum_{n=1}^{\infty}\left(a_n^2 + b_n^2\right) = A_0^2 + \frac{1}{2}\sum_{n=1}^{\infty} A_n^2 \tag{2.18}$$

EXAMPLE 2.3

Determine the Fourier series of the periodic impulse train shown in Figure 2.11. Obtain the amplitude and phase spectra.

Solution

The periodic train can be written as

$$f(t) = 10\sum_{n=-\infty}^{\infty}\delta(t - nT)$$

Using the sampling property of the impulse function,

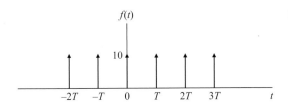

Figure 2.11. For Example 2.3.

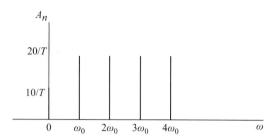

Figure 2.12. For Example 2.3; amplitude spectrum of the impulse train in Figure 2.11.

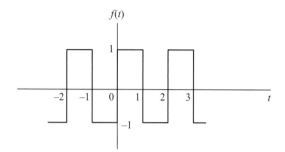

Figure 2.13. For Practice problem 2.3.

$$a_0 = \frac{1}{T} \int_0^T 10\delta(t)dt = \frac{10}{T}$$

$$a_n = \frac{2}{T} \int_0^T 10\delta(t) \cos n\omega_0 \, dt = \frac{20}{T}$$

$$b_n = \frac{2}{T} \int_0^T 10\delta(t) \sin n\omega_0 \, dt = 0$$

It is not surprising that $b_n = 0$ since the impulse function is even. From Eq. (2.14),

$$A_0 = \frac{10}{T}, \qquad A_n = \frac{20}{T}, \qquad \varphi_n = 0$$

And Eq. (2.13) becomes

$$f(t) = \frac{10}{T} \left[1 + 2\sum_{n=1}^{\infty} \cos n\omega_o t \right], \qquad \omega_o = \frac{2\pi}{T}$$

Since φ_n is zero, the phase spectrum is zero everywhere. However, the amplitude spectrum is the plot of A_n as shown in Figure 2.12.

PRACTICE PROBLEM 2.3

Find the Fourier series expansion of the square wave in Figure 2.13. Plot the amplitude and phase spectra.

(a)

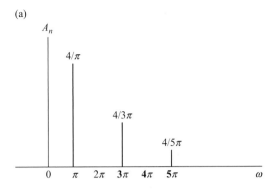

Figure 2.14. For Practice problem 2.3; amplitude and phase spectra of the function shown in Figure 2.13.

(b)

Answer: $f(t) = \frac{4}{\pi} \sum_{k=1}^{\infty} \frac{1}{n} \sin 2n\pi t,$ $n = 2k - 1$. See Figure 2.14 for the amplitude and phase spectra.

EXAMPLE 2.4

Obtain the Fourier series for the sawtooth waveform in Figure 2.15 and plot the amplitude and phase spectra.

Solution

The function can be written as

$$g(t) = \frac{2}{\pi} t, \qquad -\pi < t < \pi$$

Since $g(-t) = -g(t)$, $g(t)$ has odd symmetry so that $a_0 = a_n = 0$. The period of the function is $T = 2\pi$ so that $\omega_0 = 2\pi/T = 1$.

$$b_n = \frac{2}{T} \int_{-T/2}^{T/2} g(t) \sin n\omega_0 t \, dt = \frac{2}{2\pi} \int_{-\pi}^{\pi} \frac{2}{\pi} t \sin nt \, dt = \frac{4}{\pi^2} \int_{0}^{\pi} t \sin nt \, dt \qquad (2.4.1)$$

where we have taken advantage of the symmetry of $g(t)$ by integrating from 0 to π and multiplying by 2. But from Appendix A,

$$\int t \sin at \, dt = \frac{1}{a^2} \sin at - \frac{t}{a} \cos at \qquad (2.4.2)$$

Applying this to Eq. (2.4.1), we get

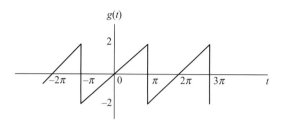

Figure 2.15. For Example 2.4.

(a)

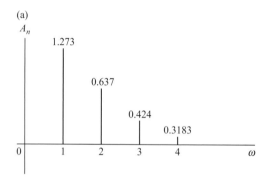

Figure 2.16. For Example 2.4.

(b)

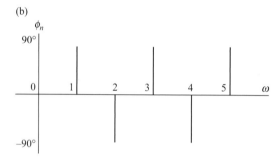

$$b_n = \frac{4}{\pi^2} \int_0^\pi t \sin nt \, dt = \frac{4}{\pi^2} \left[\frac{1}{n^2} \sin nt - \frac{t}{n} \cos nt \right]_0^\pi = \frac{4}{\pi^2} \left(0 - \frac{\pi}{n} \cos \pi n - 0 + 0 \right) = -\frac{4}{\pi n} \cos \pi n$$

Since $\cos n\pi = (-1)^n$, the Fourier series expansion of $g(t)$ is

$$g(t) = \frac{4}{\pi} \sum_{n=1}^\infty \frac{(-1)^{n+1}}{n} \sin nt$$

having only sine terms. Since $A_n = b_n$ and $\varphi_n = -90° + \alpha$, where α is due $(-1)^{n+1}$. $\alpha = 0$ when n is odd and $\alpha = 180°$ when n is even. Hence,

$$\varphi_n = \begin{cases} -90°, & n = \text{odd} \\ 90°, & n = \text{even} \end{cases}$$

The amplitude and phase spectra are shown in Figure 2.16.

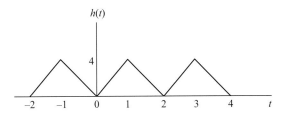

Figure 2.17. For Practice problem 2.4.

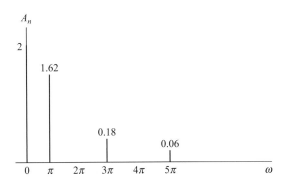

Figure 2.18. Answer for Practice problem 2.4 (not to scale).

PRACTICE PROBLEM 2.4

Determine the Fourier series of the periodic function in Figure 2.17. Sketch the amplitude and phase spectra.

Answer: $h(t) = 2 - \frac{16}{\pi^2} \sum_{k=1}^{\infty} \frac{1}{n^2} \cos n\pi t, \quad n = 2k - 1$, the phase spectrum is zero and the amplitude spectrum is shown in Figure 2.18.

2.6 EXPONENTIAL FOURIER SERIES

Since we know that a sinusoid of frequency $n\omega_0$ can be expressed in terms of exponentials $e^{jn\omega_0 t}$ and $e^{-jn\omega_0 t}$, we should intuitively expect that trigonometric Fourier series in Eqs. (2.10) and (2.13) can also be expressed in exponential form. Although it is more convenient to deal with the trigonometric Fourier series since no complex algebra is involved, the exponential form is useful for two reasons. First, it is more compact and it readily leads to the Fourier transform to be covered in the next section. Second, the trigonometric Fourier generates a one-sided spectrum, while the exponential Fourier series produces a two-sided spectrum (as we shall see), which is more convenient to use and is preferred most of the time.

A periodic signal $f(t)$ of period T and frequency $\omega = 2\pi/T$ may be represented as an exponential Fourier series as

$$f(t) = \sum_{n=-\infty}^{\infty} C_n e^{jn\omega_0 t} \tag{2.19}$$

where the complex coefficient is given by

$$C_n = \frac{1}{T} \int_0^T f(t) e^{-jn\omega_0 t} dt = |C_n| \angle \phi_n \tag{2.20}$$

where $|C_n|$ and ϕ_n are the magnitude and phase of C_n. The plots of the magnitude and phase of C_n versus $n\omega_0$ are called the *complex amplitude spectrum* and *complex phase spectrum* or simply the *exponential spectra* of $f(t)$. We should note the following:

(1) C_0 corresponds to the dc value of $f(t)$ because Eq. (2.20) is identical to Eq. (2.12a) when $n = 0$.
(2) The frequencies present are $\pm\omega_0$ (known as the *fundamental frequency*) and all integer multiples $\pm n\omega_0$ (known as the *harmonics*). The presence of the negative frequencies contradicts our common notion of frequency being the number of repetitions per unit time. The negative frequencies are due to the existence of the negative exponential term $e^{-jn\omega_0 t}$. Notice that each exponential term comes in pairs – positive and negative. The positive portion on its own does not represent a real signal but when its negative complement is added the two together form a real signal.
(3) The coefficients of the exponential Fourier series are related to those of the trigonometric Fourier series by

$$a_n - jb_n = A_n \angle \varphi_n = 2C_n = 2|C_n| \angle \phi_n \tag{2.21}$$

Thus, we can obtain the exponential Fourier series from the trigonometric series, and vice versa.

As we did for trigonometric series, we can obtain the average power P of signal $f(t)$ from the exponential series expansion. Substituting Eq. (2.19) into Eq. (2.15) gives

$$P = \frac{1}{T} \int_0^T f^2(t) dt = \frac{1}{T} \int_0^T f(t) \left[\sum_{n=-\infty}^{\infty} C_n e^{jn\omega_0 t} \right] dt$$

Interchanging the order of summation and integration yields

$$P = \sum_{n=-\infty}^{\infty} C_n \left[\frac{1}{T} \int_0^T f(t) e^{jn\omega_0 t} dt \right] = \sum_{n=-\infty}^{\infty} C_n C_n^* = \sum_{n=-\infty}^{\infty} |C_n|^2 \tag{2.22}$$

where C_n^* is the complex conjugate of C_n. Thus, we obtain another form of Parseval's theorem as:

$$P = \frac{1}{T} \int_0^T f^2(t) dt = \sum_{n=-\infty}^{\infty} |C_n|^2 \tag{2.23}$$

This is an alternative and more compact way of expressing Parseval's theorem. The *power spectrum* of signal $f(t)$ is $\sum_{n=-\infty}^{\infty} |C_n|^2$. The power spectrum shows how the total power is distributed among the dc and the harmonic components.

EXAMPLE 2.5

Determine the exponential Fourier series of the periodic impulse train of Example 2.3.

Solution

The periodic train can be written as

$$f(t) = 10 \sum_{n=-\infty}^{\infty} \delta(t - nT)$$

Using the sampling property of the impulse function,

$$C_n = \frac{1}{T} \int_0^T 10\delta(t)e^{-jn\omega_0 t} dt = \frac{10}{T}$$

Hence the Fourier series is

$$f(t) = \frac{10}{T} \sum_{n=-\infty}^{\infty} e^{jn\omega_0 t}, \quad \omega_o = \frac{2\pi}{T}$$

PRACTICE PROBLEM 2.5

Expand the rectangular pulse train shown in Figure 2.19 in exponential Fourier series.

Answer: $f(t) = 2 \sum_{n=-\infty}^{\infty} \frac{\sin \lambda}{\lambda} e^{j\lambda t}, \quad \lambda = \frac{n\pi}{5}$

EXAMPLE 2.6

Find the exponential Fourier series expansion of the periodic function

$$f(t) = e^t, \ 0 < t < 2\pi \ \text{with} \ f(t + 2\pi) = f(t).$$

Solution

Since $T = 2\pi$, $\omega = 2\pi/T = 1$. Hence

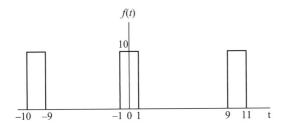

Figure 2.19. Rectangular pulse trains for Practice problem 2.5.

$$C_n = \frac{1}{T}\int_0^T f(t)e^{-jn\omega_o t}dt = \frac{1}{2\pi}\int_0^{2\pi} e^t e^{-jnt}dt = \frac{1}{2\pi}\left[\frac{1}{1-jn}e^{(1-jn)t}\right]_0^{2\pi} = \frac{1}{2\pi(1-jn)}\left[e^{2\pi}e^{-j2n\pi}-1\right]$$

Using Euler's identity,

$$e^{-j2n\pi} = \cos 2n\pi - j\sin 2n\pi = 1 - j0 = 1$$

we obtain

$$C_n = \frac{e^{2\pi}-1}{2\pi(1-jn)} = \frac{85}{1-jn}$$

Thus, the complex Fourier series is

$$f(t) = \sum_{n=-\infty}^{\infty} \frac{85}{1-jn}$$

To plot the exponential spectra, we obtain the magnitude and phase of C_n as

$$|C_n| = \frac{85}{\sqrt{1+n^2}}, \qquad \phi_n = \tan^{-1}n$$

By inserting in negative and positive values of n, we obtain the amplitude and the phase plots of C_n versus $n\omega_o = n$, as in Figure 2.20.

(a)

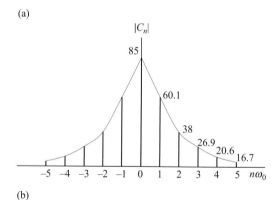

Figure 2.20. For Example 2.6: (a) amplitude spectrum; (b) phase spectrum.

(b)

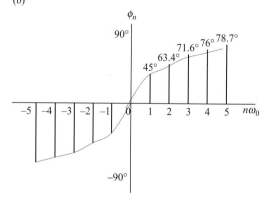

(a)

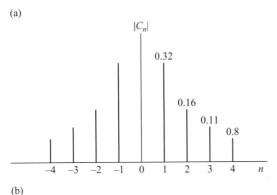

Figure 2.21. For Practice problem 2.6:
(a) amplitude spectrum; (b) phase spectrum.

(b)

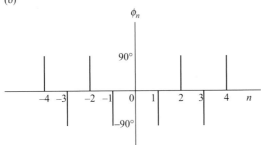

Figure 2.22. For Example 2.7

PRACTICE PROBLEM 2.6

Determine the exponential Fourier series expansion of the periodic function $f(t) = t$, $-1 < t < 1$ with $f(t + 2n) = f(t)$. Show the exponential spectra.

Answer: $f(t) = \sum_{\substack{n=-\infty \\ n \neq 0}}^{\infty} \frac{j(-1)^n}{n\pi} e^{jn\pi t}$. The exponential spectra are shown in Figure 2.21.

EXAMPLE 2.7

Determine the power of the periodic signal in Figure 2.22 and show the power spectrum.

Solution

Notice that $T = 4$ so that $\omega = 2\pi/T = \pi/2$. Hence the average power is

$$P = \frac{1}{T} \int_{-T/2}^{T/2} g^2(t)dt = \frac{1}{4} \int_{-1}^{1} 6^2 dt = \frac{36(2)}{4} = 18 \text{ W}$$

Table 2.3. **Exponential Fourier series coefficients for Example 2.9**

n	0	1	2	3	4	5
C_n	3	1.91	0	−0.636	0	0.382

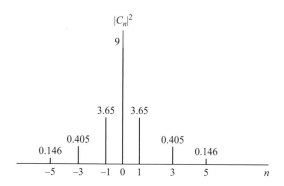

Figure 2.23. For Example 2.7; power spectrum of the signal in Figure 2.22.

$$C_n = \frac{1}{T}\int_0^T f(t)e^{-jn\omega_o t}\,dt = \frac{1}{4}\int_{-1}^1 6e^{-jn\pi t/2}\,dt = \frac{6}{4}\left[\frac{2}{-j\pi n}e^{-j\pi nt/2}\right]_{-1}^1 = \frac{6}{(-2j\pi n)}\left[e^{-jn\pi/2} - e^{jn\pi/2}\right]$$

But

$$\sin x = \frac{e^x - e^{-x}}{2j}$$

$$C_n = \frac{6}{n\pi}\sin\left(n\pi/2\right) = 3\,\mathrm{sinc}\,\left(n\pi/2\right)$$

where the sinc function is defined as $\mathrm{sinc}\,(x) = \frac{\sin x}{x}$. Table 2.3 shows some values of C_n for $n = 0, 1, 2, 3, 4, 5$. Using Parseval's theorem,

$$P = \sum_{n=-\infty}^{\infty} |C_n|^2 = C_0^2 + 2\sum_{n=1}^{\infty}|C_n|^2 = 3^2 + 2\left(1.91^2 + 0^2 + (-0.6366)^2 + 0^2 + 0.382^2 + \cdots\right)$$
$$= 17.4\text{ W}$$

which is 3.35% less than the exact value of 18 W. Greater accuracy can be achieved by taking more terms in the summation. The power spectrum is the plot of $|C_n|^2$ versus n, as in Figure 2.23.

PRACTICE PROBLEM 2.7

Obtain the power of the periodic signal in Figure 2.24. Sketch the power spectrum.

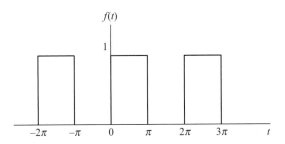

Figure 2.24. For Practice problem 2.7.

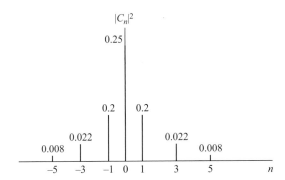

Figure 2.25. For Practice problem 2.7.

Answer: 0.5 W; see Figure 2.25 for the power spectrum.

2.7 FOURIER TRANSFORM

Fourier series analysis has limited applications because it is restricted to periodic power signals, while many signals of practical interest are nonperiodic energy signals.

However, besides their importance in many applications, Fourier series provide a foundation for Fourier transform. Fourier transform is a generalization of the complex Fourier series in the limit, as we will soon see. Fourier transform is by far the most commonly used tool in communications systems and virtually every field of science and engineering. It finds a wide range of applications in different fields such as signal processing, linear systems, electromagnetics, image analysis, filtering, spectroscopy, tomography, partial differential equations, quantum mechanics, and optics.

As a transition from Fourier series to Fourier transform, we assume a periodic signal and let its period approach infinity. Consider the exponential form of Fourier series in Eq. (2.19), namely

$$f(t) = \sum_{n=-\infty}^{\infty} c_n e^{jn\omega_o t} \tag{2.25}$$

where

$$c_n = \frac{1}{T} \int\limits_{-T/2}^{T/2} f(t)e^{-jn\omega_0 t}\,dt \tag{2.26}$$

The spacing between adjacent harmonics is

$$\Delta\omega = (n+1)\omega_0 - n\omega_0 = \omega_0 = \frac{2\pi}{T} \tag{2.27}$$

We substitute Eq. (2.26) into Eq. (2.25) to get

$$
\begin{aligned}
f(t) &= \sum_{n=-\infty}^{\infty} \left[\frac{1}{T} \int\limits_{-T/2}^{T/2} f(t)e^{-jn\omega_0 t}\,dt \right] e^{jn\omega_0 t} \\
&= \sum_{n=-\infty}^{\infty} \left[\frac{\Delta\omega}{2\pi} \int\limits_{-T/2}^{T/2} f(t)e^{-jn\omega_0 t}\,dt \right] e^{jn\omega_0 t} \\
&= \frac{1}{2\pi} \sum_{n=-\infty}^{\infty} \left[\int\limits_{-T/2}^{T/2} f(t)e^{-jn\omega_0 t}\,dt \right] \Delta\omega e^{jn\omega_0 t}
\end{aligned} \tag{2.28}
$$

As we let $T \to \infty$, the lines in the spectrum frequency come closer and closer so that $\Delta\omega$ becomes the differential frequency increment $d\omega$, the summation becomes integration, and the harmonic frequency $n\omega_0$ takes the values of frequency ω. Thus, Eq. (2.28) becomes

$$f(t) = \frac{1}{2\pi} \int\limits_{-\infty}^{\infty} \left[\int\limits_{-\infty}^{\infty} f(t)e^{-j\omega t}\,dt \right] e^{j\omega t}\,d\omega \tag{2.29}$$

The inner integral is known as the *Fourier transform* of $x(t)$ and is represented by $F(\omega)$, i.e.

$$F(\omega) = \mathcal{F}[f(t)] = \int\limits_{-\infty}^{\infty} f(t)e^{-j\omega t}\,dt \tag{2.30}$$

where $\mathcal{F}$ is the Fourier transform operator.

The **Fourier transform** of a signal $f(t)$ is the integration of the product of $f(t)$ and $e^{-j\omega t}$ over the interval from $-\infty$ to $+\infty$.

$F(\omega)$ is an integral transformation of $f(t)$ from the time domain to the frequency domain and is generally a complex function. $F(\omega)$ is known as the *spectrum* of $f(t)$. The plot of its magnitude $|F(\omega)|$ versus ω is called the *amplitude spectrum*, while that of its phase $\angle F(\omega)$ versus ω is called the *phase spectrum*. If a signal is real-valued, its magnitude spectrum is even, i.e. $|F(\omega)| = |F(-\omega)|$ and its phase spectrum is odd, i.e. $\angle F(\omega) = -\angle F(-\omega)$. Both the amplitude and the phase spectra provide the physical interpretation of the Fourier transform.

We can write Eq. (2.30) in terms of $F(\omega)$ and we obtain the *inverse Fourier transform* as

$$f(t) = \mathcal{F}^{-1}[F(\omega)] = \frac{1}{2\pi} \int_{-\infty}^{\infty} F(\omega)e^{j\omega t} d\omega \tag{2.31}$$

We say that the signal $f(t)$ and its transform $F(\omega)$ form a Fourier transform pair and denote their relationship by:

$$f(t) \quad \Leftrightarrow \quad F(\omega) \tag{2.32}$$

We will denote signals by lowercase letters, and their transforms in uppercase letters.

Since the Fourier transform is developed from the Fourier series, it follows that the conditions for its existence follow from those of the Fourier series, namely, Dirichlet conditions. Specifically, the Fourier transform $F(\omega)$ exists when the Fourier integral in Eq. (2.30) converges (i.e. exists). The Dirichlet conditions are:

1. $f(t)$ is bounded;
2. $f(t)$ has a finite number of maxima and minima;
3. $f(t)$ has a finite number of discontinuities;
4. $f(t)$ is integrable, i.e.

$$\int_{-\infty}^{\infty} |f(t)| dt < \infty \tag{2.33}$$

These are the sufficient conditions for $f(t)$ so that its Fourier transform exists. For example, functions such as $tu(t)$ (ramp function) and $f(t) = e^t, t \geq 0$ do not have Fourier transforms because they do not satisfy the conditions above. However, any signal that is a power or energy signal has a Fourier transform.

Since $F(\omega)$ is a complex-valued function, we avoid the complex algebra involved by temporarily replacing $j\omega$ with s and then replacing s with $j\omega$ at the end.

EXAMPLE 2.8
Find the Fourier transform of the following functions: (a) $\delta(t)$, (b) $e^{j\omega_0 t}$, (c) $\sin \omega_0 t$, (d) $e^{-at}u(t)$.

Solution

(a) For the impulse function,

$$F(\omega) = \mathcal{F}[\delta(t)] = \int_{-\infty}^{\infty} \delta(t)e^{-j\omega t} dt = e^{-j\omega t}\Big|_{t=0} = 1 \tag{2.8.1}$$

where the shifting property of the impulse function has been applied. Thus,

$$\mathcal{F}[\delta(t)] = 1 \tag{2.8.2}$$

This shows that the magnitude of the spectrum of the impulse function is constant; that is, all frequencies are equally represented in the impulse function.

(b) From Eq. (2.8.2)

$$\delta(t) = \mathcal{F}^{-1}[1]$$

Using the inverse Fourier transform formula in Eq. (2.31),

$$\delta(t) = \mathcal{F}^{-1}[1] = \frac{1}{2\pi} \int\limits_{-\infty}^{\infty} 1 e^{j\omega t} d\omega$$

or

$$\int\limits_{-\infty}^{\infty} e^{j\omega t} d\omega = 2\pi\delta(t) \tag{2.8.3}$$

Interchanging variables t and ω results in

$$\int\limits_{-\infty}^{\infty} e^{j\omega t} dt = 2\pi\delta(\omega) \tag{2.8.4}$$

Using this result, the Fourier transform of the given function is

$$\mathcal{F}\left[e^{j\omega t}\right] = \int\limits_{-\infty}^{\infty} e^{j\omega_o t} e^{-j\omega t} dt = \int\limits_{-\infty}^{\infty} e^{j(\omega_o - \omega)t} dt = 2\pi\delta(\omega_0 - \omega)$$

Since the impulse function is an even function, $\delta(\omega_0 - \omega) = \delta(\omega - \omega_0)$,

$$\mathcal{F}\left[e^{j\omega_0 t}\right] = 2\pi\delta(\omega - \omega_0) \tag{2.8.5}$$

By simply changing the sign of ω_0, we readily obtain

$$\mathcal{F}\left[e^{-j\omega_0 t}\right] = 2\pi\delta(\omega + \omega_0) \tag{2.8.6}$$

Also, by setting $\omega_0 = 0$,

$$\mathcal{F}[1] = 2\pi\delta(\omega) \tag{2.8.7}$$

(c) By using the result in Eqs. (2.8.5) and (2.8.6), we get

$$\mathcal{F}[\sin \omega_0 t] = \mathcal{F}\left[\frac{e^{j\omega_0 t} + e^{-j\omega_0 t}}{2j}\right]$$

$$= \frac{1}{2j}\mathcal{F}\left[e^{j\omega_0 t}\right] - \frac{1}{2j}\mathcal{F}\left[e^{-j\omega_0 t}\right] \tag{2.8.8}$$

$$= j\pi[\delta(\omega + \omega_0) - \delta(\omega - \omega_0)]$$

(d) Let $x(t) = e^{-at}u(t) = \begin{cases} e^{-at}, & t > 0 \\ 0, & t < 0 \end{cases}$

$$X(\omega) = \int_{-\infty}^{\infty} x(t)e^{-j\omega t}\,dt = \int_{0}^{\infty} e^{-at}e^{-j\omega t}\,dt = \int_{0}^{\infty} e^{-(a+j\omega)t}\,dt$$

(2.8.9)

$$\mathcal{L}[e^{-at}u(t)] = X(\omega) = \left.\frac{-1}{a+j\omega}e^{-(a+j\omega)t}\right|_{0}^{\infty} = \frac{1}{a+j\omega}$$

PRACTICE PROBLEM 2.8

Determine the Fourier transform of the following functions: (a) the rectangular pulse $\Pi(t/\tau)$, (b) $\delta(t + 3)$, (c) $2\cos\omega_0 t$.

Answers: (a) $\frac{2}{\omega}\sin\frac{\omega\tau}{2} = \tau\sin c\frac{\omega\tau}{2}$, (b) $e^{j3\omega}$, (c) $2\pi[\delta(\omega + \omega_0) - \delta(\omega - \omega_0)]$

EXAMPLE 2.9

Obtain the Fourier transform of the signal shown in Figure 2.26.

Solution

$$X(\omega) = \int_{-\infty}^{\infty} x(t)e^{-j\omega t}\,dt = \int_{-1}^{0}(-A)e^{-j\omega t}\,dt + \int_{0}^{1}Ae^{-j\omega t}\,dt$$

$$= \left.\frac{A}{j\omega}\,e^{-j\omega t}\right|_{-1}^{0} - \left.\frac{A}{j\omega}\,e^{-j\omega t}\right|_{0}^{1}$$

$$= \frac{-jA}{\omega}\left(1 - e^{j\omega} - e^{-j\omega} + 1\right)$$

$$= \frac{-j2A}{\omega}\left(1 - \cos\omega\right)$$

PRACTICE PROBLEM 2.9

Derive the Fourier transform of the triangular pulse defined as

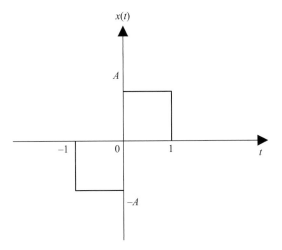

$x(t)$

A

-1 0 1 t

$-A$

Figure 2.26. For Example 2.9.

$$A(t/\tau) = \begin{cases} 1 - \dfrac{|t|}{\tau}, & |t| \leq \tau \\ 0, & |t| > \tau \end{cases}$$

Answer: $\tau \sin c^2 \left(\dfrac{\omega \tau}{2}\right)$

EXAMPLE 2.10

Find the Fourier transform of the two-sided exponential pulse shown in Figure 2.27. Sketch the transform.

Solution

Let $f(t) = e^{-a|t|} = \begin{cases} e^{at}, & t < 0 \\ e^{-at}, & t > 0 \end{cases}$

The Fourier transform is

$$F(\omega) = \int_{-\infty}^{\infty} f(t) e^{-j\omega t} dt = \int_{-\infty}^{0} e^{at} e^{-j\omega t} dt + \int_{0}^{\infty} e^{-at} e^{-j\omega t} dt$$

$$= \frac{1}{a - j\omega} + \frac{1}{a + j\omega}$$

$$= \frac{2a}{a^2 + \omega^2}$$

$F(\omega)$ is real in this case and it is sketched in Figure 2.28.

PRACTICE PROBLEM 2.10

Determine the Fourier transform of the signum function in Figure 2.29, i.e.

$$f(t) = \text{sgn}(t) = \begin{cases} 1, & t > 0 \\ -1, & t < 0 \end{cases}$$

Sketch $|F(\omega)|$.

Answer: See Figure 6.30.

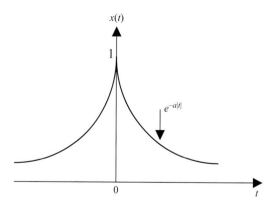

Figure 2.27. For Example 2.10.

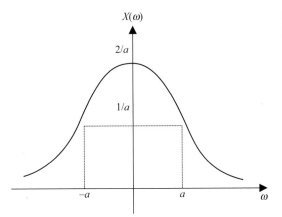

Figure 2.28. Fourier transform of $x(t)$ in Figure 2.27; for Example 2.10.

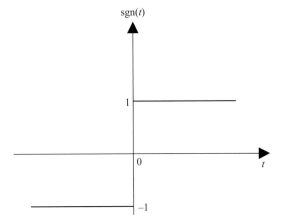

Figure 2.29. The signum function of Practice problem 2.10.

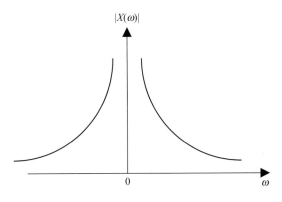

Figure 2.30. Fourier transform of the signal in Figure 2.29.

Answer: $\frac{2}{j\omega}$. See Figure 2.30.

2.8 PROPERTIES OF THE FOURIER TRANSFORM

In this section, we will develop some of the important properties of the Fourier transform and show how they are used in finding the transforms of complicated functions from the transforms of simple functions. For each of the properties, we follow this same pattern: we first state it, derive it, and then illustrate it with an example.

2.8.1 Linearity

Since integrals are linear operators, the linearity property holds for the Fourier transform just as it holds for Laplace transform. If $F_1(\omega)$ and $F_2(\omega)$ are the Fourier transforms of $f_1(t)$ and $f_2(t)$ respectively, then

$$\boxed{\mathcal{F}[a_1 f_1(t) + a_2 f_2(t)] = a_1 F_1(\omega) + a_2 F_2(\omega)} \tag{2.34}$$

where a_1 and a_2 are constants. This property states that the Fourier transform of a linear combination of functions is the same linear combination of the transform of the individual functions. By definition,

$$
\begin{aligned}
\mathcal{F}[a_1 f_1(t) + a_2 f_2(t)] &= \int_{-\infty}^{\infty} [a_1 f_1(t) + a_2 f_2(t)] e^{-j\omega t} dt \\
&= \int_{-\infty}^{\infty} a_1 f_1(t) e^{-j\omega t} dt + \int_{-\infty}^{\infty} a_2 f_2(t) e^{-j\omega t} dt \\
&= a_1 F_1(\omega) + a_2 F_2(\omega)
\end{aligned}
\tag{2.35}
$$

This can be extended to a linear combination of an arbitrary number of signals.

For example, $\cos \omega_0 t = \frac{1}{2}(e^{j\omega_0 t} + e^{-j\omega_0 t})$. Using the linearity property,

$$
\begin{aligned}
\mathcal{F}[\cos \omega_0 t] &= \frac{1}{2}\left[\mathcal{F}(e^{j\omega_0 t}) + \mathcal{F}(e^{-j\omega_0 t})\right] \\
&= \pi[\delta(\omega - \omega_0) + \delta(\omega + \omega_0)]
\end{aligned}
\tag{2.36}
$$

where we have applied Eqs. (2.8.5) and (2.8.6) in Example 2.8.

2.8.2 Time scaling

Let $F(\omega) = \mathcal{F}[f(t)]$ and a be a real constant. Then

$$\boxed{\mathcal{F}[f(at)] = \frac{1}{|a|} F\left(\frac{\omega}{a}\right)} \tag{2.37}$$

resulting in a new frequency ω/a. Equation (2.37) implies that expansion in one domains leads to compression in the other domain and vice versa. To establish the time-scaling property, we note that by definition,

$$\mathcal{F}[f(at)] = \int_{-\infty}^{\infty} f(at)e^{-j\omega t}\,dt$$

Letting $\lambda = at$, $d\lambda = adt$ so that

$$\mathcal{F}[f(at)] = \int_{-\infty}^{\infty} f(\lambda)e^{-j\omega\lambda/a}\frac{d\lambda}{a} = \frac{1}{a}F\left(\frac{\omega}{a}\right) \tag{2.38}$$

Notice that if $a = -1$, then Eq. (2.37) becomes

$$\mathcal{F}[f(-t)] = F(-\omega) = F^*(\omega) \tag{2.39}$$

where the asterisk stands for the complex conjugate. This is known as time reversal.
For example, for the rectangular pulse $p(t) = A\Pi(t/\tau)$ in Practice problem 2.8,

$$\mathcal{F}[p(t)] = A\tau\sin c\frac{\omega\tau}{2} \tag{2.40}$$

Using Eq. (2.37),

$$\mathcal{F}[p(2t)] = \frac{A\tau}{2}\sin c\frac{\omega\tau}{4} \tag{2.41}$$

The frequency scaling property states

$$\frac{1}{|a|}\mathcal{F}[f(t/a)] = F(a\omega) \tag{2.42}$$

2.8.3 Time shifting

If $F(\omega) = \mathcal{F}[f(t)]$ and t_o is a constant, then

$$\boxed{\mathcal{F}[f(t - t_o)] = e^{-j\omega t_o}F(\omega)} \tag{2.43}$$

This implies that a delay or time shift in the time domain implies a phase shift in the frequency domain. To find the Fourier transform of a shifted signal, we multiply the Fourier transform of the original signal by $e^{-j\omega t_o}$. Only the phase is affected by time shifting; the magnitude does not change. To derive this property, we note that by definition

$$\mathcal{F}[f(t - t_o)] = \int_{-\infty}^{\infty} f(t - t_o)e^{-j\omega t}\,dt \tag{2.44}$$

Let $\lambda = t - t_o$, $d\lambda = dt$, and $t = \lambda + t_o$, so that

$$\mathcal{F}[f(t - t_o)] = \int_{-\infty}^{\infty} f(\lambda)e^{-j\omega(\lambda + t_o)}\,d\lambda$$

$$= e^{-j\omega t_o}\int_{-\infty}^{\infty} f(\lambda)e^{-j\omega\lambda}\,d\lambda = e^{-j\omega t_o}X(\omega) \tag{2.45}$$

By following similar steps,

$$\mathcal{F}[f(t + t_o)] = e^{j\omega t_o} F(\omega)$$

For example, from Example 2.8(d),

$$\mathcal{F}[e^{-at} u(t)] = \frac{1}{a + j\omega} \tag{2.46}$$

We obtain the transform of $x(t) = e^{-a(t-3)} u(t - 3)$ as

$$\mathcal{F}\left[e^{-a(t-3)} u(t - 3)\right] = \frac{e^{-j3\omega}}{a + j\omega} \tag{2.47}$$

2.8.4 Frequency shifting

This is the dual of the time-shifting property. This property states that if $X(\omega) = \mathcal{F}[x(t)]$ and ω_0 is constant, then

$$\boxed{\mathcal{F}[f(t) e^{j\omega_0 t}] = F(\omega - \omega_0)} \tag{2.48}$$

This means that a shifting in the frequency domain is equivalent to a phase shift in the time domain. By definition,

$$\mathcal{F}[f(t) e^{j\omega_0 t}] = \int_{-\infty}^{\infty} f(t) e^{j\omega_0 t} e^{-j\omega t} dt$$

$$= \int_{-\infty}^{\infty} f(t) e^{-j(\omega - \omega_0)t} dt = F(\omega - \omega_0) \tag{2.49}$$

For example, since $\cos \omega_0 t = \frac{1}{2}(e^{j\omega_0 t} + e^{-j\omega_0 t})$. Applying the property in Eq. (2.48) yields

$$\mathcal{F}[f(t) \cos \omega_0 t] = \frac{1}{2} \mathcal{F}\left[f(t) e^{j\omega_0 t}\right] + \frac{1}{2} \mathcal{F}\left[f(t) e^{-j\omega_0 t}\right]$$

$$= \frac{1}{2} F(\omega - \omega_0) + \frac{1}{2} F(\omega + \omega_0) \tag{2.50}$$

This is an important result in modulation. For example, if the amplitude spectrum of $f(t)$ is as shown in Figure 2.31(a), then the corresponding amplitude spectrum of $f(t) \cos \omega_0 t$ will be as shown in Figure 2.31(b).

2.8.5 Time differentiation

If $F(\omega) = \mathcal{F}[f(t)]$, then the time differential property gives

$$\boxed{\mathcal{F}[f'(t)] = j\omega F(\omega)} \tag{2.51}$$

(a)

(b)

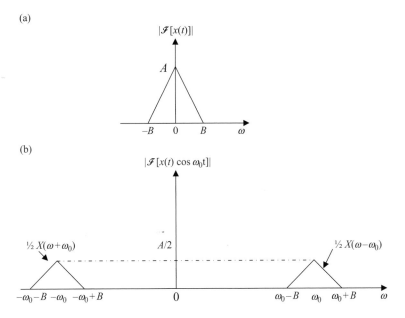

Figure 2.31. Amplitude spectra of: (a) signal $f(t)$; (b) modulated signal $x(t) \cos \omega_0 t$.

This states that the transform of the derivative of $f(t)$ is obtained by multiplying its transform $F(\omega)$ by $j\omega$. By definition,

$$f(t) = \mathcal{F}^{-1}[F(\omega)] = \frac{1}{2\pi} \int_{-\infty}^{\infty} F(\omega)e^{j\omega t} d\omega \tag{2.52}$$

Taking the derivative of both sides with respect to t gives

$$\frac{df(t)}{dt} = \frac{j\omega}{2\pi} \int_{-\infty}^{\infty} F(\omega)e^{j\omega t} d\omega = j\omega \mathcal{F}^{-1}[F(\omega)]$$

or

$$\mathcal{F}[f'(t)] = j\omega F(\omega) \tag{2.53}$$

Repeated applications of Eq. (2.51) give the Fourier transform of the nth derivative of $x(t)$ as

$$\mathcal{F}\left[f^{(n)}(t)\right] = (j\omega)^n F(\omega) \tag{2.54}$$

For example, if $f(t) = e^{-at}u(t)$, then

$$f'(t) = -ae^{-at}u(t) + \delta(t) = -af(t) + \delta(t) \tag{2.55}$$

where the $\delta(t)$ accounts for the discontinuity at $t = 0$. Taking the Fourier transform of the first and the last terms, we obtain

$$j\omega F(\omega) = -aF(\omega) + 1 \quad \Rightarrow \quad F(\omega) = \frac{1}{a + j\omega} \tag{2.56}$$

which agrees with what we got in Example 2.8(d).

2.8.6 Frequency differentiation

This property states that if $F(\omega) = \mathcal{F}[f(t)]$, then

$$\boxed{\mathcal{F}[(-jt)^n f(t)] = \frac{d^n}{d\omega^n} F(\omega)} \tag{2.57}$$

This property is also called multiplication by a power of t. We establish this by using the basic definition of the Fourier transform.

$$\begin{aligned}
\frac{d^n}{d\omega^n} X(\omega) &= \frac{d^n}{d\omega^n} \left(\int_{-\infty}^{\infty} f(t) e^{-j\omega t} dt \right) = \int_{-\infty}^{\infty} f(t) \frac{d^n}{d\omega^n} e^{-j\omega t} dt \\
&= \int_{-\infty}^{\infty} f(t)(-jt)^n e^{-j\omega t} dt = \int_{-\infty}^{\infty} (-jt)^n f(t) e^{-j\omega t} dt \\
&= \mathcal{F}((-jt)^n f(t))
\end{aligned} \tag{2.58}$$

For example, from Practice problem 2.8,

$$\mathcal{L}[\Pi(t/\tau)] = \tau \sin c \frac{\omega\tau}{2} \tag{2.59}$$

Letting $n = 1$

$$\begin{aligned}
\mathcal{L}[-jt\Pi(t/\tau)] &= \frac{d}{d\omega} \tau \sin c \frac{\omega\tau}{2} = \tau \frac{d}{d\omega} \left(\frac{\sin \omega\tau/2}{\omega\tau/2} \right) \\
&= \tau \frac{\omega\tau/2[\tau/2 \cos(\omega\tau/2)] - \tau/2 \sin(\omega\tau/2)}{(\omega\tau/2)^2} \\
&= \frac{\omega\tau/2 \cos(\omega\tau/2) - \sin(\omega\tau/2)}{\omega^2/2}
\end{aligned} \tag{2.60}$$

2.8.7 Time integration

If $F(\omega) = \mathcal{F}[f(t)]$, then

$$\boxed{\mathcal{F}\left[\int_{-\infty}^{t} f(t) dt \right] = \frac{F(\omega)}{j\omega} + \pi F(0)\delta(\omega)} \tag{2.61}$$

This states that the transform of the integral of $x(t)$ is obtained by dividing the transform of $x(t)$ by $j\omega$ plus the impulse term that reflects the dc component $X(0)$. If we replace ω by 0 in Eq. (2.30),

$$F(0) = \int_{-\infty}^{\infty} f(t)dt \tag{2.62}$$

indicating that the dc component is zero when the integral of $x(t)$ over all time vanishes. The time integration property in Eq. (2.61) will be proved later when we consider the convolution property.

For example, we know from Example 2.8(a) that $F[\delta(t)] = 1$ and that integrating the impulse function gives the unit step function $u(t)$. By applying Eq. (2.61),

$$\mathcal{F}[u(t)] = \mathcal{F}\left[\int_{-\infty}^{t} \delta(t)dt\right] = \frac{1}{j\omega} + \pi\delta(\omega) \tag{2.63}$$

2.8.8 Duality

The duality property states that if $F(\omega)$ is the Fourier transform of $f(t)$, then the Fourier transform of $f(t)$ is $2\pi f(-\omega)$; i.e.

$$\mathcal{F}[f(t)] = F(\omega) \quad \Rightarrow \quad \mathcal{F}[F(t)] = 2\pi f(-\omega) \tag{2.64}$$

This expresses the fact that the Fourier transform pairs are symmetric. To derive the property, we recall from Eq. (2.31) that

$$f(t) = \mathcal{F}^{-1}[F(\omega)] = \frac{1}{2\pi}\int_{-\infty}^{\infty} F(\omega)e^{j\omega t}d\omega \tag{2.65}$$

or

$$2\pi f(t) = \int_{-\infty}^{\infty} F(\omega)e^{j\omega t}d\omega \tag{2.66}$$

Replacing t with $-t$ gives

$$2\pi f(-t) = \int_{-\infty}^{\infty} F(\omega)e^{-j\omega t}d\omega \tag{2.67}$$

If we interchange t and ω, we obtain

$$2\pi f(-\omega) = \int_{-\infty}^{\infty} F(t)e^{-j\omega t}dt = \mathcal{F}[F(t)] \tag{2.68}$$

as expected.

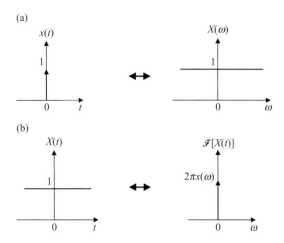

Figure 2.32. A typical illustration of the duality property of the Fourier transform: (a) transform of impulse; (b) transform of unit dc level.

For example, if $f(t) = e^{-|t|}$, then

$$F(\omega) = \frac{2}{\omega^2 + 1} \tag{2.69}$$

By the duality property, the Fourier transform of $F(t) = \frac{2}{t^2+1}$ is

$$2\pi f(-\omega) = 2\pi f(\omega) = 2\pi e^{-|\omega|} \tag{2.70}$$

Figure 2.32 illustrates another example of the duality property. If $f(t) = \delta(t)$ so that $F(\omega) = 1$, as in Figure 2.32(a), then the Fourier transform of $f(t) = 1$ is $2\pi f(\omega)$ as in Figure 2.32(b).

2.8.9 Convolution

If $x(t)$ and $h(t)$ are two signals, their convolution $y(t)$ is given by the convolution integral

$$y(t) = h(t) * x(t) = \int_{-\infty}^{\infty} h(\tau)x(t-\tau)d\tau \tag{2.71}$$

If $X(\omega)$, $H(\omega)$, and $Y(\omega)$ are the Fourier transforms of $x(t)$, $h(t)$, and $y(t)$ respectively, then

$$Y(\omega) = \mathcal{F}[h(t) * x(t)] = H(\omega)X(\omega) \tag{2.72}$$

which states that convolution of the time-domain signals corresponds to multiplying their Fourier transforms.

To derive the convolution property, we take the Fourier transform of both sides of Eq. (2.71) and obtain

$$Y(\omega) = \int_{-\infty}^{\infty} \left[\int_{-\infty}^{\infty} h(\tau)x(t-\tau)d\tau \right] e^{-j\omega t} dt \tag{2.73}$$

Table 2.4. **Properties of the Fourier transform**

Property	$f(t)$	$F(\omega)$
Linearity	$a_1 f_1(t) + a_2 f_2(t)$	$a_1 F_1(\omega) + a_2 F_2(\omega)$
Scaling	$f(at)$	$\frac{1}{\|a\|} F\left(\frac{\omega}{a}\right)$
Time shift	$f(t-a)u(t-a)$	$e^{-j\omega a} F(\omega)$
Frequency shift	$e^{j\omega_o t} f(t)$	$F(\omega - \omega_o)$
Modulation	$\cos(\omega_o t) f(t)$	$\frac{1}{2}[F(\omega + \omega_o) + F(\omega - \omega_o)]$
Time differentiation	$\frac{df}{dt}$	$j\omega F(\omega)$
	$\frac{d^n f}{dt^n}$	$(j\omega)^n F(\omega)$
Time integration	$\int_{-\infty}^{t} f(t) dt$	$\frac{F(\omega)}{j\omega} + \pi F(0)\delta(\omega)$
Frequency differentiation	$t^n f(t)$	$(j)^n \frac{d^n}{d\omega^n} F(\omega)$
Time reversal	$f(-t)$	$F(-\omega)$ or $F^*(\omega)$
Duality	$F(t)$	$2\pi f(-\omega)$
Time convolution	$f_1(t) * f_2(t)$	$F_1(\omega) F_2(\omega)$
Frequency convolution	$f_1(t) f_2(t)$	$\frac{1}{2\pi} F_1(\omega) * F_2(\omega)$

If we change the order of integration and factor out $h(\tau)$ since it does not depend on t, we get

$$Y(\omega) = \int_{-\infty}^{\infty} h(\tau) \left[\int_{-\infty}^{\infty} x(t-\tau) e^{-j\omega t} dt \right] d\tau$$

The inner integral can be simplified by letting $\lambda = t - \tau$ so that $t = \lambda + \tau$ and $dt = d\lambda$.

$$Y(\omega) = \int_{-\infty}^{\infty} h(\tau) \left[\int_{-\infty}^{\infty} x(\lambda) e^{-j\omega(\lambda + \tau)} d\lambda \right] d\tau$$

$$= \int_{-\infty}^{\infty} h(\tau) e^{-j\omega\tau} d\tau \int_{-\infty}^{\infty} x(\lambda) e^{-j\omega\lambda} d\lambda = H(\omega) X(\omega)$$

(2.74)

as expected.

These properties of the Fourier transform are listed in Table 2.4. The transform pairs of some common functions are presented in Table 2.5.

Table 2.5. **Fourier transform pairs**

$f(t)$	$F(\omega)$	Power or energy signal
$\delta(t)$	1	Power
1	$2\pi\delta(\omega)$	Power
$u(t)$	$\pi\delta(\omega)+\frac{1}{j\omega}$	Power
$u(t+\tau)-u(t-\tau)$	$2\frac{\sin\omega\tau}{\omega}$	Energy
$|t|$	$\frac{-2}{\omega^2}$	Neither
$\operatorname{sgn}(t)$	$\frac{2}{j\omega}$	Power
$e^{-at}u(t)$	$\frac{1}{a+j\omega}$	Energy
$e^{at}u(-t)$	$\frac{1}{a-j\omega}$	Energy
$t^n e^{-at}u(t)$	$\frac{n!}{(a+j\omega)^{n+1}}$	Neither
$e^{-a|t|}$	$\frac{2a}{a^2+\omega^2}$	Energy
$e^{j\omega_o t}$	$2\pi\delta(\omega-\omega_o)$	Power
$\sin\omega_o t$	$j\pi[\delta(\omega+\omega_o)-\delta(\omega-\omega_o)]$	Power
$\cos\omega_o t$	$\pi[\delta(\omega+\omega_o)+\delta(\omega-\omega_o)]$	Power
$e^{-at}\sin\omega_o t\,u(t)$	$\frac{\omega_o}{(a+j\omega)^2+\omega_o^2}$	Energy
$e^{-at}\cos\omega_o t\,u(t)$	$\frac{a+j\omega}{(a+j\omega)^2+\omega_o^2}$	Energy
$\Pi\left(\frac{t}{\tau}\right)=\begin{cases}1, & |t|<\tau/2\\0, & |t|>\tau/2\end{cases}$	$\tau\operatorname{sinc}\left(\frac{\omega\tau}{2}\right)$	Energy
$\Delta\left(\frac{t}{\tau}\right)=\begin{cases}1-|t|/\tau, & |t|<\tau\\0, & |t|>\tau\end{cases}$	$\tau\operatorname{sinc}^2\left(\frac{\omega\tau}{2}\right)$	Energy
$e^{-a^2 t^2}$	$e^{-\omega^2/4a^2}$	Energy
$\sum_{n=-\infty}^{\infty}f(t-nT)$	$\omega_o\sum_{n=-\infty}^{\infty}F(n\omega_o)(\omega-n\omega_o),\ \omega_o=\frac{2\pi}{T}$	Power
$\sum_{n=-\infty}^{\infty}\delta(t-nT)$	$\omega_o\sum_{n=-\infty}^{\infty}\delta(\omega-n\omega_o),\ \omega_o=\frac{2\pi}{T}$	Power

EXAMPLE 2.11

A signal $f(t)$ has a Fourier transform given by

$$F(\omega) = \frac{5(1 + j\omega)}{8 - \omega^2 + 6j\omega}$$

Without finding $f(t)$, find the Fourier transform of the following.

(a) $f(t - 3)$
(b) $f(4t)$
(c) $e^{-j2t}f(t)$
(d) $f(-2t)$

Solution

We apply the appropriate property for each case.

(a) $\mathcal{F}[f(t - 3)] = e^{-j\omega 3}F(\omega) = \dfrac{5(1 + j\omega)e^{-j\omega 3}}{8 - \omega^2 + j6\omega}$

(b) $\mathcal{F}[f(4t)] = \frac{1}{4}F\left(\frac{\omega}{4}\right) = \dfrac{\frac{5}{4}(1 + j\omega/4)}{8 - \omega^2/16 + j6\omega/4} = \dfrac{5(4 + j\omega)}{128 - \omega^2 + j24\omega}$

(c) $\mathcal{F}[e^{-j2t}f(t)] = F(\omega + 2) = \dfrac{5[1 + j(\omega + 2)]}{8 - (\omega + 2)^2 + 6j(\omega + 2)} = \dfrac{5(1 + j\omega + j2)}{4 - \omega^2 - 4\omega + 6j\omega + j12}$

(d) $\mathcal{F}[f(-2t)] = \frac{1}{2}F\left(\frac{\omega}{-2}\right) = \dfrac{\frac{5}{2}(1 - j\omega/2)}{8 - \frac{\omega^2}{4} - \frac{6j\omega}{2}} = \dfrac{5(2 - j\omega)}{32 - \omega^2 - 12j\omega}$

PRACTICE PROBLEM 2.11

A signal $x(t)$ has the Fourier transform

$$F(\omega) = \frac{9}{9 + \omega^2}$$

Find the Fourier transform of the following signals:

(a) $y(t) = f(2t - 1)$
(b) $z(t) = df(2t)/dt$
(c) $h(t) = \int_{-\infty}^{t} f(\lambda)d\lambda$

Answers: (a) $\dfrac{18e^{-j\omega/2}}{36 + \omega^2}$, (b) $\dfrac{18j\omega}{36 + \omega^2}$, (c) $\pi\delta(\omega) + \dfrac{9}{j\omega(9 + \omega^2)}$

EXAMPLE 2.12

Determine the Fourier transform of the signal in Figure 2.33.

Solution

Although the Fourier transform of $f(t)$ can be found directly using Eq. (2.30), it is much easier to find it using the derivative property. Taking the first derivative of $g(t)$ produces the signal in Figure 2.34(a). Taking the second derivative gives us the signal in Figure 2.34(b). From this,

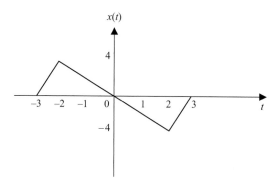

(a)

Figure 2.33. For Example 2.12.

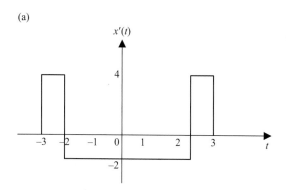

(b)

Figure 2.34. For Example 2.12: (a) first derivative of $x(t)$; (b) second derivative of $x(t)$.

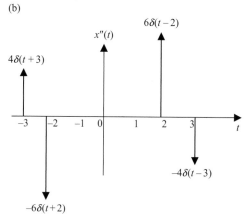

$$f''(t) = 4\delta(t+3) - 6\delta(t+2) + 6\delta(t-2) - 4\delta(t-3)$$

Taking the Fourier transform of each term, we obtain

$$(j\omega)^2 F(\omega) = 4e^{j3\omega} - 6e^{j2\omega} + 6e^{-j2\omega} - 4e^{-j3\omega}$$
$$-\omega^2 F(\omega) = 4(e^{j3\omega} - e^{-j3\omega}) + 6(e^{j2\omega} - e^{-j2\omega})$$
$$= j8\sin 3\omega - j12\sin 2\omega$$
$$F(\omega) = \frac{j}{\omega^2}(12\sin 2\omega - 8\sin 3\omega)$$

PRACTICE PROBLEM 2.12

Determine the Fourier transform of the function in Figure 2.35.

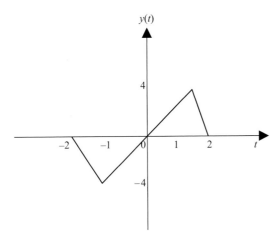

Figure 2.35 For Practice problem 2.12.

Answer: $Y(\omega) = \frac{j10}{\omega^2}(\sin 2\omega - 2\sin \omega)$

EXAMPLE 2.13

Find the inverse Fourier transform of:

(a) $G(\omega) = \dfrac{10j\omega}{(-j\omega + 2)(j\omega + 3)}$

(b) $Y(\omega) = \dfrac{\delta(\omega)}{(j\omega + 1)(j\omega + 2)}$

Solution

(a) To avoid complex algebra, let $s = j\omega$. Using partial fractions,

$$G(s) = \frac{10s}{(2 - s)(3 + s)} = \frac{-10s}{(s - 2)(s + 3)} = \frac{A}{s - 2} + \frac{B}{s + 3}, \quad s = j\omega$$

$$A = (s - 2)G(s)\Big|_{s=2} = \frac{-10(2)}{2 + 3} = -4$$

$$B = (s + 3)G(s)\Big|_{s=-3} = \frac{-10(-3)}{-3 - 2} = -6$$

$$G(\omega) = \frac{-4}{j\omega - 2} - \frac{6}{j\omega + 3}$$

Taking the inverse Fourier transform of each term,

$$g(t) = -4e^{2t}u(-t) - 6e^{-3t}u(t)$$

(b) Because of the delta function, we use Eq. (2.31) to find the inverse.

$$y(t) = \frac{1}{2\pi}\int_{-\infty}^{\infty} \frac{\delta(\omega)e^{j\omega t}\,d\omega}{(2 + j\omega)(j\omega + 1)} = \frac{1}{2\pi}\frac{e^{j\omega t}}{(2 + j\omega)(j\omega + 1)}\Big|_{\omega=0} = \frac{1}{2\pi}\frac{1}{2} = \frac{1}{4\pi}$$

where the sifting property has been applied.

PRACTICE PROBLEM 2.13

Obtain the inverse Fourier transform of:

(a) $F(\omega) = \dfrac{e^{-j2\omega}}{1+j\omega}$

(b) $G(\omega) = \dfrac{\pi\delta(\omega)}{(5+j\omega)(2+j\omega)}$

Answers: (a) $f(t) = e^{-(t-2)}u(t-2)$, (b) $g(t) = 0.05$

2.9 APPLICATIONS – FILTERS

Fourier analysis has several applications in communications. Such applications include filtering, sampling, and amplitude modulation. In this section, we limit ourselves to just filtering. Filtering is the process of separating a desired signal from unwanted signal (noise). It is a universal tool used in many engineering fields such as electronics, communication, and signal processing as well as in physics, biology, astronomy, economics, and finance. Communication systems in particular require extensive applications of filters.

A device or system that is capable of performing filtering is called a *filter*.

> A **filter** is a circuit or system that passes certain frequencies of the input signal but rejects or attenuates other frequencies.

Filters have been used in practical applications for more than eight decades. Filter technology feeds related areas such as equalizers, impedance matching networks, transformers, shaping networks, power dividers, attenuators, and directional couplers, and continually provides practicing engineers with opportunities to innovate and experiment.

As a linear system, a filter has an input $x(t)$, an output $y(t)$, and an input response $h(t)$. The three are related by the convolution integral, namely,

$$y(t) = h(t) * x(t) = \int_0^\infty x(\lambda)h(t-\lambda)d\lambda \tag{2.75}$$

$$\text{or} \quad Y(\omega) = \mathcal{F}[h(t) * x(t)] = H(\omega)X(\omega) \tag{2.76}$$

The transfer function of the filter is

$$H(\omega) = \frac{Y(\omega)}{X(\omega)} = |H(\omega)|\angle\theta \tag{2.77}$$

where $|H(\omega)|$ is the magnitude of H (also known as the amplitude response) and θ is the phase of H since H is generally complex.

As a filtering device, a filter is characterized by its *stopband* and *passband*. The passband of a filter is the frequency range that the filter passes with little or no attenuation, while the stopband is the range of frequencies that the filter does not pass (attenuates or eliminates).

As shown in Figure 2.36, there are four types of filter:

1. A *low-pass filter* passes low frequencies and stops high frequencies, as shown ideally in Figure 2.36(a), i.e

$$|H(\omega)| = \begin{cases} 1, & -B \leq \omega \leq B \\ 0, & \text{otherwise} \end{cases} \tag{2.78}$$

2. A *high-pass filter* passes high frequencies and rejects low frequencies, as shown ideally in Figure 2.36(b).

$$|H(\omega)| = \begin{cases} 0, & -B \leq \omega \leq B \\ 1, & \text{otherwise} \end{cases} \tag{2.79}$$

3. A *bandpass filter* passes frequencies within a frequency band and blocks or attenuates frequencies outside the band, as shown ideally in Figure 2.36(c)

$$|H(\omega)| = \begin{cases} 1, & B_1 \leq |\omega| \leq B_2 \\ 0, & \text{otherwise} \end{cases} \tag{2.80}$$

4. A *band-stop filter* passes frequencies outside a frequency band and blocks or attenuates frequencies within the band, as shown ideally in Figure 2.36(d).

$$|H(\omega)| = \begin{cases} 0 & B_1 \leq |\omega| \leq B_2 \\ 1, & \text{otherwise} \end{cases} \tag{2.81}$$

A filter is said to be ideal if it has a perfectly flat response within the desired frequency range and zero response outside that range. As is evident from Eqs. (2.78) to (2.81) as well as Figure 2.38, the magnitude of the transfer function of ideal filters $|H(\omega)| = 1$ in the passband and $|H(\omega)| = 0$ in the stopband. Unfortunately, ideal filters cannot be built with practical components such as resistors, inductors, and capacitors. This can be proved by taking the inverse Fourier transform of the ideal filters; we obtain impulse responses $h(t)$ which are noncausal and therefore not physically realizable. For this reason, the ideal filters are known as *unrealizable filters*. Physically realizable filters have an amplitude response $|H(\omega)|$ that varies gradually without abrupt transitions between passband and stopbands as in Figure 2.38. However, realizable filters whose characteristics approach those of ideal filters do exist.

Although to cover all types of practically realizable filters would require a whole book, we will attempt to discuss a standard filter type to gain some insight into how we can make realizable filters approach the behavior of ideal filters. Standard classes of filters include

(a)

(b)

(c)

(d)

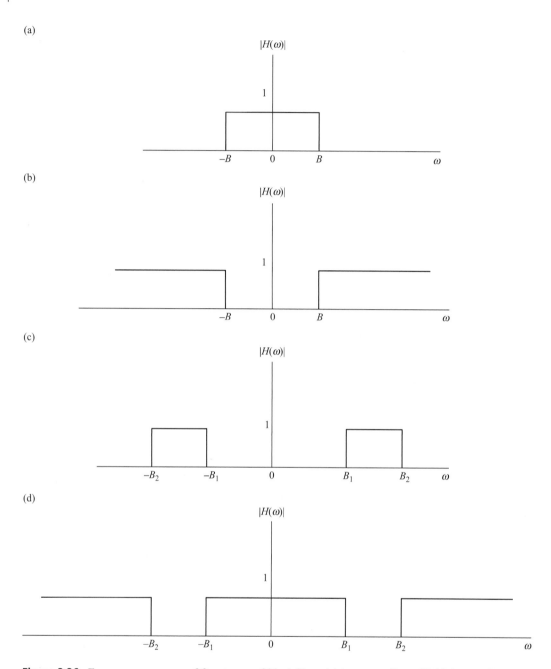

Figure 2.36. Frequency response of four types of ideal filter: (a) low-pass filter; (b) high-pass filter; (c) bandpass filter; (d) band-stop filter.

Butterworth, Chebyshev, elliptic, and Bessel filters. We consider only Butterworth low-pass filters, which belong to the simplest class of filters. We consider only low-pass filters because we can construct high-pass, bandpass, and band-stop filters from any low-pass filters by frequency transformation.

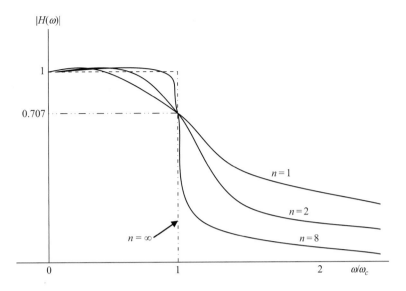

Figure 2.37. Magnitude response of Butterworth filters.

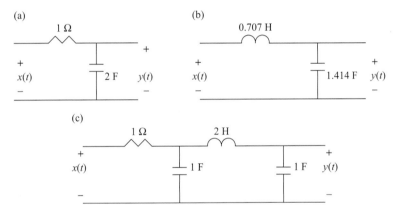

Figure 2.38. Typical *RLC* circuit realizations of Butterworth filters with $\omega_c = 1$: (a) first-order; (b) second-order; (c) third-order.

Butterworth filters are commonly used to meet specific design specifications and their characteristics are readily available and extensively tabulated. They are characterized by the fact that the square of the magnitude of the frequency response $H(\omega)$ is of the form

$$|H(\omega)|^2 = \frac{1}{1 + \left(\frac{\omega}{\omega_c}\right)^{2n}} \tag{2.82}$$

where n is the order of the filter or the order of the differential equation that will describe the transfer function in Eq. (2.69); n also corresponds to the number of storage elements (inductors and capacitors) required to implement the filter. The parameter $\omega_c = 2\pi f_c$ is the cutoff frequency,

Table 2.6. Coefficients of Butterworth polynomial

n	a_0	a_1	a_2	a_3	a_4	a_5	a_6
1	1						
2	1	1.414	1				
3	1	2	2	1			
4	1	2.613	3.414	2.613	1		
5	1	3.236	5.236	5.236	3.236	1	
6	1	3.864	7.464	9.141	7.464	3.864	1

which is the frequency at which the magnitude of the frequency response is $1/\sqrt{2}$ times its value at dc, i.e. $|H(\omega_c)| = |H(0)|/\sqrt{2}$. In other words, ω_c is the frequency $|H(\omega)|$ in dB which is down by 3 dB on the Bode plot. Figure 2.37 illustrates a plot of $|H(\omega)|$ for various values of n. It is evident from the plots that the Butterworth characteristic approaches that of the ideal filter when $n \to \infty$. Butterworth filters are known as having a *maximally flat* frequency response (the flattest possible curve) because the first $2n - 1$ derivatives of $|H(\omega)|$ are zero at the origin (dc or $\omega = 0$) for any given n.

Although Eq. (2.69) specifies the response magnitude of Butterworth filters, it does not provide how to construct or realize the filters. To do this requires a transfer function of the form

$$H(s) = \frac{K}{(s - p_0)(s - p_1)(s - p_2) \ldots (s - p_n)} \qquad (2.83)$$

where $s = j\omega$ and p_0 to p_n are the poles of the filter and K is a constant. Rather than having the denominator of the Butterworth transfer function in factored form as in Eq. (2.70), we can multiply the factors and obtain

$$H(s) = \frac{K}{a_0 s^n + a_1 \omega_c s^{n-1} + a_2 \omega_c^2 s^{n-2} + \cdots + a_{n-1} \omega_c^{n-1} s + a_n \omega_c^n} \qquad (2.84)$$

The coefficients a_0 to a_n (known as the coefficients of Butterworth polynomials or the denominator of $H(s)$) are listed in Table 2.6 for $n = 1$ to $n = 6$. The constant K can be determined by realizing that $H(s = 0) = 1$, i.e. the dc gain is unity.

Figure 2.38 shows typical realizations of the first-, second-, and third-order Butterworth filters. Notice that the first-order Butterworth filter is identical to an RC low-pass filter and would not achieve a "good enough" approximation to the ideal filter. The approximation improves and frequency response becomes "flat enough" in the passband as the order n increases by adding more storage elements.

To determine the order of a Butterworth filter, it is usually specified that the stopband begins at $\omega = \omega_s$ with a minimum attenuation of δ. From Eq. (2.82), we need at least

$$\delta^2 = \frac{1}{1 + \left(\frac{\omega_s}{\omega_c}\right)^{2n}} \quad \rightarrow \quad \left(\frac{\omega_s}{\omega_c}\right)^{2n} \geq \frac{1}{\delta^2} - 1$$

Taking the logarithm of both sides leads to

$$n \geq \frac{\log_{10}\left(\frac{1}{\delta^2} - 1\right)}{2 \log_{10}\left(\frac{\omega_s}{\omega_c}\right)} \tag{2.85}$$

For example, if $\delta = 0.001$ so that the stopband is at least $20 \log_{10}(0.001) = -60$ dB in magnitude and the stopband begins at $\omega = \omega_s = 3\omega_c$, then

$$n \geq \frac{\log_{10}(999999)}{2 \log_{10}(3)} = 6.288$$

Since n must be integer, $n = 7$ or higher should be selected.

EXAMPLE 2.14

A Butterworth filter is designed to have a gain of -40 dB at $\omega = 3\omega_c$. What must the order of the filter be? Obtain its transfer function.

Solution

$$-40 \text{ dB} = 20 \log_{10}|H| \quad \rightarrow \quad |H| = 10^{-40/20} = 0.01.$$

From Eq. (2.82),

$$|H(\omega)|^2 = \frac{1}{1 + \left(\frac{\omega}{\omega_c}\right)^{2n}} = (0.01)^2 \quad \rightarrow \quad 1 + 3^{2n} = 1000$$

or $3^{2n} = 999$

Taking the logarithm of both sides gives

$$2n \log_{10} 3 = \log_{10} 999 \quad \rightarrow \quad n = 3.1434$$

Since n must be an integer, a fourth-order filter is required. From Eq. (2.84), when $n = 4$

$$H(s) = \frac{K}{a_0 s^4 + a_1 \omega_c s^3 + a_2 \omega_c^2 s^2 + a_3 \omega_c^3 s + a_4 \omega_c^4}$$

From Table 2.4, we obtain the coefficients of the Butterworth filter. Since $a_4 = 1$,

$$H(0) = 1 = K/\omega_c^4 \quad \rightarrow \quad K = \omega_c^4$$

Hence,

$$H(s) = \frac{\omega_c^4}{s^4 + 2.613\omega_c s^3 + 3.414\omega_c^2 s^2 + 2.613\omega_c^3 s + \omega_c^4}$$

PRACTICE PROBLEM 2.14

If a third-order Butterworth filter is designed to have a gain of -20 dB at 50 MHz, determine the cutoff frequency of the filter and the corresponding transfer function.

Answer: 10.81 MHz, $H(s) = \frac{\omega_c^3}{s^3 + 2\omega_c s^2 + 2\omega_c^2 s + \omega_c^3}$, $\omega_c = 6.791 \times 10^7$ rad/s

EXAMPLE 2.15

The circuit in Figure 2.39 is to be designed as a second-order Butterworth filter with a cutoff frequency of 10 rad/s. Assuming $R = 1\ \Omega$, find L and C.

Solution

The given circuit in Figure 2.39 is second-order because two storage elements are involved. Using current division,

$$I_o = \frac{1/sC}{1/sC + R + sL} I = \frac{I}{1 + sC(R + sL)}$$

$$V = I_o R = \frac{RI}{1 + sRC + s^2 LC}$$

Hence the transfer function of the circuit is

$$H(s) = \frac{V(s)}{I(s)} = \frac{R}{1 + sRC + s^2 LC} = \frac{R/LC}{s^2 + sR/L + 1/LC} \tag{2.15.1}$$

From Eq. (2.83) and Table 2.6, the transfer function for the second-order Butterworth filter is

$$H(s) = \frac{K}{s^2 + 1.414\omega_c s + \omega_c^2} \tag{2.15.2}$$

Comparing Eqs. (2.15.1) and (2.15.2), we notice that

$$K = R/LC, \quad \omega_c^2 = 1/LC, \quad 1.414\omega_c = R/L$$

Given that $R = 1\ \Omega$ and $\omega_c = 10$ rad/s,

$$L = \frac{R}{1.414\omega_c} = \frac{1}{14.14} = 70.72\ \text{mH}$$

$$C = \frac{1}{\omega_c^2 L} = \frac{14.14}{100} = 141.4\ \text{mF}$$

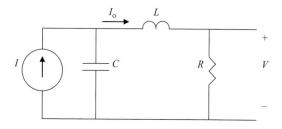

Figure 2.39. For Example 2.15.

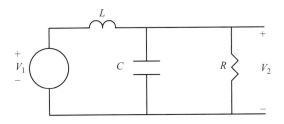

Figure 2.40. For Practice problem 2.15.

PRACTICE PROBLEM 2.15

Given the *RLC* circuit in Figure 2.40 in which $R = 1\ \Omega$, $\omega_c = 100$ rad/s, find L and C such that $H = V_2/V_1$ produces a Butterworth frequency response.

Answer: $L = 1.414$ H and $C = 7.07$ mF

2.10 COMPUTATION USING MATLAB

MATLAB is a software package that is used throughout this book. It is particularly useful for signal analysis. A review of MATLAB is provided in Appendix B for a beginner. This section shows how to use the software to numerically perform most of the operations we had in this chapter. Those operations include plotting, Fourier analysis, and filtering. MATLAB has the **fft** command for the discrete fast Fourier transform (FFT).

2.10.1 Plotting a signal

The MATLAB command **plot** can be used to plot $x(t)$. For example, to plot

$$x(t) = 2e^{-t} + 4\cos(3t - \pi/6), \qquad 1 < t < 2$$

we use the following MATLAB statements:

```
» t=1:0.001:2;
» x=2*exp(-t) + 4*cos(3*t - pi/6);
» plot(t,x)
```

where an increment or step size 0.001 is selected. In MATLAB, *t* and *x* are taken as vectors and must be of the same size in order to plot them. MATLAB has no command for finding the Fourier transform $F(\omega)$, but you can use the command **plot** to plot $F(\omega)$ once you get it.

2.10.2 Fourier series

MATLAB does not provide a command for getting the Fourier series expansion of a signal. But once the Fourier series expansion is obtained, MATLAB can be used to plot it for a finite

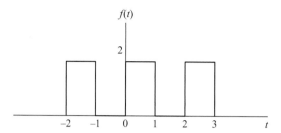

$f(t)$

Figure 2.41. Rectangular pulse train.

number of harmonics or check whether the partial sums of the Fourier series approach the exact signal. To illustrate this, consider the rectangular pulse train in Figure 2.41. Comparing that signal with the signal in the first entry of Table 2.2 shows the signal in Figure 2.41 is only a raised version of the signal in the table. It is raised by A, where $A = 1$ and $T = 2$ or $\omega = 2\pi/T = \pi$. Thus, the Fourier series expansion has a dc value of A and is

$$f(t) = 1 + \frac{4}{\pi}\sum_{n=1}^{\infty}\frac{1}{k}\sin k\pi t, \quad k = 2n - 1 \tag{2.86}$$

Keep in mind that the Fourier series must be truncated for computational reasons and only a partial sum is possible even with computers. Let the harmonics be summed from $n = 1$ to $n = N$, where $N = 5$, and suppose we want to plot $f(t)$ for $-2 < t < 2$, the MATLAB commands for generating the partial sum (or truncated series)

$$f_N(t) = 1 + \frac{4}{\pi}\sum_{n=1}^{N}\frac{1}{k}\sin k\pi t, \quad k = 2n - 1 \tag{2.87}$$

are given below:

```
N=5;
t=-2:0.001:2;
  f0=1.0; % dc component
  fN=f0*ones(size(t));
  for n=1:N
    k=2*n-1;
    mag=4/pi;
    arg=k*pi*t;
    fN = fN + mag*sin(arg)/k;
end
plot(t,fN)
```

The plot is shown in Figure 2.42 for $N = 5$. If we increase the value of N to $N = 20$, the plot becomes that shown in Figure 2.43. Notice the partial sum oscillates above and below the actual value of $f(t)$. At the neighborhood of the points of discontinuity ($t = 0, \pm1, \pm2, \ldots$), there is overshoot and damped oscillation. In fact, an overshoot of about 9 percent of the peak value is always present, regardless of the number of terms used to approximate $f(t)$. This is known as the *Gibbs phenomenon*.

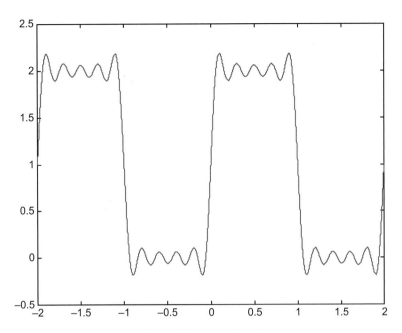

Figure 2.42. Plot of partial sum $f_N(t)$ when $N = 5$.

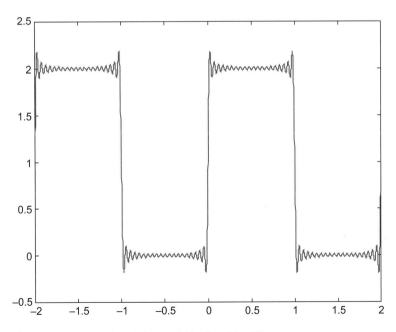

Figure 2.43. Plot of partial sum $f_N(t)$ when $N = 20$.

We can perform the computation of the partial sum for exponential Fourier series. If we consider the same rectangular pulse train in Figure 2.41, it can readily be shown that the truncated series is

$$f_N(t) = 1 + \sum_{\substack{n=-N \\ n \neq 0}}^{N} C_n e^{jn\omega_o t}, \quad \omega_o = \pi, \quad C_n = \frac{j}{n\pi} \left[e^{-jn\pi} - 1 \right] \qquad (2.88)$$

To ease programming, this can be written as

$$f_N(t) = 1 + \sum_{n=1}^{N} C_n e^{jn\pi t} + \sum_{n=1}^{N} C_{-n} e^{-jn\pi t} \qquad (2.89)$$

where C_{-n} is the complex conjugate of C_n. The commands for computing the partial sum in Eq. (2.89) are given below. By making $N = 5$ and 20, we obtain similar results to those shown in Figures 2.42 and 2.43.

```
N=20;
t=-2:0.001:2;
 f0=1.0; % dc component
 fN=f0*ones(size(t));
 for n=1:N
    mag1= j*(exp(-j*n*pi) -1)/(pi*n);
    mag2= conj(mag1);
    arg=n*pi*t;
    fN = fN + mag1*exp(j*arg) + mag2*exp(-j*arg);
end
plot(t,fN)
```

2.10.3 Filtering

In section 2.9, we noticed that it is sometimes necessary to be able to find the poles of the transfer function $H(s)$ of a filter. MATLAB can be used to find the roots of a polynomial by using the command **roots**. For example, if the transfer function of a system is

$$H(s) = \frac{s + 4}{s^3 + 6s^2 + 11s + 6} \qquad (2.90)$$

We can find the poles of $H(s)$ or roots of $s^3 + 6s^2 + 11s + 6 = 0$ by entering

```
» roots([1 6 11 6])
```

or

```
» den = [1 6 11 6]; % denominator of H(s)
» roots(den)
```

Either way, the roots are provided as -1, -2, and -3. In factored form, $H(s)$ becomes

$$H(s) = \frac{s + 4}{(s + 1)(s + 2)(s + 3)}$$

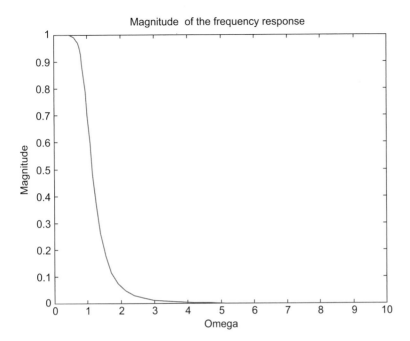

Figure 2.44. Magnitude Bode plot of the frequency response of the third-order Butterworth filter.

The command **buttap** can be used to find the zeros and poles of the nth-order Butterworth filter. For example, the following sequence of MATLAB statements produces the magnitude frequency response of the fourth-order Butterworth filter. The plot of the response is provided in Figure 2.44.

```
» [z,p,k]=buttap(4); % returns the zeros, poles, and constant k of the 4th-order
Butterworth filter
» num=k*poly(z); % forms the numerator
» den=poly(p); % forms the denominator
» [mag,phase,w]=bode(num,den); % returns magnitude, phase (in degrees), and fre-
quency vector w (automatically)
» plot(w,mag) % plots the magnitude verse w
» title('Magnitude of the frequency response')
» xlabel('Omega')
» ylabel('Magnitude')
```

Rather than using linear scale for the magnitude, we could use log scale (in dB) by replacing the **plot**(w,mag) statement with

```
» semilogx(w,20*log10(abs(mag)))
```

Summary

1. A signal is a time-varying function representing messages or information. Signals may be classified as continuous or discrete, analog or digital, periodic or aperiodic, energy or power.

2. A system is a functional relationship between the input $x(t)$ and output $y(t)$. Systems may be classified as linear or nonlinear, continuous or discrete, time-varying or time-invariant, causal or noncausal, analog or digital.

3. Spectral analysis is an inestimable mathematical tool for studying communication systems. It deals with the description of signals in the frequency domain using Fourier series for periodic signals and Fourier transform for nonperiodic signals.

4. The Fourier series of a periodic function is a summation of harmonics of a fundamental frequency. Any periodic function satisfying Dirichlet conditions can be expressed in terms of Fourier series in any of these three forms:

$$f(t) = \underbrace{a_0}_{\text{dc}} + \underbrace{\sum_{n=1}^{\infty} (a_n \cos n\omega_0 + b_n \sin n\omega_0)}_{\text{ac}} \quad \text{(quadrature form)}$$

$$a_0 = \frac{1}{T} \int_0^T f(t)dt, \quad a_n = \frac{1}{T} \int_0^T f(t) \cos n\omega_o t \, dt$$

$$b_n = \frac{1}{T} \int_0^T f(t) \sin n\omega_o t \, dt$$

$$f(t) = A_0 + \sum_{n=1}^{\infty} A_n \cos (n\omega_0 + \varphi_n) \quad \text{(amplitude – phase form)}$$

$$A_0 = a_o, \quad A_n = \sqrt{a_n^2 + b_n^2}, \quad \varphi_n = -\tan^{-1} \frac{b_n}{a_n}$$

$$f(t) = \sum_{n=-\infty}^{\infty} C_n e^{jn\omega_0 t} \quad \text{(exponential form)}$$

$$C_n = \frac{1}{T} \int_0^T f(t) e^{-jn\omega_0 t} dt = |C_n| \angle \phi_n$$

5. One form of Parseval's theorem (for periodic signals) states that the total average power of a signal is the sum of the average powers of its harmonic components, i.e.

$$P = \frac{1}{T} \int_0^T f^2(t)dt = \sum_{n=-\infty}^{\infty} |c_n|^2$$

6. The Fourier transform $F(\omega)$ is the frequency-domain representation of $f(t)$,

$$F(\omega) = \int_{-\infty}^{\infty} f(t)e^{-j\omega t}dt$$

7. The inverse Fourier transform is

$$f(t) = \frac{1}{2\pi} \int_{-\infty}^{\infty} F(\omega)e^{j\omega t}d\omega$$

8. Important Fourier transform properties and pairs are summarized in Tables 2.4 and 2.5.
9. Another form of Parseval's theorem (for energy signals) states that

$$E = \int_{-\infty}^{\infty} f^2(t)dt = \frac{1}{2\pi} \int_{-\infty}^{\infty} |F(\omega)|^2 d\omega$$

10. Filters are devices used for removing unwanted frequency components from a signal. They are classified according to their suppressed frequency bands as low-pass, high-pass, bandpass, and band-stop.
11. Ideal filters pass all frequency components of the input within their passband and reject completely all frequency components outside the passband.
12. Butterworth filters are standard or prototype filters that approximate some aspects of ideal filters by compromising the others.
13. MATLAB is a powerful tool for signal analysis. It is used in this chapter to plot, find partial sum of a Fourier series, and design filters.

Review questions

2.1 A signal can be both a power signal and an energy signal.
(a) True.   (b) False.

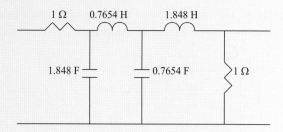

Figure 2.45. For Review question 2.10.

2.2 If the input $x(t)$ and output $y(t)$ of a system are related as $y(t) = 10x(t - 2)$, then the system is:

    (a) Time-varying.   (b) Time-invariant.   (c) Causal.   (d) Noncausal.

2.3 Parseval's theorem implies superposition of average power.

    (a) True.   (b) False.

2.4 Which of the following signals is NOT a power signal?

    (a) 3.   (b) $u(t)$.   (c) $\cos 5t$.   (d) $e^{-2|t|}$.

2.5 Which of the following signals are energy signals?

    (a) 10.   (b) $\sin 4tu(t)$.   (c) $\delta(t)$.   (d) $e^{-2t}u(t)$.

2.6 If $x(t) = 10 + 8 \cos t + 4 \cos 3t + 2 \cos 5t + \ldots$, the frequency of the sixth harmonic is:

    (a) 12.   (b) 11.   (c) 9.   (d) 6.

2.7 Which of these functions does not have a Fourier transform?

    (a) $e^t u(-t)$.   (b) $te^t u(t)$.   (c) $1/t$.   (d) $|t|u(t)$.

2.8 The inverse Fourier transform of $\delta(\omega)$ is

    (a) $\delta(t)$.   (b) $u(t)$.   (c) 1.   (d) $1/2\pi$.

2.9 What kind of Butterworth filters are discussed in this chapter?

    (a) Low-pass.   (b) High-pass.   (c) Bandpass.   (d) Band-stop.

2.10 What is the order of the Butterworth filter shown in Figure 2.45?

    (a) 3.   (b) 4.   (c) 5.   (d) 6.

Answers: 2.1 b, 2.2 b,c, 2.3 a, 2.4 d, 2.5 b,d, 2.6 d, 2.7 c, 2.8 d, 2.9 a, 2.10 b

Problems

Sections 2.2 and 2.3 Classifications and operations on signals

2.1 Define each of the following terms:

    (a) an analog signal

    (b) a digital signal

    (c) a continuous-time signal

    (d) a discrete-time signal

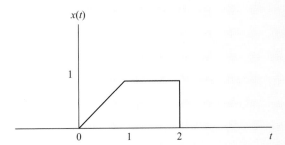

Figure 2.46. For Problem 2.5.

2.2 Give a brief description of each of the following terms:
   (a) a periodic signal
   (b) an aperiodic signal
   (c) a power signal
   (d) an energy signal

2.3 Sketch the following signals.
   (a) $x_1(t) = 3u(t-1) - u(t-2)$
   (b) $x_2(t) = 2\Pi(10t)$
   (c) $x_3(t) = 5\Delta(t/4)$

2.4 Give a sketch of each of the following signals.
   (a) $y_1(t) = 2 \,\text{sinc}\,(\pi t/3)$
   (b) $y_2(t) = \Delta(t) - \Pi(t-2)$
   (c) $y_3(t) = 4\,\text{sgn}(t+2)$

2.5 Given that $x(t)$ is shown in Figure 2.46, sketch the following signals.
   (a) $x_1(t) = x(-t)$
   (b) $x_2(t) = x(2+t)$
   (c) $x_3(t) = 2x(t) + 1$
   (d) $x_4(t) = x(2t)$
   (e) $x_5(t) = x(t/4)$

2.6 Determine whether the following signals are power or energy signals or neither.
   (a) $x(t) = e^{-t}$ (exponential)
   (b) $y(t) = r(t) = tu(t)$ (ramp)
   (c) $z(t) = \Delta(t)$ (triangular pulse)

2.7 Is it possible to generate a power signal in a lab? Explain.

2.8 If $f(t)$ is a power signal with average power P, find the average power of $g(t) = af(bt + c)$, where a, b, and c are constants.

Section 2.4 Classifications of systems

2.9 Give a brief description of the following terms.
   (a) a linear system
   (b) a nonlinear system
   (c) a continuous-time system
   (d) a discrete-time system

2.10 Define the following terms.
    (a) a time-varying system
    (b) a time-invariant system
    (c) a causal system
    (d) a non-causal system
    (e) an analog system
    (f) a digital system

2.11 For each of the systems described below, $x(t)$ is the input signal and $y(t)$ is the output. Specify whether each system is linear or nonlinear.
    (a) $y(t) = 10 + 2x(t)$
    (b) $y(t) = x(t) + 2x^2(t)$
    (c) $y(t) = 3tx(t)$

2.12 Repeat Problem 2.11 for the following cases.
    (a) $\int_{-\infty}^{t} x(4\lambda)d\lambda$
    (b) $y(t) = \ln [x(t)]$
    (c) $y(t) = \sin(t)\, x(t)$

Section 2.5 Trigonometric Fourier series

2.13 Determine the Fourier series expansion of the signal in Figure 2.47.

2.14 Find the Fourier series of the half-wave rectified cosine function shown in Figure 2.48.

2.15 Obtain the Fourier series for the periodic signal in Figure 2.49.

2.16 For the signal in Figure 2.50, determine the trigonometric Fourier series. Evaluate $f(t)$ at $t = 2$ using the first three non-zero harmonics.

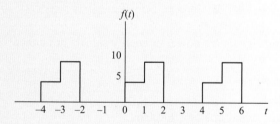

Figure 2.47. For Problem 2.13.

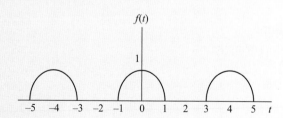

Figure 2.48. For Problem 2.14.

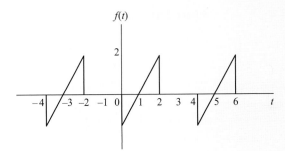

f(t)

Figure 2.49. For Problem 2.15.

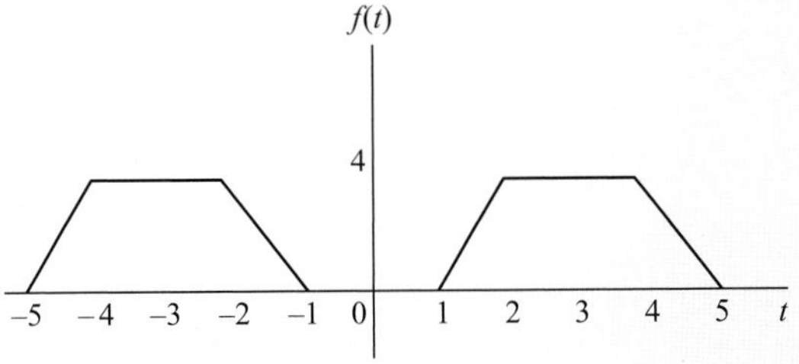

f(t)

Figure 2.50. For Problem 2.16.

2.17 A periodic signal is expressed as

$$f(t) = 4 + 2\cos{(t + 15°)} - 0.5\cos{(3t + 20°)} + 0.25\sin{(5t + 25°)}$$

Sketch the magnitude and phase spectra.

2.18 Given that

$$f(t) = \sum_{\substack{n=1 \\ n=\text{odd}}}^{\infty} \left(\frac{20}{n^2\pi^2}\cos{2nt} - \frac{3}{n\pi}\sin{2nt} \right)$$

plot the first five terms of the amplitude and phase spectra for the signal.

2.19 An amplitude modulated (AM) waveform is given by

$$f(t) = [40 - 20\sin{(2\pi t + \pi/6)}]\cos{5\pi t}$$

Show that *f(t)* can be represented as

$$f(t) = a_1\cos{(\omega_1 t + \theta_1)} + a_1\cos{(\omega_2 t + \theta_2)} + a_3\cos{(\omega_3 t + \theta_3)}$$

and determine $a_1, a_2, a_3, \omega_1, \omega_2, \omega_2, \theta_1, \theta_2,$ and θ_3.

2.20 Determine the Fourier series expansion of the signal defined over its period as:

$$f(t) = \begin{cases} 4t, & 0 < t < 1 \\ 4, & 1 < t < 3 \\ 8 - 4t, & 3 < t < 4 \end{cases}$$

2.21 A periodic signal $f(t) = 2t/\pi,\ -\pi/2 < t < \pi/2$ with $f(t \pm \pi) = f(t)$.
   (a) Find its Fourier series expansion.
   (b) Calculate the fraction of its power which is contained in the first four harmonics.

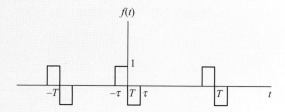

Figure 2.51. For Problem 2.23.

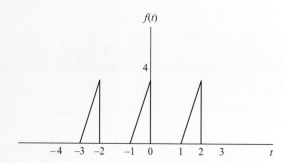

Figure 2.52. For Problem 2.24.

Section 2.6 Exponential Fourier series

2.22 Obtain the exponential Fourier series for $f(t) = t^2$, $-\pi < t < \pi$, $f(t + 2\pi n) = f(t)$.

2.23 For the triangular pulse in Figure 2.51, find the exponential Fourier series.

2.24 Find the exponential Fourier series for the clipped sawtooth waveform in Figure 2.52.

2.25 The periodic signal $f(t)$ is represented by

$$ f(t) = \sum_{n=\infty}^{\infty} C_n e^{jn\omega_o t} $$

Determine the coefficients C_n' of each of the following signals.
(a) $g(t) = f(t - 2)$
(b) $h(t) = 2\frac{df(t)}{dt}$
(c) $y(t) = \frac{d^2 f}{dt^2} - \frac{df}{dt}$

Sections 2.7 and 2.8 Fourier transform and its properties

2.26 Determine the Fourier transform of the following signals.
(a) $x(t) = e^{-t} \sin \pi t$
(b) $y(t) = \frac{1}{3}[\delta(t + 1/3) + \delta(t - 1/3)]$
(c) $z(t) = e^{-t} \mathrm{sgn}(t)$

2.27 Determine the Fourier transform of the signals in Figure 2.53.

2.28 Find the Fourier transform of the signals in Figure 2.54.

2.29 Determine the Fourier transform of the sawtooth pulse shown in Figure 2.55.

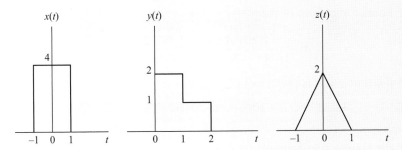

Figure 2.53. For Problem 2.27.

(a)   (b)   **Figure 2.54.** For Problem 2.28.

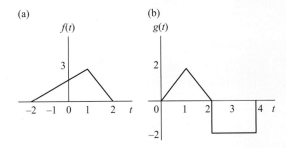

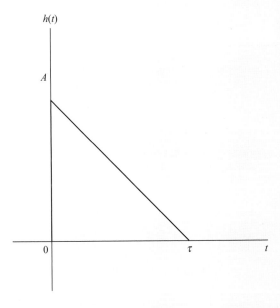

Figure 2.55. For Problem 2.29.

2.30 Obtain the Fourier transform of these signals.

(a) $f(t) = (1 + m \cos \alpha t) \cos \beta t, \quad -\infty < t < \infty$

(b) $g(t) = \begin{cases} \sin t, & 0 < t < \pi \\ 0, & \text{otherwise} \end{cases}$

2.31   Find the Fourier transform of the pulse

$$p(t) = \begin{cases} \cos \pi t/\tau, & |t| < \tau/2 \\ 0, & \text{otherwise} \end{cases}$$

2.32   The raised cosine pulse is given by

$$r(t) = 10(1 + \cos \pi t)\Pi(t/2)$$

(a) sketch $r(t)$,

(b) find the Fourier transform of $r(t)$.

2.33   Find the inverse Fourier transform of the following:

(a) $F_1(\omega) = \cos(\pi\omega/4)$

(b) $F_2(\omega) = \frac{e^{-j\omega 2}}{4+j\omega}$

(c) $F_3(\omega) = \frac{\omega^2+2}{\omega^4+3\omega^2+2}$

2.34   Determine the inverse Fourier transform of the spectrum in Figure 2.56.

2.35   Obtain $f(t)$ corresponding to the $F(\omega)$ shown in Figure 2.57.

2.36   Determine the inverse Fourier transform of the signal whose spectrum is shown in Figure 2.58.

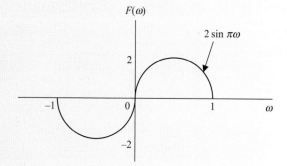

Figure 2.56. For Problem 2.34.

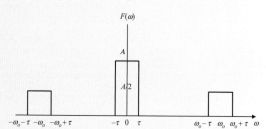

Figure 2.57. For Problem 2.35.

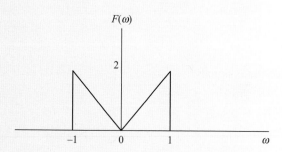

Figure 2.58. For Problem 2.36.

2.37 Given that the Fourier transform of $g(t)$ is

$$G(\omega) = \frac{20}{(1+j\omega)}$$

find the Fourier transform of the following:
(a) $g(-2t)$
(b) $(1+t)g(1+t)u(1+t)$
(c) $t\frac{dg(t)}{dt}$
(d) $g(t)\cos \pi t$

2.38 The Fourier transform of $x(t)$ is

$$X(\omega) = \frac{4+j\omega}{-\omega^2 + j2\omega + 3}$$

Find the Fourier transform of these signals
(a) $x(t)e^{-j2t}$
(b) $x(t)\sin \pi(t-1)$
(c) $x(t)*\delta(t-2)$
(d) $\displaystyle\int_{-\infty}^{t} x(\tau)d\tau$

2.39 Use Parseval's theorem to evaluate the following integrals.
(a) $\displaystyle\int_{-\infty}^{\infty} \frac{\sin^2 t}{t^2}\, dt$
(b) $\displaystyle\int_{-\infty}^{\infty} \frac{dt}{(t^2+4)^4}\, dt$

2.40 Use Parseval's theorem to determine the energy of the following signals.
(a) $x(t) = e^{-3t}u(t)$
(b) $y(t) = \Pi(t/4)$

2.41 If $f(t) = 5\Pi(t/2)$ and $g(t) = f(t+2) + f(t-2)$, determine $G(\omega)$.

Section 2.9 Applications – filters

2.42 Determine the frequency response $V_o(\omega)/V_i(\omega)$ of the circuit in Figure 2.59. What type of ideal filter does the circuit represent?

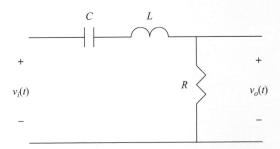

Figure 2.59. For Problem 2.42.

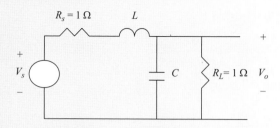

Figure 2.60. For Problem 2.46.

2.43 The transfer function of a low-pass filter has

$$H(\omega) = \begin{cases} 1, & |\omega| \leq B \\ 0, & \text{otherwise} \end{cases}$$

If the signal $x(t) = 10e^{-100\pi t}u(t)$ is applied to the filter, determine the value of B in rad/s that causes only one-third of the energy of $x(t)$ to be passed.

2.44 Find the impulse response $h(t)$ of a third-order Butterworth filter with $\omega_c = 1$ rad/s.

2.45 Determine $H(s)$ such that

$$|H(\omega)|^2 = \frac{1}{1 + \omega^6}$$

2.46 Given the RLC circuit in Figure 2.60 in which $R = 1$ Ω and $\omega_c = 10$ rad/s, find L and C such that $H = V_o/V_s$ produces a Butterworth frequency response.

Section 2.10 Computation using MATLAB

2.47 Use MATLAB to plot the following signals.
    (a) $x(t) = t^2 - 2t + 3,$ $-2 < t < 2$
    (b) $y(t) = 4\cos(2\pi t - 12°) + 3\sin 2\pi t,$ $-\pi < t < \pi$
    (c) $z(t) = 10(1 - e^{-2t}),$ $0 < t < 5$

2.48 Use MATLAB to plot:
    (a) $|F(\omega)| = \frac{1}{\sqrt{4+\omega^2}},$ $-5 < \omega < 5$
    (b) $G(\omega) = 10\,\text{sinc}^2(5\omega),$ $-10 < \omega < 10$

2.49 Use MATLAB to plot the magnitude and phase of

$$F(\omega) = 10je^{-j\pi\omega}\frac{\sin \pi\omega}{1 - \omega^2}$$

for $-5 < \omega < 5$.

2.50 Write a MATLAB script to plot the partial sum of the trigonometric Fourier series of the signal in Problem 2.13. Take $N = 15$.

2.51 Develop a MATLAB program to plot the partial sum of the trigonometric Fourier series in Problem 2.14. Take $N = 25$.

2.52 Plot a 31-term partial sum of $f(t)$ given in Problem 2.20 using MATLAB.

2.53 The response of the fifth-order Butterworth filter is

$$H(s) = \frac{1}{s^5 + 3.236s^4 + 5.236s^3 + 5.236s^2 + 3.236s + 1}$$

Use MATLAB to find the poles of $H(s)$.

2.54 Use MATLAB to plot the magnitude frequency response of the sixth-order Butterworth filter. Use a semilog scale.

3 | Amplitude modulation

To climb steep hills requires a slow pace at first.

WILLIAM SHAKESPEARE

HISTORICAL NOTE – The information age in full gallop

Samuel F. B. Morse (1791–1872), was born into a world devoid of the conveniences of modern communication – no phones, radio, television, or the internet! Unthinkable now, but that was the case not so long ago. In the early 1830s, the fastest means of communication was the postal system operating via railways, horse riders, and carriages. It took weeks for a letter to cross the continent from the US east coast to the west coast, and vice versa. That postal system embodied ages of progress since the dawn of man, including the invention of the wheel, writing, paper, pen, the printing press, the locomotive engine, etc. Then in 1838 Samuel Morse successfully demonstrated his practical electrical telegraph system. Messages could now be sent across the continent in a matter of minutes. At that time it seemed almost magical. As if to add to the magic, a rapid succession of inventions and improvements followed in its wake. Major among these were the roll film camera, wireless telegraphy, electronic amplifying tubes, and the phonograph.

Alexander Graham Bell (1847–1922), patented the electric telephone in 1876. This was a spectacular device that permitted two-way conversation and transmitted music and sounds of all kinds. Communications was exploding. A spate of inventions, including the radio, the television, and the computer followed. Older technologies quickly became obsolete as they were superseded by newer and superior innovations. Fast forward to the 21st century; communications technology is on a sprint. Witness the wonder of the internet, HDTV, and the smart cell phone. Modern electronic communication is indispensable to the individual and to our society. Our age is witnessing the information age in full gallop!

As students of the principles of communication, how do we fit into the picture? Communication is a very valuable and exciting field of applied science. Communication principles provide a powerful means of serving human needs and enabling modern technologies. Careers in communication are very rewarding. The hope for future advances in the field lies in those who master the fundamentals. Perhaps you will be one of those future giants in the field. Remember, all that life requires of each of us is that we try our best.

3.1   INTRODUCTION

A communication system is used for transmission of information-bearing signals (message signals) over varying distances. Examples of message signals are electrical signal outputs of source transducers, such as microphones and video cameras. Most message signals are *baseband signals*.

> **Baseband signals** are low-pass signals occupying a range of frequencies from zero to the signal bandwidth.

Baseband communication involves the transmission of baseband signals. It is beset with serious limitations. Because of their overlapping spectra, only one baseband signal can be transmitted over a given communication medium. Baseband signals are not suitable for radio communication. Consider the problem of broadcasting an audio signal of 10 kHz bandwidth over a radio channel. The efficiency of antenna radiation decreases from a maximum at an antenna length of $\lambda/2$, to a minimum acceptable level at an antenna length of $\lambda/10$, where λ, the wavelength of the electromagnetic wave, is related to its frequency f and to its velocity c by

$$c = \lambda f = 3 \times 10^8 \text{ m/s} \tag{3.1}$$

From the above formula, the 10 kHz microphone signal corresponds to a wavelength of 30 km. It requires an infeasible antenna height of 3 km! The serious limitations of baseband communication are easily overcome through the technique of *modulation*.

> **Modulation** is the translation of the spectrum of message signals to higher-frequency bands to facilitate efficient transmission through communication media, or simultaneous transmission of multiple signals over the same medium, or both.

Consider again the 10 kHz microphone signal. Suppose it is translated to a frequency band centered on 1 MHz. The required $\lambda/10$ antenna height of 30 m is quite feasible for a radio antenna mast. The relatively high-frequency 1 MHz signal is called the *carrier*, the 10 kHz microphone signal is the *message signal* or the *modulating signal*, and the carrier with the message signal superimposed on it is the *modulated signal*.

Suppose there are three message signals m_1, m_2, and m_3, each of 10 kHz bandwidth. They can be respectively modulated using carriers of 1 MHz, 1.02 MHz, and 1.04 MHz, to occupy adjacent and non-overlapping frequency bands. Each can be recovered with a receiver whose input filter has a pass band corresponding to only the desired modulated signal. Thus modulation permits the simultaneous transmission of different message signals through the same communication channel.

Modulation is achieved by varying (modulating) a parameter of the carrier, in such a way that a one-to-one correspondence exists between that parameter and the message signal. The carrier is usually a sinusoid, but this need not always be the case. In effect, it is the high-frequency carrier that is transmitted, but one of its parameters (amplitude, frequency, or phase) carries the information in the message signal, resulting in what is aptly termed *carrier modulation*. Modulation does a good job of matching the transmitted signal to the characteristics of the channel, but the modulated signal is usually above the frequency range of human perception and is otherwise unusable without further processing at the receiver. *Demodulation*, the inverse of modulation, must be performed at the receiver to return the received signal to its original form prior to modulation in the transmitter.

Communication systems are broadly classified into analog and digital systems, depending on whether the transmitted signal is analog or digital. Analog communication techniques include *amplitude modulation, frequency modulation, and phase modulation*. These names correspond respectively to the parameter of the carrier that is modulated: amplitude, frequency, and phase. This chapter is devoted to amplitude modulation. In it we shall study various amplitude modulation and demodulation techniques, and their applications.

3.2 AMPLITUDE MODULATION (AM)

In amplitude modulation, the amplitude of a carrier is varied linearly with the message signal. The carrier is much higher in frequency than the baseband message signal $m(t)$. Consider a sinusoidal carrier $c(t) = A_c \cos(\omega_c t + \theta)$. Of the three parameters of the carrier, only the amplitude A_c is varied in AM. The constant phase of the carrier can be set to zero. Thus the carrier may be more simply represented as $c(t) = A_c \cos \omega_c t$. The amplitude-modulated signal is given by

$$\phi_{AM}(t) = [A_c + m(t)] \cos \omega_c t = A_c \cos \omega_c t + m(t) \cos \omega_c t \tag{3.2}$$

Figure 3.1 illustrates the amplitude modulation technique. Figure 3.1(a) shows the baseband modulating signal, while Figure 3.1(b) shows the high-frequency sinusoidal carrier. The linear variation of the modulated carrier amplitude with the message signal is shown in Figure 3.1(c) and (d). If the amplitude of the carrier equals or exceeds that of the message signal, $A_c + m(t)$ is always non-negative. This is the usual case for AM shown in Figure 3.1(c), in which the AM signal envelope, $E(t) = A_c + m(t)$, is a true replica of the message signal. If the carrier is smaller in amplitude than the message signal, $A_c + m(t)$ does assume both positive and negative values. This results in the AM signal of Figure 3.1(d) in which the envelope $E(t) = |A_c + m(t)|$ is a rectified version of $A_c + m(t)$. In this case, the envelope of the modulated signal is a distorted version of the message signal, and the carrier is said to be over-modulated. *Over-modulation* and the resulting *envelope distortion* are undesirable.

> **Demodulation** is the inverse of modulation. It is the recovery of the original message signal from the modulated signal.

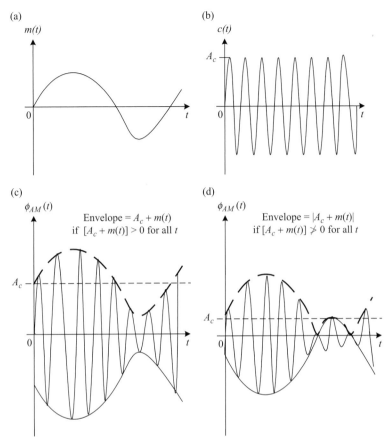

Figure 3.1. Illustration of amplitude modulation. (a) Baseband (message) signal; (b) sinusoidal carrier; (c) amplitude-modulated waveform with an undistorted envelope; (d) amplitude-modulated waveform with a distorted envelope.

Following modulation at the transmitter, it is necessary to demodulate or *detect* the original message signal at the receiver. For AM, demodulation is easily accomplished by recovering the envelope from the modulated signal through a process known as *envelope detection*. Because envelope detection is simple and cheap relative to the alternative, it is the preferred demodulation technique in AM. Envelope detection is feasible only if envelope distortion did not occur during modulation. Thus, envelope detection requires that the envelope $E(t)$ satisfy the relationship

$$E(t) = [A_c + m(t)] \geq 0 \text{ for all } t \tag{3.3}$$

The above criterion can also be expressed in terms of the *modulation index*.

The **modulation index** μ is the ratio of the peak value of the message signal to the amplitude of the carrier.

Let the peak value of the modulating signal be m_p, then the modulation index is given by

$$\boxed{\mu = \frac{m_p}{A_c}} \tag{3.4}$$

The modulation index can also be expressed as a percentage. If $m_p = A_c$, then $\mu = 1$, resulting in *full modulation* or *100% modulation*. Over-modulation and envelope distortion correspond to $\mu > 1$. In terms of the modulation index μ the requirement for envelope detection is

$$0 \le \mu \le 1 \tag{3.5}$$

Although the limiting case of $\mu = 1$ meets the mathematical requirement for envelope detection, it should be avoided. As we should be well aware of, noise is inevitable in any communication system. If $\mu = 1$, over-modulation will result when noise increases the peak value of the modulated signal beyond that due to the carrier and the message signal. For the same reason, high values of modulation index close to 100% should also be avoided.

Tone modulation

In tone modulation the message signal is a single-frequency sinusoid. As such, tone modulation does not result in transmission of information, but it provides a very valuable means of analyzing communication systems. Consider an AM signal for which the message signal $m(t) = A_m \cos \omega_m t$ has amplitude A_m and frequency ω_m. The amplitude A_m is the same as m_p, the peak value of the message signal as defined earlier. The modulation index is given by $\mu = \frac{m_p}{A_c} = \frac{A_m}{A_c}$. Thus the expression for the AM signal given in Eq. (3.2) can be re-written as

$$\phi_{AM}(t) = [A_c + A_m \cos \omega_m t] \cos \omega_c t = A_c \left[1 + \frac{A_m}{A_c} \cos \omega_m t\right] \cos \omega_c t$$

Replacing $\frac{A_m}{A_c}$ with μ gives

$$\phi_{AM}(t) = A_c[1 + \mu \cos \omega_m t] \cos \omega_c t \tag{3.6}$$

EXAMPLE 3.1

For an amplitude-modulated signal, the carrier is given by $c(t) = 10 \cos \omega_c t$, and the modulating signal is given by $m(t) = A_m \cos \omega_m t$, where $\omega_c \gg \omega_m$. Find the modulation index, state whether or not the criterion for envelope detection is met, and sketch the amplitude-modulated waveform for $A_m = 5$, $A_m = 10$, and $A_m = 15$.

Solution

(a) $\mu = \frac{A_m}{A_c} = \frac{5}{10} = 0.5$, or ($\mu = 50\%$). This meets the envelope detection criterion.
(b) $\mu = \frac{A_m}{A_c} = \frac{10}{10} = 1$, or ($\mu = 100\%$). This barely meets the envelope detection criterion.
(c) $\mu = \frac{A_m}{A_c} = \frac{15}{10} = 1.5$ or ($\mu = 150\%$). This does not meet the criterion.

The corresponding waveforms are sketched in Figure 3.2.

(a)

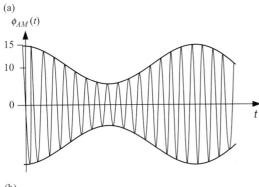

(b)

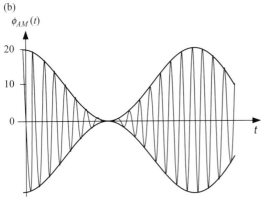

(c)

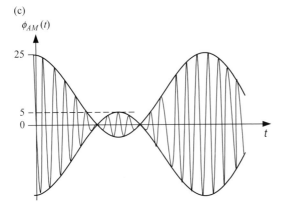

Figure 3.2. Waveforms for Example 3.1 for different values of modulation index: (a) modulation index = 0.5; (b) modulation index = 1; (c) modulation index = 1.5.

PRACTICE PROBLEM 3.1

For an amplitude-modulated waveform, the carrier is given by $c(t) = 5 \cos \omega_c t$, and the modulating signal is as shown in Figure 3.3. The carrier frequency is much higher than that of the triangular wave shown in Figure 3.3. Find the modulation index and sketch the amplitude-modulated waveform for $m_p = 3$, $m_p = 5$, and $m_p = 7$.

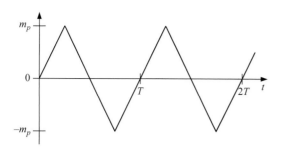

Figure 3.3. Modulating signal for Practice problem 3.1

Answer: $\mu = 0.6$, $\mu = 1.0$, and $\mu = 1.4$, respectively. These values of μ meet, barely meet, and do not meet the requirement for envelope detection, respectively. The sketches are similar to those of Figure 3.2 except that the envelope is triangular rather than sinusoidal.

EXAMPLE 3.2

An AM waveform with tone modulation as displayed on an oscilloscope is shown in Figure 3.4. Find the modulation index and an expression for the modulated signal.

Solution

The carrier is a sine wave since it is sinusoidal and has a zero crossing at $t = 0$. The modulating signal is also a sine wave since the AM envelope is sinusoidal and has a value at $t = 0$ which is equal to its mean value. A sine wave is just a cosine wave with a 90° phase shift and vice versa. Thus the AM waveform shown can be expressed as

$$\phi_{AM}(t) = [A_c + A_m \sin \omega_m t] \sin \omega_c t \tag{3.7}$$

Let $A(t)$ be the time-varying amplitude of the carrier, and let $A_{\max}$ and $A_{\min}$ be its maximum and minimum values. Then

$$A_{\max} = A_c + A_m \tag{3.8}$$

$$A_{\min} = A_c - A_m \tag{3.9}$$

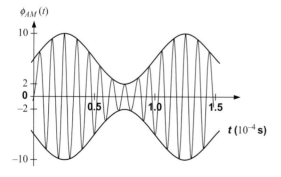

Figure 3.4. AM waveform displayed on the oscilloscope, for Example 3.2.

Solving for the carrier and the modulating signal amplitudes from the last two equations,

$$A_c = \frac{1}{2}[A_{max} + A_{min}] \tag{3.10}$$

$$A_m = \frac{1}{2}[A_{max} - A_{min}] \tag{3.11}$$

$$\mu = \frac{A_m}{A_c} = \frac{A_{max} - A_{min}}{A_{max} + A_{min}} \tag{3.12}$$

From the given waveform, $A_{max} = 10$ and $A_{min} = 2$. Thus,

$$\mu = \frac{A_{max} - A_{min}}{A_{max} + A_{min}} = \frac{10 - 2}{10 + 2} = 0.667$$

$$A_c = \frac{1}{2}[A_{max} + A_{min}] = 6$$

$$A_m = \frac{1}{2}[A_{max} - A_{min}] = 4$$

Thus $\quad \phi_{AM}(t) = [6 + 4\sin\omega_m t]\sin\omega_c t$

PRACTICE PROBLEM 3.25

The envelope of a tone-modulated AM waveform displayed on an oscilloscope has a peak value of 13 and a minimum value of 3. The carrier is a cosine wave, while the message signal is a sine wave. Find the modulation index, and give the expression for the AM waveform.

Answer: $\mu = 0.625$

$$\phi_{AM}(t) = [8 + 5\sin(\omega_m t)]\cos(\omega_c t)$$

3.2.1 Spectrum of amplitude-modulated signals

Consider a message signal $m(t)$ and the AM signal $\phi_{AM}(t) = A_c\cos\omega_c t + m(t)\cos\omega_c t$. Let $M(\omega)$ be the Fourier transform of $m(t)$ and $\Phi_{AM}(\omega)$ be the Fourier transform of $\phi_{AM}(t)$. Taking the Fourier transform of Eq. (3.2),

$$\Phi_{AM}(\omega) = \frac{1}{2}[M(\omega + \omega_c) + M(\omega - \omega_c)] + \pi A_c[\delta(\omega + \omega_c) + \delta(\omega - \omega_c)] \tag{3.13}$$

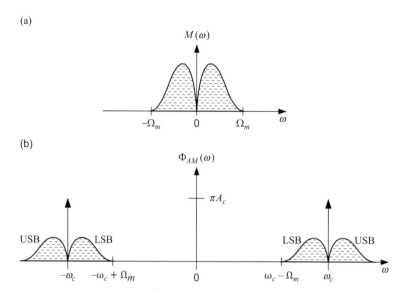

(a)

$M(\omega)$

$-\Omega_m$ 0 Ω_m ω

(b)

$\Phi_{AM}(\omega)$

πA_c

USB   LSB   LSB   USB

$-\omega_c$ $-\omega_c + \Omega_m$ 0 $\omega_c - \Omega_m$ ω_c ω

Figure 3.5. Spectrum of amplitude-modulated signals: (a) message signal spectrum; (b) AM signal spectrum.

$M(\omega)$ and $\Phi_{AM}(\omega)$ are both shown in Figure 3.5. Note that $\Phi_{AM}(\omega)$ contains replicas of $M(\omega)$, scaled in amplitude by a factor of $\frac{1}{2}$, and shifted to the right and to the left by $\pm\omega_c$. It also contains the frequency-domain representation of the carrier as the two impulse functions located at $\pm\omega_c$, each with an amplitude of πA_c. The replicas of $M(\omega)$ consist of the *upper sideband* (USB) and the *lower sideband* (LSB), depending on whether the frequency band occupied is above or below ω_c. The bandwidth of $\Phi_{AM}(\omega)$ is double that of $M(\omega)$. AM is aptly referred to as a double sideband plus carrier system. Each of the sidebands contains all the information necessary for recovery of the message signal.

The bandwidth of $M(\omega)$ is $\Omega_m = 2\pi B$ radians/sec, where B is its natural frequency bandwidth in Hz. This is equal to the bandwidth of each of the two sidebands. From an examination of the AM spectrum, it is clear that if an undistorted version of the message signal is to be recovered from the modulated signal, the positive and negative frequency versions of the LSB should not overlap at the origin. This corresponds to the requirement that $\omega_c - \Omega_m \geq 0$, or equivalently, $f_c - B \geq 0$, where f_c is the carrier frequency in Hz. This gives the minimum carrier frequency for amplitude modulation as $f_c \geq B$.

To achieve higher antenna radiation efficiency or a better matching of the modulated signal to the characteristics of the channel, the carrier frequency is usually much higher than the above minimum requirement. Consider commercial AM broadcasting. Each AM channel has a bandwidth of 10 kHz. The bandwidth of an AM signal is double that of the message signal, so the message signal bandwidth is 5 kHz. AM channels are spaced 10 kHz apart from 550 to 1600 kHz. Thus, for an AM station broadcasting at 1000 kHz, the carrier frequency is 200 times the bandwidth of the message signal.

As a *double sideband system*, AM is wasteful in bandwidth since only one sideband is necessary for transmitting the information in the message signal. As will be shown next, it is also inefficient with regard to the transmitted power, since it transmits a large carrier amplitude together with the sidebands, but the carrier does not convey information.

3.2.2 Power efficiency in amplitude modulation

Consider the amplitude-modulated signal $\phi_{AM}(t) = A_c \cos \omega_c t + m(t) \cos \omega_c t$. The carrier is $A_c \cos \omega_c t$, while $m(t) \cos \omega_c t$ represents the sidebands. The power in the carrier is $P_c = \frac{A_c^2}{2}$. Let P_s be the power in the sidebands and P_m the power in the message signal.

Then

$$P_s = \lim_{T \to \infty} \frac{1}{T} \int_{T/2}^{T/2} m^2(t) \cos^2(\omega_c t) dt = \frac{1}{2} \lim_{T \to \infty} \frac{1}{T} \int_{-T/2}^{T/2} m^2(t)[1 + \cos(2\omega_c t)] dt$$

But $\int_{T/2}^{T/2} m^2(t) \cos(2\omega_c t) dt = 0$

Therefore,

$$P_s = \frac{1}{2} \lim_{T \to \infty} \frac{1}{T} \int_{T/2}^{T/2} m^2(t) dt \tag{3.14}$$

The right-hand side of the last equation is half the power P_m in the message signal. Thus,

$$P_s = \frac{1}{2} P_m \tag{3.15}$$

For AM, the power in the information-bearing part, or the useful power, is the power in the sidebands. The efficiency of a modulation technique with regard to power consumption is denoted by its *power efficiency η*.

The **power efficiency** of a modulated signal is the ratio or percentage of the power in the information-bearing part relative to the total power in the modulated signal.

Thus

$$\eta = \frac{\text{useful power}}{\text{total power}} = \frac{\text{sideband power}}{\text{total power}}$$

i.e.

$$\eta = \frac{P_s}{P_c + P_s} = \frac{\frac{1}{2} P_m}{P_c + \frac{1}{2} P_m} \tag{3.16}$$

But $P_c = \frac{A_c^2}{2}$, thus

$$\boxed{\eta = \frac{P_m}{A_c^2 + P_m}} \tag{3.17}$$

Consider the case of tone modulation for which $m(t) = A_m \cos \omega_m t = \mu A_c \cos \omega_m t$.

$$P_m = \frac{A_m^2}{2} = \frac{(\mu A_c)^2}{2} \qquad (3.18)$$

Thus

$$\eta = \frac{A_m^2}{2A_c^2 + A_m^2} = \frac{\mu^2}{2 + \mu^2} \qquad (3.19)$$

From the preceding equation, the power efficiency η for the case of tone modulation will have a maximum value of $1/3$ when μ has its maximum value of 1 that barely satisfies the envelope detection criterion.

EXAMPLE 3.3

For an amplitude-modulated signal, $\mu = 0.5$, find the power efficiency η if the modulating signal is (a) a sinusoid, (b) the triangular signal shown in Figure 3.6.

Solution

(a) This is a case of tone modulation.

$$\eta = \frac{\mu^2}{2 + \mu^2} = \frac{(0.5)^2}{2 + (0.5)^2} = 0.111 \text{ or } 11.1\%$$

(b) Because of the symmetry of the modulating signal, its power P_m can be found by integrating over only a quarter of the triangular waveform.

$$m(t) = \frac{4A}{T}t, \ \ 0 \le t \le \frac{T}{4}$$

$$P_m = \frac{4}{T}\int_0^{T/4} m^2(t)dt = \frac{4}{T}\int_0^{T/4} \frac{16A^2}{T^2}t^2 dt = \frac{64A^2}{3T^3}t^3\Big|_0^{T/4} = \frac{64A^2}{3T^3}\left[\frac{T}{4}\right]^2 = \frac{A^2}{3}$$

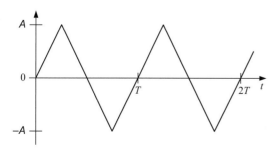

Figure 3.6. Modulating signal for Example 3.3 and Practice problem 3.3

The carrier amplitude is $A_c = 2A$ since $\mu = 0.5$.

$$\therefore \ \eta = \frac{P_m}{A_c^2 + P_m} = \frac{\frac{A^2}{3}}{(2A)^2 + \frac{A^2}{3}} = \frac{\frac{1}{3}}{4 + \frac{1}{3}} = 0.0769 \text{ or } 7.69\%$$

PRACTICE PROBLEM 3.3

Find the power efficiency η for an AM wave form with $\mu = 0.25$, if the modulating signal is (a) a sinusoid, (b) the triangular signal shown in Figure 3.6.

Answers: (a) $\eta = 0.0303$ or 3.03%. (b) $\eta = 0.02041$ or 2.041%

Poor power efficiency is a serious drawback of the AM system. The preceding Example and Practice problem show that AM power efficiency varies with the message signal type, and that it becomes progressively smaller as μ decreases. Typical audio signals tend to have pronounced and widely spaced peaks, but are low in amplitude most of the time. Avoiding over-modulation at the high peaks lowers the average modulation index and exacerbates the power efficiency problem. One way in which AM systems increase average modulation index and power efficiency without risking over-modulation is by passing the audio message signal through a peak-limiter prior to modulation.

3.2.3 Generation of AM signals

Nonlinear and switching AM modulators are discussed in this section. Figure 3.7(a) illustrates the principle of operation of a nonlinear or a switching AM modulator. In each modulator, the sum of the carrier and the modulating signal is passed through a nonlinear or a switching device. A bandpass filter is used to select the fundamental (AM signal) component from the output of the nonlinear/switching device. Figure 3.7(b) shows a circuit which can operate as either a nonlinear or a switching AM modulator. The diode input voltage is the sum of the carrier and the modulating signal. The diode can operate as a nonlinear device or a switching device, depending on whether its input voltage is smaller or much larger than the diode turn-on voltage V_{on}.

Nonlinear AM modulator

Consider the diode in Figure 3.7(b). The diode input voltage $v_D(t)$ is given by $v_D(t) = m(t) + A_c \cos \omega_c t$. If $v_D(t)$ is less than the diode turn-on voltage V_{on}, the diode I–V characteristic is nonlinear. Consequently the diode current $i_D(t)$ and the voltage drop $x(t)$ it produces across the resistance R are nonlinear functions of $v_D(t)$. Assuming that only the first two terms of the nonlinearity of $i_D(t)$ are significant and letting b_1, b_2, a_1, and a_2 be constants,

$$i_D(t) = b_1 v_D(t) + b_2 v_D^2(t) \tag{3.20}$$

$$\begin{aligned} x(t) = i_D(t)R &= b_1 R v_D(t) + b_2 R v_D^2(t) \\ &= a_1 v_D(t) + a_2 v_D^2(t) \end{aligned} \tag{3.21}$$

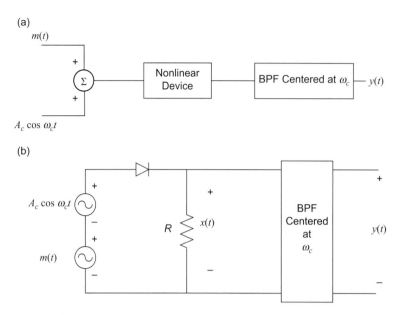

Figure 3.7. Nonlinear or switching AM generator: (a) block diagram; (b) nonlinear or switching AM generator circuit employing a diode as the nonlinear device.

Thus $x(t)$ can be expressed as

$$x(t) = a_1[m(t) + A_c \cos \omega_c t] + a_2[m(t) + A_c \cos \omega_c t]^2$$
$$= a_1 m(t) + a_2 m^2(t) + a_2 A_c^2 \cos^2 \omega_c t + A_c[a_1 + 2a_2 m(t)] \cos \omega_c t$$

The input $x(t)$ into the bandpass filter and the filter output $y(t)$ are respectively:

$$x(t) = a_1 m(t) + a_2 m^2(t) + \frac{a_2 A_c^2}{2}[\cos 2\omega_c t + 1] + A_c a_1 \left[1 + \frac{2a_2 m(t)}{a_1}\right] \cos \omega_c t \qquad (3.22)$$

$$y(t) = \phi_{AM}(t) = A_c a_1 \left[1 + \frac{2a_2 m(t)}{a_1}\right] \cos \omega_c t \qquad (3.23)$$

To satisfy the criterion for envelope detection, $\frac{2a_2|m(t)|}{a_1} < 1$.

Switching AM modulator

As previously indicated, the diode input voltage is $v_D(t) = m(t) + A_c \cos \omega_c t$ in Figure 3.7(b). The circuit will behave as a switching modulator only if the value of $v_D(t)$ is such as to either turn the diode fully on ($|v_D(t)| \gg V_{on}$), or else turn the diode fully off ($|v_D(t)| \leq 0$). Because the carrier amplitude exceeds message signal amplitude in AM, $|v_D(t)| \gg V_{on}$ is achieved by using a large amplitude carrier ($A_c \gg V_{on}$). Thus the diode will be fully on during the positive half cycle, and fully off during the negative half cycle of the carrier. Consequently, during each positive cycle of the carrier $x(t) = v_D(t)$, but during each negative half cycle $x(t) = 0$. In effect, $x(t)$ is equivalent to the product of $v_D(t)$ and the pulse train $p(t)$ shown in Figure 3.8.

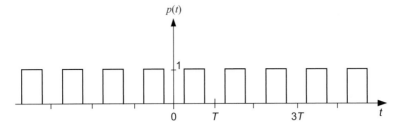

Figure 3.8. Switching pulse train.

Thus $x(t) = [m(t) + A_c \cos \omega_c t]p(t)$. Because $p(t)$ is a periodic signal similar to the rectangular pulse train of Table 2.2, its Fourier series expression can be obtained from the corresponding expression in that table, but with $A = 1$, $\tau = \frac{1}{2}T$, and $\omega_c = \frac{2\pi}{T}$. Since $\sin \frac{n\pi\tau}{T} = \sin \frac{n\pi}{2} = 0$ for even values of n, the Fourier series for $p(t)$ contains only odd powers of n and is given by

$$p(t) = \frac{1}{2} + \frac{2}{\pi}\left[\cos \omega_c t - \frac{1}{3}\cos 3\omega_c t + \frac{1}{5}\cos 5\omega_c t - \cdots\right] \qquad (3.24)$$

Thus,

$$x(t) = [m(t) + A_c \cos \omega_c t]\left[\frac{1}{2} + \frac{2}{\pi}\left(\cos \omega_c t - \frac{1}{3}\cos 3\omega_c t + \frac{1}{5}\cos 5\omega_c t - \cdots\right)\right]$$

$$= \frac{1}{2}\left[A_c \cos \omega_c t + \frac{4}{\pi}m(t)\cos \omega_c t\right] + \text{other terms not centered at } \omega_c \qquad (3.25)$$

The other terms referred to in this equation include dc, unmodulated, and higher-frequency modulation terms. These terms are suppressed by the bandpass filter centered at ω_c to yield the amplitude-modulated signal

$$y(t) = \phi_{AM}(t) = \frac{1}{2}\left[A_c \cos \omega_c t + \frac{4}{\pi}m(t)\cos \omega_c t\right] \qquad (3.26)$$

3.2.4 Demodulation of AM signals

An important advantage of amplitude modulation is that its demodulation by non-coherent methods is possible.

Coherent (or **synchronous**) **demodulation** (or **detection**) is a method of recovering the message signal from the modulated signal that requires a carrier at the receiver, which matches the transmitted carrier in phase and frequency.

Although possible, coherent demodulation of AM signals is rare because of its great complexity and high cost. Two noncoherent methods of demodulating AM signals employ the *envelope detector* and the *rectifier detector.*

AM envelope detector

The envelope detector and its principle of operation are illustrated in Figure 3.9. The input voltage in the circuit of part (a) is the AM signal $v_i(t) = [A_c + m(t)] \cos \omega_c t$. In the absence of the circuit components to the right of the resistor R, the voltage across R is just a half-wave rectified version of the AM waveform. This waveform is smoothed by the capacitor C, which charges to the peak value of the modulated carrier $v_i(t)$ during the positive half cycles. The charging time constant is $r_D C$, where r_D is the very small resistance of the diode in forward bias. Consequently, $r_D C$ is very small, and the charging curves hug the positive half cycles up to their peaks. As $v_i(t)$ decreases below the capacitor voltage $v_c(t)$, the diode turns off, and the capacitor discharges through R at a rate determined by the time constant RC. By suitable choice of the time constant, each discharge curve meets the succeeding positive half cycle near its peak. Thus $v_c(t) \approx |A_c + m(t)|$, and $v_c(t)$ gives a good approximation to the envelope of $v_i(t)$. Typically, this envelope is much smoother than is depicted in Figure 3.9(b), because the carrier is much higher in frequency relative to the message signal than is depicted in the illustration. As shown in Figure 3.9(c), this envelope is the message signal plus a dc component, which is the carrier amplitude. R_1 and C_1 constitute a dc-blocking device that removes the dc component.

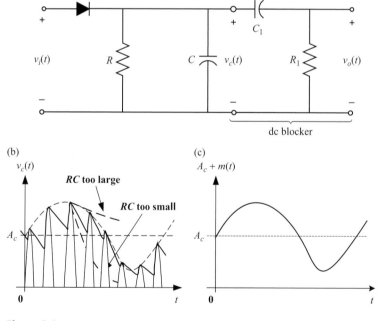

Figure 3.9. Envelope detection of AM signals. (a) Envelope detector plus dc blocker; (b) voltage across the capacitor; (c) desired envelope.

The output signal $v_o(t)$ will be a good approximation to the message signal $m(t)$, provided the AM signal was not over-modulated and its envelope was not distorted.

Consider an AM signal with carrier frequency f_c and message signal bandwidth B. To ensure a smooth envelope, the RC time constant should be much greater than the period of the carrier, otherwise the capacitor discharge curves would meet the succeeding positive half cycles of the carrier at rather low voltages. This requires that $RC \gg 1/f_c$. Also, RC should not be too large, otherwise the discharge curve would miss some carrier signal peaks when the envelope is decreasing in magnitude. This requires that $RC \ll 1/B$. Thus, RC should be chosen to lie in the range $\frac{1}{B} \gg RC \gg \frac{1}{f_c}$, or equivalently,

$$B \ll \frac{1}{RC} \ll f_c \tag{3.27}$$

An RC time constant within the above range is easily achievable since for AM the message signal bandwidth is $B = 5$ kHz, and f_c ranges from 550 kHz to 1600 kHz. An optimum choice of the RC time constant for a good tracking of the AM envelope is the geometric mean of its lower and upper limits:

$$RC = \sqrt{\frac{1}{Bf_c}} \tag{3.28}$$

EXAMPLE 3.4

The input signal to the envelope detector in a commercial AM radio receiver is given by $\phi_{AM}(t) = \left[10 + 4\sin\left(10\pi \times 10^3 t\right)\right]\cos\left(9.1\pi \times 10^5 t\right)$. The envelope detector uses a 10 nF smoothing capacitor. Specify an optimum value for the resistor in parallel with the smoothing capacitor for a good tracking of the AM envelope.

Solution

B = bandwidth of message signal = $\left(10\pi \times 10^3\right)/2\pi = 5$ kHz. f_c = carrier frequency = $\left(9.1\pi \times 10^5\right)/2\pi = 455$ kHz.

$$B \ll \frac{1}{RC} \ll f_c \Rightarrow 0.5 \times 10^4 \ll \frac{1}{RC} \ll 45.5 \times 10^4.$$

Choosing the optimum value of the RC time constant as the geometric mean of its possible minimum and maximum values,

$$\frac{1}{RC} = \sqrt{Bf_c} = 10^4\sqrt{0.5 \times 45.5} = 4.77 \times 10^4$$

$$\therefore R = \frac{1}{C\left(4.77 \times 10^4\right)} = \frac{1}{10 \times 10^{-9}\left(4.77 \times 10^4\right)} = 2.096 \text{ k}\Omega$$

PRACTICE PROBLEM 3.4

The AM signal $\phi_{AM}(t) = \left[5 + 2\sin\left(8\pi \times 10^3 t\right)\right]\cos\left(10^6\pi t\right)$ is the input signal into an envelope detector. Specify an optimum value for the resistor in parallel with a 5 nF smoothing capacitor for a smooth tracking of the AM envelope.

Answer: $R = 4.472$ kΩ

AM rectifier detector

The AM rectifier detector is illustrated in Figure 3.10. The input signal $v_i(t)$ is the AM signal. The half-wave rectified waveform $v_R(t)$ shown in Figure 3.10(a) is applied to a low-pass filter rather than to a smoothing capacitor as in the AM envelope detector. Note that $v_R(t) = v_i(t)$ during positive half cycles of the carrier, and $v_R(t) = 0$ during negative half cycles of the carrier. Consequently, $v_R(t) = v_i(t)p(t)$, where $p(t)$ is the periodic square wave signal given by Eq. (3.24) and depicted in Figure 3.8. Thus $v_R(t)$, which is shown in Figure 3.10(b), is given by

$$v_R(t) = v_i(t)p(t) = \left[(A_c + m(t))\cos\omega_c t\right]p(t)$$

$$= [A_c + m(t)]\cos\omega_c t\left[\frac{1}{2} + \frac{2}{\pi}\left(\cos\omega_c t - \frac{1}{3}\cos 3\omega_c t + \frac{1}{5}\cos 5\omega_c t - \cdots\right)\right]$$

(a)

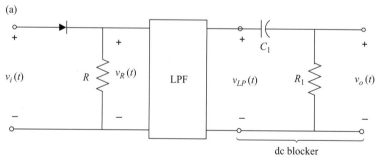

(b)   (c)

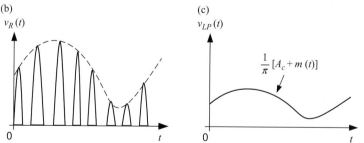

Figure 3.10. Rectifier detection of AM signals. (a) Rectifier detector plus dc blocker; (b) rectified AM signal; (c) low filter output.

Thus $v_R(t) = [A_c + m(t)] \left[\dfrac{1}{\pi} + \dfrac{1}{2} \cos \omega_c t + \dfrac{2}{3\pi} \cos 2\omega_c t + \text{higher frequency terms} \right].$

The output of the low-pass filter is $\frac{1}{\pi}[A_c + m(t)]$. The dc component of this signal is removed by the dc blocker formed by R_1 and C_1 to yield the demodulated signal $v_o(t) = m(t)/\pi$, which is a replica of the message signal. Since $p(t)$ contains the carrier and its harmonics, this technique utilizes the carrier in the received AM signal for demodulation. It is in effect a synchronous demodulation technique, although it does not require that a carrier be locally generated at the receiver.

3.3 DOUBLE SIDEBAND-SUPPRESSED CARRIER (DSB-SC) AM

In DSB-SC, the modulated signal does not contain a carrier component, therefore it has a power efficiency of 100%. DSB-SC modulation is illustrated in Figure 3.11. The block diagram of Figure 3.11(a) shows that the modulated signal $\phi_{DSB}(t)$ is just the product of the carrier $A_c \cos(\omega_c t + \theta_c)$ and the message signal $m(t)$. To simplify the relevant expressions, let the constant carrier amplitude A_c be equal to unity and its constant phase θ_c be zero. Then the carrier is simply $\cos \omega_c t$. Thus the modulated signal and its spectrum are respectively

$$\boxed{\phi_{DSB}(t) \quad = \quad m(t) \cos \omega_c t} \tag{3.29}$$

$$\boxed{\Phi_{DSB}(\omega) \quad = \quad \frac{1}{2}[M(\omega - \omega_c) + M(\omega + \omega_c)]} \tag{3.30}$$

The modulated waveform is shown in Figure 3.11(c), and its spectrum is shown in Figure 3.11(e). The DSB-SC spectrum consists of an upper sideband and a lower sideband. Like AM, it has a bandwidth of $2\Omega_m = 4\pi B$, which is twice the bandwidth of the message signal, but, unlike the AM spectrum, it does not contain carrier impulses at $\pm\omega_c$.

Synchronous demodulation of DSB-SC signals

Demodulation of DSB-SC signals involves shifting back the sidebands from $\pm\omega_c$ to the frequency origin, and passing the result through a low-pass filter. This procedure is depicted in the block diagram of Figure 3.12(a). To simplify subsequent mathematical expressions, the local carrier at the demodulator has an amplitude of 2, but it is identical in frequency and phase to the carrier used in the modulator. Hence, this is a coherent (synchronous) detector. The signal $x(t)$ and its spectrum $X(\omega)$ are given respectively by

$$x(t) = 2m(t) \cos^2 \omega_c t = m(t) + m(t) \cos 2\omega_c t \tag{3.31}$$

$$\therefore X(\omega) = M(\omega) + \frac{1}{2}[M(\omega - 2\omega_c) + M(\omega + 2\omega_c)] \tag{3.32}$$

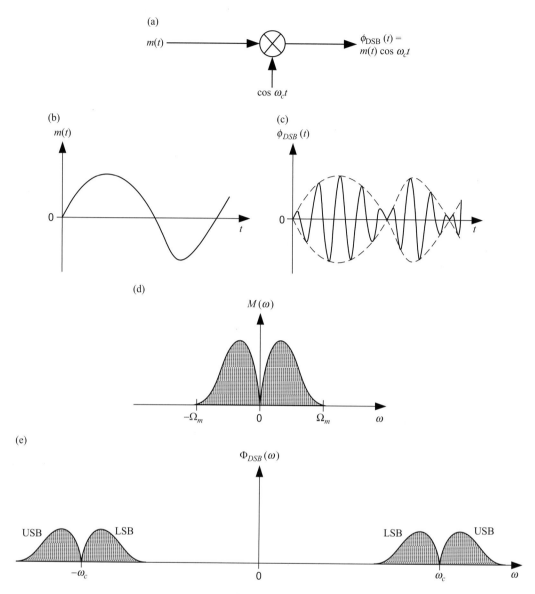

Figure 3.11. DSB-SC modulation. (a) Block diagram; (b) message signal; (c) DSB-SC waveform; (d) message signal spectrum; (e) DSB-SC signal spectrum.

Figure 3.12(b) and (c) shows $X(\omega)$ and the low-pass filter output which is the detector output $Y_d(\omega) = M(\omega)$. To prevent overlap of this message signal and the LSBs of its replicas centered at $\pm 2\omega_c$, Ω_m must not exceed $2\omega_c - \Omega_m$. Thus $\omega_c \geq \Omega_m$ for successful recovery of the message signal from the DSB-SC signal.

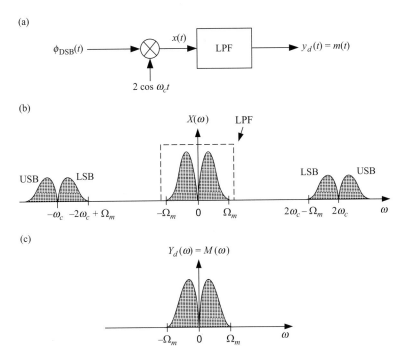

Figure 3.12. Synchronous demodulation of DSB-SC AM. (a) Block diagram of demodulator; (b) spectrum before low-pass filtering; (c) spectrum of recovered message signal.

EXAMPLE 3.5

In the DSB modulator of Figure 3.11(a), the carrier is $A_c \cos \omega_c t$ and the message signal is $m(t) = A_m \cos \omega_m t$. Find an expression for $\Phi_{DSB}(\omega)$ and sketch it.

Solution

$$\phi_{DSB}(t) = A_c A_m \cos \omega_c t \cos \omega_m t$$

$$= \frac{A_c A_m}{2} \left[\cos (\omega_c - \omega_m)t + \cos (\omega_c + \omega_m)t \right]$$

Taking the Fourier transform of $\phi_{DSB}(t)$,

$$\Phi_{DSB}(\omega) = \frac{\pi A_c A_m}{2} \{ [\delta(\omega - \omega_c + \omega_m)t + \delta(\omega + \omega_c - \omega_m)t]$$

$$+ [\delta(\omega - \omega_c - \omega_m)t + \delta(\omega + \omega_c + \omega_m)t] \}$$

The sketch for $\Phi_{DSB}(\omega)$ is shown in Figure 3.13. It consists of two impulses at $\pm(\omega_c + \omega_m)$ representing the upper sideband, and two impulses at $\pm(\omega_c - \omega_m)$ representing the lower sideband. Note that there are no carrier components in the spectrum.

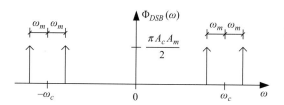

Figure 3.13. Spectrum for DSB signal of Example 3.5.

PRACTICE PROBLEM 3.5

In the DSB-SC modulator of Figure 3.11(a), the carrier is $\cos \omega_c t$ and the message signal is $m(t) = 6 \cos \omega_m t + 4 \sin 2\omega_m t$, where $\omega_c \gg \omega_m$. Find an expression for $\Phi_U(\omega)$, the upper sideband of the modulated signal $\phi_{DSB}(t)$, and for $\Phi_L(\omega)$, the lower sideband of $\phi_{DSB}(t)$.

Answer:

$$\Phi_U(\omega) = 3\pi[\delta(\omega + \omega_c + \omega_m) + \delta(\omega - \omega_c - \omega_m)]$$
$$+ j2\pi[\delta(\omega + \omega_c + 2\omega_m) + \delta(\omega - \omega_c - 2\omega_m)]$$

$$\Phi_L(\omega) = 3\pi[\delta(\omega + \omega_c - \omega_m) + \delta(\omega - \omega_c + \omega_m)]$$
$$+ j2\pi[\delta(\omega + \omega_c - 2\omega_m) + \delta(\omega - \omega_c + 2\omega_m)]$$

EXAMPLE 3.6

For a DSB-SC modulator the message signal is $m(t) = 8 \cos \omega_m t + 4 \cos 2\omega_m t$ and the carrier is $\sin \omega_c t$, where $\omega_m = 10\pi \times 10^3$ rad/s and $\omega_c = 40\pi \times 10^3$ rad/s. The DSB-SC signal is demodulated with a synchronous demodulator in which the local carrier is $\sin \omega_c t$ and the low-pass filter bandwidth is $15\pi \times 10^3$ rad/s. Find an expression for $x(t)$, the input to the low-pass filter and for $y_d(t)$, the output signal of the demodulator.

Solution

$$\phi_{DSB}(t) = (8 \cos \omega_m t + 4 \cos 2\omega_m t) \sin \omega_c t$$
$$= 4[\sin (\omega_c + \omega_m)t + \sin (\omega_c - \omega_m)t] + 2[\sin (\omega_c + 2\omega_m)t + \sin (\omega_c - 2\omega_m)t]$$
$$x(t) = \phi_{DSB} \sin \omega_c t = 4[\sin (\omega_c + \omega_m)t + \sin (\omega_c - \omega_m)t] \sin \omega_c t$$
$$+ 2[\sin (\omega_c + 2\omega_m)t + \sin (\omega_c - 2\omega_m)t] \sin \omega_c t$$
$$x(t) = 2[\cos (\omega_m)t - \cos (2\omega_c + \omega_m)t + \cos (\omega_m)t - \cos (2\omega_c - \omega_m)t]$$
$$+ \cos 2\omega_m t - \cos (2\omega_c + 2\omega_m)t + \cos 2\omega_m t - \cos (2\omega_c - 2\omega_m)t$$
$$x(t) = 4 \cos \omega_m t + 2 \cos 2\omega_m t$$
$$- \cos (2\omega_c - 2\omega_m)t - 2 \cos (2\omega_c - \omega_m)t - 2 \cos (2\omega_c + \omega_m)t - \cos (2\omega_c + 2\omega_m)t.$$

But $\omega_c = 4\omega_m$ and the low-pass filter bandwidth is $1.5\omega_m$, so

$$x(t) = 4 \cos \omega_m t + 2 \cos 2\omega_m t - \cos 6\omega_m t - 2 \cos 7\omega_m t - 2 \cos 9\omega_m t - \cos 10\omega_m t$$

The output signal is $y_d(t) = 4\cos\omega_m t$.

The first two terms in $x(t)$ should constitute the demodulated message signal. Because the low-pass filter bandwidth is smaller than the message signal bandwidth, the message signal component at $2\omega_m$ is blocked by the filter. It is crucial that the low-pass filter bandwidth be appropriate for the message signal.

PRACTICE PROBLEM 3.6

In the DSB-SC modulator of Figure 3.11(a), the carrier is $\cos\omega_c t$ and the message signal is $m(t) = 4\cos\omega_m t + 2\sin 3\omega_m t$, where $\omega_m = 5\pi \times 10^3$ rad/s and $\omega_c = 2\pi \times 10^4$ rad/s. The modulated signal is used as the input signal of the demodulator of Figure 3.12(a), in which the carrier is $2\cos\omega_c t$ and the low-pass filter bandwidth is 20 kHz. Find an expression for $x(t)$ of Figure 3.12(a) and for $y_d(t)$, the output signal of the demodulator.

Answer:

$$x(t) = 4\cos\omega_m t + 2\cos 7\omega_m t + 2\cos 9\omega_m t + 2\sin 3\omega_m t - \sin 5\omega_m t + \sin 11\omega_m t.$$

The low-pass filter bandwidth is $8\omega_m$, so the demodulator output is

$$y_d(t) = 4\cos\omega_m t + 2\sin 3\omega_m t - \sin 5\omega_m t + 2\cos 7\omega_m t.$$

Because the bandwidth of the low-pass filter exceeds that of the message signal, the components at $5\omega_m$ and $7\omega_m$, which are not part of the message signal, are present in the demodulator output. This again underscores the need for a careful choice of the low-pass filter bandwidth.

3.3.1 DSB-SC modulators

DSB-SC modulation and demodulation as illustrated in the block diagrams of Figure 3.11 and Figure 3.12 are both very simple conceptually. However, circuit implementation of the signal multiplication they involve is complex. Obtaining a local carrier at the receiver that is synchronous with the transmitted carrier is even more complex. Consequently, DSB-SC is too complex and expensive for use in commercial radio broadcasting. It is mostly employed in point-to-point communication where few receivers are needed, and where its high power efficiency is a great advantage in the long distances often involved.

Product DSB-SC modulators

Product modulators implement analog multiplication of signals electronically. One example is based on the variable transconductance multiplier, which is available in integrated circuit (IC) form. One input signal is applied to a differential amplifier with a gain proportional to its

(a)

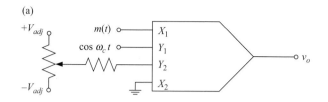

(b)

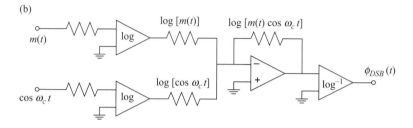

Figure 3.14. Product modulators for DSB-SC AM. (a) Variable transducer modulator; (b) log/antilog modulator.

variable transconductance. The other input signal is applied to a voltage-to-current converter, whose output current determines the transconductance of the differential amplifier. Thus, the output of the differential amplifier is proportional to the product of the two input signals. Such a product modulator employing AD534, an analog IC, by Analog Devices is shown in Figure 3.14(a). For this IC, the maximum value for each signal is 10 V. Let the input signals be $X_1 = m(t)$ and $Y_1 = \cos \omega_c t$. The output signal is $v_0 = \frac{1}{10} X_1 Y_1$ to keep v_o below the supply voltages $\pm V_{cc}$, which is usually about ± 15 V. Possible component mismatches can result in non-zero v_o when either of the input signals is zero. Such an error can be nulled with a small adjustable voltage within the ± 30 mV range applied to the Y_2 input as shown in the figure.

Another example of a product modulator is the logarithmic product modulator shown in Figure 3.14(b). It utilizes two input log amplifiers, a summing amplifier, and an output antilog amplifier. When the two input signals are $m(t)$ and $\cos \omega_c t$, it is clear from the signal flow that the output is the modulated signal $\phi_{DSB}(t) = m(t) \cos \omega_c t$.

Nonlinear DSB-SC modulators

Two nonlinear AM modulators can be used to implement a nonlinear DSB-SC modulator. Such an implementation, which utilizes two nonlinear AM modulators of the type shown in Figure 3.7, is shown in Figure 3.15. Note that the input signals to one of the AM modulators are $m(t)$ and $\cos \omega_c t$, while the inputs to the other AM modulator are $-m(t)$ and $\cos \omega_c t$. Applying the results of Eq. (3.22) to $x_1(t)$ and $x_2(t)$ of Figure 3.15,

$$x_1(t) = a_1 m(t) + a_2 m^2(t) + \frac{a_2 A_c^2}{2} [\cos 2\omega_c t + 1] + A_c a_1 \left[1 + \frac{2a_2 m(t)}{a_1}\right] \cos \omega_c t$$

$$x_2(t) = -a_1 m(t) + a_2 m^2(t) + \frac{a_2 A_c^2}{2} [\cos 2\omega_c t + 1] + A_c a_1 \left[1 - \frac{2a_2 m(t)}{a_1}\right] \cos \omega_c t$$

$$\therefore \quad x(t) = x_1(t) - x_2(t) = 2a_1 m(t) + 4A_c a_2 m(t) \cos \omega_c t \tag{3.33}$$

After passing $x(t)$ through the bandpass filter, the filter output is the DSB-SC signal

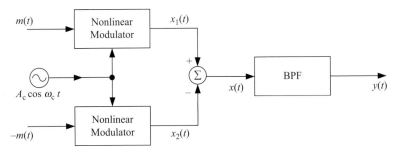

Figure 3.15. Nonlinear DSB-SC modulator.

$$y(t) = \phi_{DSB}(t) = 4A_c a_2 m(t) \cos \omega_c t \tag{3.34}$$

Note that the input signal to the bandpass filter $x(t)$ does not contain a carrier component. Hence, the modulator is said to be *balanced* with respect to the carrier.

A **balanced modulator** is one whose unfiltered output does not contain either a carrier component or a message signal component. It is balanced with respect to the component missing from the unfiltered output. If both the carrier and message signal components are missing from the unfiltered output, the modulator is said to be **double balanced**.

Of what significance is the balance of a modulator? If the above modulator were not balanced with respect to the carrier, the input to the bandpass filter centered at ω_c would contain a carrier component at that frequency. Thus the filter would not suppress the carrier, and the filter output would be an AM signal instead of the desired DSB-SC signal. In general, the more balanced a modulator is in a suppressed carrier system, the easier the filtering required to obtain the desired modulated signal.

Switching DSB-SC modulators

Three switching modulators which employ diodes as switching elements are discussed below. As explained earlier, the diodes act as switching elements when the amplitude of the carrier is much greater than the diode turn-on voltage V_{on}.

Series-bridge modulator

The series-bridge modulator is shown in Figure 3.16(a). When the carrier $\cos \omega_c t$ is positive at a, all the diodes are on, $V_b = V_d$, and so $x(t) = m(t)$. During the negative half cycle of the carrier, all the diodes are off, making $x(t) = 0$. Thus, the input signal to the bandpass filter is $x(t) = m(t)P(t)$, where $p(t)$ is the square pulse waveform shown in Figure 3.16(c), which has the same frequency as the carrier $\cos \omega_c t$. It will be shown that input signal to the bandpass filter is also $x(t) = m(t)P(t)$ in the shunt-bridge modulator to be discussed next. Hence, further analysis is deferred until after the discussion of the operation of that modulator.

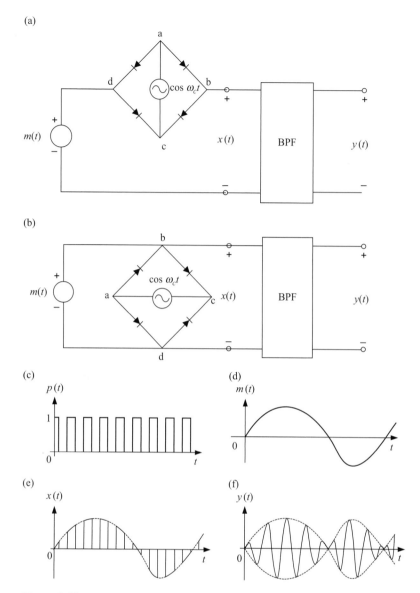

Figure 3.16. Switching DSB-SC modulators. (a) Series-bridge modulator; (b) shunt-bridge modulator; (c) switching pulse train; (d) message signal; (e) product of pulse train and message signal; (f) DSB-SC modulated output signal.

Shunt-bridge modulator

The shunt-bridge modulator is shown in Figure 3.16(b). When the carrier is positive at a, all the diodes are on, a short circuit exists between b and d, consequently $x(t) = 0$. During the negative half cycle of the carrier, all the diodes are off, an open circuit exists between b and d, making $x(t) = m(t)$. Thus, the bandpass filter input signal is $x(t) = m(t)P(t)$, as in the series-bridge modulator. Hence the following analysis suffices for both modulators.

Employing the expression for $p(t)$ given in Eq. (3.24),

$$x(t) = m(t)\left\{\frac{1}{2} + \frac{2}{\pi}\left[\cos\omega_c t - \frac{1}{3}\cos 3\omega_c t + \frac{1}{5}\cos 5\omega_c t - \cdots\right]\right\}$$

$$= \frac{1}{2}m(t) + \frac{2}{\pi}\left[m(t)\cos\omega_c t - \frac{1}{3}m(t)\cos 3\omega_c t + \frac{1}{5}m(t)\cos 5\omega_c t - \cdots\right] \qquad (3.35)$$

Note that $x(t)$ contains a message signal term but not a carrier term. Thus, both modulators are *balanced with respect to the carrier*. The waveforms for $m(t)$, $p(t)$, $x(t)$, and $y(t)$, are shown in Figure 3.16. The output of the bandpass filter centered at ω_c is the DSB-SC signal $y(t)$, and is given by

$$y(t) = \phi_{DSB} = \frac{2}{\pi}m(t)\cos\omega_c t \qquad (3.36)$$

Ring modulator

The ring modulator is illustrated in Figure 3.17(a). The center-tapped transformers can be assumed to have a 1:1 transformation ratio. To underscore the fact that the switching signal should be a periodic signal of the same frequency as the carrier, but not necessarily a sinusoid, a periodic pulse waveform similar to $p_1(t)$ of Figure 3.17(b) is employed here in place of a sinusoidal carrier. When the pulse amplitude is positive at a, diodes D_1 and D_2 conduct, while diodes D_3 and D_4 are off, so $x(t) = m(t)$. During the negative half cycle of the periodic pulse waveform, diodes D_1 and D_2 are off while diodes D_3 and D_4 conduct, the direction of current through the output transformer is reversed, resulting in $x(t) = -m(t)$. Consequently the signal at the input of the bandpass filter can be expressed as $x(t) = m(t)p_1(t)$. Note that $p(t)$ is a unipolar signal, whereas $p_1(t)$ is a bipolar signal. By expressing $p_1(t)$ in terms of $p(t)$, its Fourier series can be obtained from the Fourier series for $p(t)$ given in Eq. (3.24).

$$p_1(t) = 2\left[p(t) - \frac{1}{2}\right] = 2p(t) - 1$$

$$= 2\left\{\frac{1}{2} + \frac{2}{\pi}\left[\cos\omega_c t - \frac{1}{3}\cos 3\omega_c t + \frac{1}{5}\cos 5\omega_c t - \cdots\right]\right\} - 1$$

$$\therefore\ p_1(t) = \frac{4}{\pi}\left[\cos\omega_c t - \frac{1}{3}\cos 3\omega_c t + \frac{1}{5}\cos 5\omega_c t - \cdots\right] \qquad (3.37)$$

The input signal into the bandpass filter $x(t) = m(t)p(t)$ is

$$x(t) = \frac{4}{\pi}\left[m(t)\cos\omega_c t - \frac{1}{3}m(t)\cos 3\omega_c t + \frac{1}{5}m(t)\cos 5\omega_c t - \cdots\right] \qquad (3.38)$$

The output of the bandpass filter centered at ω_c is the DSB-SC signal

$$y(t) = \phi_{DSB}(t) = \frac{4}{\pi}m(t)\cos\omega_c t \qquad (3.39)$$

(a)

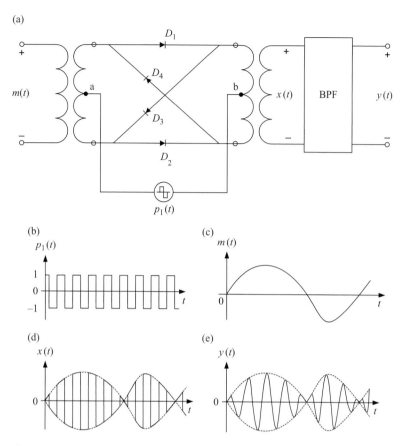

Figure 3.17. (a) Ring modulator; (b) switching pulse train; (c) message signal; (d) product of pulse train and message signal; (e) DSB-SC modulated output signal.

The waveforms for $m(t)$, $p_1(t)$, $x(t)$, and $y(t)$ are shown in Figure 3.17. Note that the input to the bandpass filter $x(t)$ does not contain any carrier or message signal components. Hence the ring modulator is a *double balanced modulator*.

3.3.2 Frequency mixer

Frequency mixing, also known as *frequency conversion or heterodyning*, consists of changing the carrier frequency of a modulated signal to new carrier frequency. It plays a very important role in radio receivers in which it is used to convert the carrier frequency of any channel to which the receiver is tuned to the *intermediate carrier frequency*. A demodulator with characteristics matched to the intermediate frequency is then used for demodulating any received channel.

The principle of operation of a frequency mixer is illustrated in the block diagram of Figure 3.18(a). The input modulated signal is $\phi(t) = m(t) \cos \omega_c t$, and the desired output of the mixer at the intermediate frequency ω_I is $\phi_I(t) = m(t) \cos \omega_I t$. The local oscillator (local to

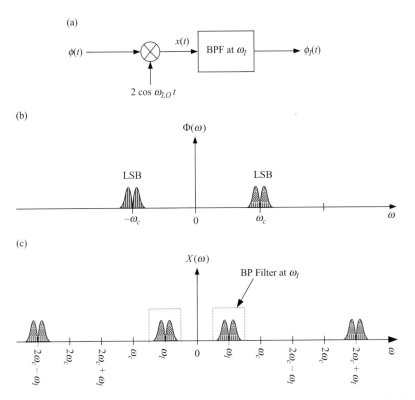

Figure 3.18. Frequency translation. (a) Block diagram; (b) spectrum of modulated input signal; (c) Spectrum of product of input modulated signal and local oscillator signal.

the receiver) signal is $\cos(\omega_{LO}t)$, where ω_{LO} is chosen so that either $\omega_{LO} = \omega_c - \omega_I$ or else $\omega_{LO} = \omega_c + \omega_I$. The multiplier output is $x(t) = 2\phi(t)\cos\omega_{LO}t = 2m(t)\cos\omega_{LO}t\cos\omega_c t$.

If $\omega_{LO} = \omega_c - \omega_I$, then $x(t) = m(t)[\cos(\omega_c - \omega_I - \omega_c)t + \cos(\omega_c - \omega_I + \omega_c)t]$. Because the cosine is an even function of its argument, $\cos(-\omega_I t) = \cos(\omega_I t)$ and

$$x(t) = m(t)\cos\omega_I t + m(t)\cos(2\omega_c - \omega_I)t \qquad (3.40)$$

If $\omega_{LO} = \omega_c + \omega_I$, then $x(t) = m(t)[\cos(\omega_c + \omega_I - \omega_c)t + \cos(\omega_c + \omega_I + \omega_c)t]$, giving

$$x(t) = m(t)\cos\omega_I t + m(t)\cos(2\omega_c + \omega_I)t \qquad (3.41)$$

In either case, $x(t)$ consists of two modulated signals, one at the intermediate carrier frequency ω_I and the other at either the higher frequency $2\omega_c - \omega_I$, or the higher frequency $2\omega_c + \omega_I$. The output of the bandpass filter centered at ω_I is the desired modulated signal at the intermediate frequency. A superheterodyning mixer is one in which $\omega_{LO} = \omega_c + \omega_I$, whereas a *heterodyning* mixer is one in which $\omega_{LO} = \omega_c - \omega_I$. Example 3.8 below shows that superheterodyning results in a lower tuning range for frequency-converter local oscillators. Consequently, super-heterodyning is preferred in radio receivers since the difficulty of component realization increases with tuning range. The spectrum for a superheterodyne frequency mixer is illustrated in Figure 3.18(c).

EXAMPLE 3.7

Derive the relationship between ω_{LO} and ω_c such that centering the bandpass filter of the frequency converter of Figure 3.18 at $\omega_{LO} - \omega_c$ will ensure that ω_I is always greater than ω_c.

Solution

Bandpass filter centered at $\omega_{LO} - \omega_c \Rightarrow \omega_I = \omega_{LO} - \omega_c$.

For, $\omega_I > \omega_c$. Let $\omega_I = \omega_c + x$, where x can be arbitrarily small.

Then $\omega_I = \omega_{LO} - \omega_c = \omega_c + x \Rightarrow \omega_{LO} = 2\omega_c + x$

Thus, ω_{LO} and ω_c must satisfy: $\omega_{LO} > 2\omega_c$

PRACTICE PROBLEM 3.7

Derive the relationship between ω_{LO} and ω_c such that centering the bandpass filter of the frequency converter of Figure 3.18 at $\omega_{LO} - \omega_c$ will ensure that ω_I is always less than ω_c.

Answer: $\omega_{LO} < 2\omega_c$

EXAMPLE 3.8

For a frequency converter the carrier frequency of the output signal is 500 kHz and the carrier frequency of the AM input signal ranges from 600 kHz to 1700 kHz. Find the tuning ratio of the local oscillator $\frac{f_{LO,max}}{f_{LO,min}}$, if the frequency of the local oscillator is given by (a) $f_I = f_{LO} - f_c$ and (b) $f_I = f_c - f_{LO}$.

Solution

(a) $f_I = f_{LO} - f_c \Rightarrow f_{LO} = f_c + f_I \Rightarrow$ superheterodyning

$$\therefore \quad \frac{f_{LO,max}}{f_{LO,min}} = \frac{f_{c,max} + f_I}{f_{c,min} + f_I} = \frac{1700 + 500}{600 + 500} = 2.$$

(b) $f_I = f_c - f_{LO} \Rightarrow f_{LO} = f_c - f_I \Rightarrow$ heterodyning

$$\therefore \quad \frac{f_{LO,max}}{f_{LO,min}} = \frac{f_{c,max} - f_I}{f_{c,min} - f_I} = \frac{1700 - 500}{600 - 500} = 12.$$

PRACTICE PROBLEM 3.8

Repeat Example 3.8 if the carrier frequency of the frequency-converter output signal is 425 kHz, and the carrier frequency of the input signal ranges from 500 kHz to 1500 kHz.

Solution: (a) $\frac{f_{LO,max}}{f_{LO,min}} = 2.081$. (b) $\frac{f_{LO,max}}{f_{LO,min}} = 14.33$

3.4 QUADRATURE AMPLITUDE MODULATION (QAM)

Both DSB-SC AM and conventional AM are wasteful of bandwidth; they utilize twice the bandwidth of the message signal. QAM overcomes this shortcoming by transmitting two DSB-SC signals in the bandwidth of one DSB-SC signal. Interference between the two modulated signals is prevented by using two carriers which have the same frequency but are in phase quadrature, and hence are orthogonal to each other. The QAM modulator and demodulator are illustrated in the block diagrams of Figure 3.19. The channel with a carrier of $\cos \omega_c t$ is the *in-phase channel* (*I*-channel), while the channel with a carrier of $\cos (\omega_c t - 90°) = \sin \omega t$ is the *quadrature channel* (*Q*-channel). The QAM signal is easily obtained from the QAM modulator block diagram of Figure 3.19(a) as

$$\boxed{\phi_{QAM}(t) \quad = \quad m_1(t) \cos \omega_c t + m_2(t) \sin \omega_c t} \qquad (3.42)$$

The carriers employed in the transmitter and the receiver are synchronous with each other, since each has a frequency of ω_c and a phase angle of zero. Thus, QAM receivers employ synchronous (coherent) detection. The signals $z_1(t)$ and $z_2(t)$ in the demodulator of Figure 3.19(b) are given respectively by

$$z_1(t) = 2 \cos \omega_c t \left[\phi_{QAM}(t)\right] = 2 \cos \omega_c t[m_1(t) \cos \omega_c t + m_2(t) \sin \omega_c t]$$
$$= 2m_1(t) \cos^2 \omega_c t + 2m_2(t) \cos \omega_c t \sin \omega_c t$$
$$= m_1(t) + m_1(t) \cos 2\omega_c t + m_2(t) \sin 2\omega_c t$$

and

$$z_2(t) = 2 \sin \omega_c t \left[\phi_{QAM}(t)\right] = 2 \sin \omega_c t[m_1(t) \cos \omega_c t + m_2(t) \sin \omega_c t]$$
$$= 2m_2(t) \sin^2 \omega_c t + 2m_1(t) \sin \omega_c t \cos \omega_c t$$
$$= m_2(t) - m_2(t) \cos 2\omega_c t + m_1(t) \sin 2\omega_c t$$

After passing $z_1(t)$ and $z_2(t)$ through the low-pass filters, the respective outputs of the *I*-channel and the *Q*-channel are $y_1(t) = m_1(t)$ and $y_2(t) = m_2(t)$, which are the original message signals. Unfortunately, phase and frequency errors in receiver local carriers cause serious problems in coherent detection of QAM. As illustrated in the examples below, these problems are more serious in QAM than in DSB-SC. In addition to the signal attenuation and distortion they cause in both DSB-SC and QAM receivers, local carrier phase and frequency errors also cause *co-channel interference* in QAM receivers.

Channels which share the same bandwidth, such as the *I*-channel and the *Q*-channel in QAM, are known as **co-channels**, and interference between them is known as **co-channel interference**.

(a)

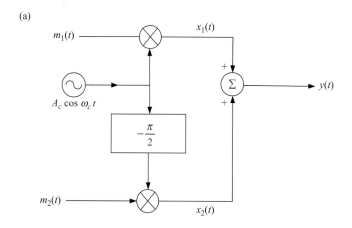

(b)

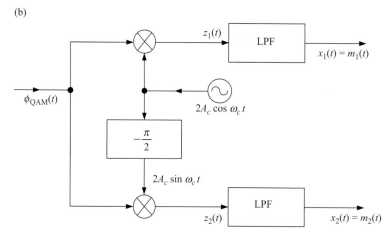

Figure 3.19. Quadrature amplitude modulation. (a) Block diagram of QAM modulator; (b) block diagram of QAM demodulator.

EXAMPLE 3.9

The local carrier in a DSB-SC receiver and a QAM receiver is $2\cos(\omega_c t + \alpha)$, while the carrier at each transmitter is $\cos\omega_c t$. Derive expressions for the demodulated output signals for the receivers, and compare the effects of phase error in synchronous detection of DSB-SC and QAM.

Solution

For the DSB-SC receiver of Figure 3.12,

$$x(t) = 2\cos(\omega_c t + \alpha)\phi_{DSB}(t) = 2m(t)\cos\omega_c t\cos(\omega_c t + \alpha)$$
$$= m(t)\cos\alpha + m(t)\cos(2\omega_c t + \alpha)$$

The low-pass filter suppresses the term centered at $2\omega_c$, yielding the output

$$y(t) = m(t)\cos\alpha \tag{3.43}$$

For the QAM receiver, $z_1(t)$ and $z_2(t)$ are given respectively by

$$
\begin{aligned}
z_1(t) &= 2\cos\left(\omega_c t + \alpha\right)\left[\phi_{QAM}(t)\right] = 2\cos\left(\omega_c t + \alpha\right)[m_1(t)\cos\omega_c t + m_2(t)\sin\omega_c t] \\
&= 2m_1(t)\cos\omega_c t\cos\left(\omega_c t + \alpha\right) + 2m_2(t)\sin\omega_c t\cos\left(\omega_c t + \alpha\right) \\
&= m_1(t)\cos\alpha + m_1(t)\cos\left(2\omega_c t + \alpha\right) - m_2(t)\sin\alpha + m_2(t)\sin\left(2\omega_c t + \alpha\right)
\end{aligned}
$$

and

$$
\begin{aligned}
z_2(t) &= 2\sin\left(\omega_c t + \alpha\right)\left[\phi_{QAM}(t)\right] = 2\sin\left(\omega_c t + \alpha\right)[m_1(t)\cos\omega_c t + m_2(t)\sin\omega_c t] \\
&= 2m_1(t)\cos\omega_c t\sin\left(\omega_c t + \alpha\right) + 2m_2(t)\sin\omega_c t\sin\left(\omega_c t + \alpha\right) \\
&= m_1(t)\sin\alpha + m_1(t)\sin\left(2\omega_c t + \alpha\right) + m_2(t)\cos\alpha - m_2(t)\cos\left(2\omega_c t + \alpha\right)
\end{aligned}
$$

The low-pass filters suppress the terms centered at $2\omega_c$, yielding respectively,

$$
y_1(t) = m_1(t)\cos\alpha - m_2(t)\sin\alpha \tag{3.44}
$$

$$
y_2(t) = m_2(t)\cos\alpha + m_1(t)\sin\alpha \tag{3.45}
$$

Like the demodulated output of the DSB-SC receiver, the first term in each of the two QAM receiver outputs is the message signal attenuated by a factor of $\cos\alpha$. The second terms in the QAM receiver outputs are portions of undesired channel message signals which appear at the desired channel output. These represent *co-channel interference*.

PRACTICE PROBLEM 3.9

The local carrier with a small phase error in a DSB-SC receiver and a QAM receiver is $2\cos\left(\omega_c t + 10°\right)$, while the carrier at each transmitter is $\cos\omega_c t$. Derive expressions for the receiver outputs.

Answer: DSB-SC receiver output: $y(t) = 0.985m(t)$
QAM receiver outputs: $y_1(t) = 0.985m_1(t) - 0.174m_2(t)$

$$
y_2(t) = 0.985m_2(t) + 0.174m_2(t)
$$

EXAMPLE 3.10

Compare the effect of a small frequency error in the local carrier for a DSB-SC receiver and a QAM receiver by deriving expressions for the receiver outputs. The carrier at the transmitter is $\cos\omega_c t$ and the carrier at the receiver is $2\cos\left(\omega_c + \Delta\omega\right)t$.

Solution

For the DSB-SC receiver, the input into the low-pass filter is given by

$$
\begin{aligned}
x(t) &= \phi_{DSB}(t)[2\cos\left(\omega_c + \Delta\omega\right)t] = 2m(t)\cos\omega_c t\cos\left(\omega_c + \Delta\omega\right)t \\
&= m(t)\cos\Delta\omega t + m(t)\cos\left(2\omega_c + \Delta\omega\right)t
\end{aligned}
$$

The low-pass filter suppresses the component centered at $2\omega_c$. The demodulator output is

$$y(t) = m(t) \cos \Delta\omega t \qquad (3.46)$$

For the QAM receiver, the inputs to the two low-pass filters are

$$
\begin{aligned}
z_1(t) &= 2\cos(\omega_c + \Delta\omega)t \left[\phi_{QAM}(t)\right] = 2\cos(\omega_c + \Delta\omega)t[m_1(t)\cos\omega_c t + m_2(t)\sin\omega_c t] \\
&= 2m_1(t)\cos\omega_c t \cos(\omega_c + \Delta\omega)t + 2m_2(t)\sin\omega_c t \cos(\omega_c + \Delta\omega)t \\
&= m_1(t)\cos\Delta\omega t + m_1(t)\cos(2\omega_c + \Delta\omega)t - m_2(t)\sin\Delta\omega t + m_2(t)\sin(2\omega_c + \Delta\omega)t
\end{aligned}
$$

and

$$
\begin{aligned}
z_2(t) &= 2\sin(\omega_c + \Delta\omega)t \left[\phi_{QAM}(t)\right] = 2\sin(\omega_c + \Delta\omega)t[m_1(t)\cos\omega_c t + m_2(t)\sin\omega_c t] \\
&= 2m_1(t)\cos\omega_c t \sin(\omega_c + \Delta\omega)t + 2m_2(t)\sin\omega_c t \sin(\omega_c + \Delta\omega)t \\
&= m_1(t)\sin\Delta\omega t + m_1(t)\sin(2\omega_c + \Delta\omega)t + m_2(t)\cos\Delta\omega t - m_2(t)\cos(2\omega_c + \Delta\omega)t
\end{aligned}
$$

After low-pass filtering, the two channel outputs are

$$y_1(t) = m_1(t)\cos\Delta\omega t - m_2(t)\sin\Delta\omega t \qquad (3.47)$$

$$y_2(t) = m_2(t)\cos\Delta\omega t + m_1(t)\sin\Delta\omega t \qquad (3.48)$$

The first term in each of the two QAM receiver outputs is similar to the output of the DSB-SC receiver. Each is the recovered message signal multiplied by a very low-frequency sinusoid. The resulting distortion is a slow and sinusoidal attenuation of the recovered message signal. At zeros of $\cos\Delta\omega t$, the recovered message signal is completely nulled out, but just at that time, the second term, the co-channel interference which is proportional to $\sin\Delta\omega t$, is at its maximum.

PRACTICE PROBLEM 3.10

The local carrier in a QAM receiver has both a phase error and a frequency error. The carrier at the transmitter is given by $\cos\omega_c t$, while the carrier at the receiver is given by $2\cos[(\omega_c + \Delta\omega)t + \alpha]$. Derive expressions for the demodulated output signals of the I-channel and the Q-channel.

Answer:

$$y_1(t) = m_1(t)\cos(\Delta\omega t + \alpha) - m_2(t)\sin(\Delta\omega t + \alpha)$$

$$y_2(t) = m_2(t)\cos(\Delta\omega t + \alpha) + m_1(t)\sin(\Delta\omega t + \alpha)$$

3.5   SINGLE SIDEBAND (SSB) AM

As indicated earlier, each of the sidebands in a double sideband system carries all the information in the message signal. Thus each sideband is sufficient for recovery of the message

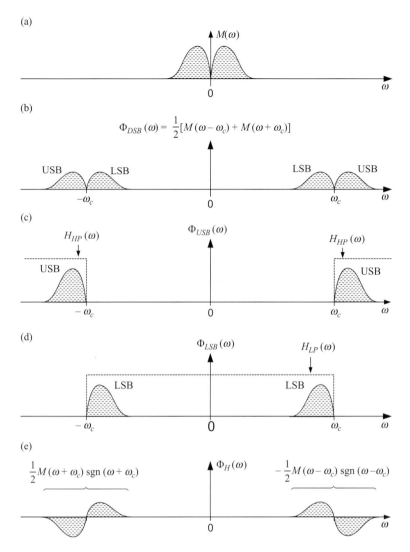

Figure 3.20. Illustration of signal sideband AM. (a) Message signal spectrum; (b) DSB-SC spectrum; (c) upper sideband version of SSB AM; (d) lower sideband version of SSB AM; (e) $\Phi_H(\omega)$.

signal. Single sideband (SSB) AM avoids waste of bandwidth by transmitting only one of the sidebands. A frequency-domain illustration of SSB AM is shown in Figure 3.20. The upper sideband version $\Phi_{USB}(\omega)$ shown in part (c) can be obtained by passing the DSB-SC signal of part (b) through a high-pass filter, while the lower sideband version, $\Phi_{LSB}(\omega)$, of part (d) can be obtained by passing $\Phi_{DSB}(\omega)$ through a low-pass filter. The spectrum $\Phi_H(\omega)$ shown in part (e) is obtained by performing some mathematical operations on the DSB spectrum. The upper and lower sideband spectra $\Phi_{USB}(\omega)$ and $\Phi_{LSB}(\omega)$ can be generated from linear combinations of $\Phi_{DSB}(\omega)$ and $\Phi_H(\omega)$. The ultimate objective is to obtain time-domain expressions for the upper and lower sideband versions of SSB.

3.5.1 The Hilbert transform and the time-domain expression for SSB

The Hilbert transform is crucial in the derivation of the time-domain expression for SSB. The spectrum $\Phi_H(\omega)$ is helpful in introducing the Hilbert transform. $\Phi_{DSB}(\omega)$ and the spectrum $\Phi_H(\omega)$ of Figure 3.20(e) can respectively be expressed as

$$\Phi_{DSB}(\omega) = \frac{1}{2}[M(\omega + \omega_c) + M(\omega - \omega_c)]$$

$$\Phi_H(\omega) = \frac{1}{2}[M(\omega + \omega_c)]\text{sgn}(\omega + \omega_c) - \frac{1}{2}[M(\omega - \omega_c)]\text{sgn}(\omega - \omega_c)$$

Note that $\text{sgn}(\omega + \omega_c)$ and $\text{sgn}(\omega - \omega_c)$ which appear in the expression for $\Phi_H(\omega)$ are frequency-shifted versions of the signum function $\text{sgn}(\omega)$, which is $+1$ for $\omega > 0$ and -1 for $\omega < 0$. The upper and lower sidebands $\Phi_{USB}(\omega)$ and $\Phi_{LSB}(\omega)$ can be obtained by combining $\Phi_{DSB}(\omega)$ and $\Phi_H(\omega)$ as follows:

$$\Phi_{USB}(\omega) = [\Phi_{DSB}(\omega) - \Phi_H(\omega)] \tag{3.49}$$

$$\Phi_{LSB}(\omega) = [\Phi_{DSB}(\omega) + \Phi_H(\omega)] \tag{3.50}$$

But what exactly is the signal $\Phi_H(\omega)$? The Fourier transform of the time-domain signum function was found in Example 2.14 to be $\text{sgn}(t) \Leftrightarrow \frac{2}{j\omega}$. Applying the duality property of the Fourier transform, and noting that the signum function is an odd function of its argument, yields

$$\frac{2}{jt} \Leftrightarrow 2\pi\,\text{sgn}(-\omega) = -2\pi\,\text{sgn}(\omega)$$

$$\therefore \quad \frac{1}{\pi t} \Leftrightarrow -j\,\text{sgn}(\omega) = \begin{cases} j, & \omega < 0 \\ -j, & \omega > 0 \end{cases} \tag{3.51}$$

The above implies that $-jM(\omega)\text{sgn}(\omega)$ is obtained by passing $M(\omega)$ through a pure phase filter which shifts the phase of its positive frequency part by $-90°$ and that of its negative frequency part by $+90°$. Such an operation is known as *Hilbert transformation* or *quadrature filtering*.

Hilbert transformation or *quadrature filtering* is an all-pass filtering operation which leaves the magnitude spectrum of the input signal unchanged, but causes a phase shift of $-90°$ for its positive frequency spectrum, and a phase shift of $+90°$ for its negative frequency spectrum.

Let $m_h(t)$ and $M_h(\omega)$ be the Hilbert transform of $m(t)$ in time and frequency domains. Then

$$m(t) * \frac{1}{\pi t} = m_h(t) \Leftrightarrow M_h(\omega) = -jM(\omega)\text{sgn}(\omega) \tag{3.52}$$

Thus the Hilbert transform of $m(t)$ in the time domain is the convolution

$$\boxed{m_h(t) \quad = \quad \frac{1}{\pi} \int_{-\infty}^{\infty} \frac{m(\lambda)}{t - \lambda} d\lambda} \tag{3.53}$$

To recover $M(\omega)$ from $M_h(\omega)$, or to perform *inverse Hilbert transformation in the frequency domain*, multiply $M_h(\omega)$ by $j \operatorname{sgn}(\omega)$ to obtain

$$M_h(\omega)[j \operatorname{sgn}(\omega)] = \{M(\omega)[-j \operatorname{sgn}(\omega)]\}[j \operatorname{sgn}(\omega)] = M(\omega) \tag{3.54}$$

Conversely, to perform *inverse Hilbert transformation in the time domain*, convolve $m_h(t)$ with $F^{-1}[j \operatorname{sgn}(\omega)] = -\frac{1}{\pi t}$ to obtain

$$\boxed{m(t) \quad = \quad -\frac{1}{\pi} \int_{-\infty}^{\infty} \frac{m_h(\lambda)}{t - \lambda} d\lambda} \tag{3.55}$$

Armed with the Hilbert transform, we are now in a position to derive the time-domain expression for $\Phi_H(\omega)$ of Figure 3.20(e) and hence the time-domain expression for the SSB signal. Since $-\frac{1}{j\pi t} \Leftrightarrow \operatorname{sgn}(\omega)$, it is clear from the frequency-shifting property that $-\frac{1}{j\pi t} e^{j\omega_c t} \Leftrightarrow \operatorname{sgn}(\omega - \omega_c)$ and that $-\frac{1}{j\pi t} e^{-j\omega_c t} \Leftrightarrow \operatorname{sgn}(\omega + \omega_c)$. Thus

$$m_h(t) e^{j\omega_c t} = \left[\frac{1}{\pi t} * m(t)\right] e^{j\omega_c t} \Leftrightarrow -jM(\omega - \omega_c)\operatorname{sgn}(\omega - \omega_c) = M_h(\omega - \omega_c)$$

$$m_h(t) e^{-j\omega_c t} = \left[\frac{1}{\pi t} * m(t)\right] e^{-j\omega_c t} \Leftrightarrow -jM(\omega + \omega_c)\operatorname{sgn}(\omega + \omega_c) = M_h(\omega + \omega_c)$$

But as previously obtained from Figure 3.20, $\Phi_H(\omega)$ is given by

$$\Phi_H(\omega) = \frac{1}{2}[M(\omega + \omega_c)]\operatorname{sgn}(\omega + \omega_c) - \frac{1}{2}[M(\omega - \omega_c)]\operatorname{sgn}(\omega - \omega_c)$$

Thus the frequency- and time-domain expressions for $\Phi_H(\omega)$ are given respectively by

$$\Phi_H(\omega) = \frac{1}{j2}[M_h(\omega - \omega_c) - M_h(\omega + \omega_c)] \tag{3.56}$$

$$\phi_h(t) = \frac{1}{j2} m_h(t) \left[e^{+j\omega_c t} - e^{-j\omega_c t}\right] = m_h(t) \sin \omega_c t \tag{3.57}$$

Clearly, the frequency-domain SSB expressions of Eq. (3.49) and Eq. (3.50) correspond in the time domain to

$$\phi_{USB}(t) = m(t) \cos \omega_c t - m_h(t) \sin \omega_c t \tag{3.58}$$

$$\phi_{LSB}(t) = m(t) \cos \omega_c t + m_h(t) \sin \omega_c t \tag{3.59}$$

The above two equations can be combined into the single SSB equation below in which the $-$ applies to the USB version and the $+$ applies to the LSB version:

$$\boxed{\phi_{SSB}(t) \;=\; m(t)\cos\omega_c t \mp m_h(t)\sin\omega_c t} \tag{3.60}$$

EXAMPLE 3.11

For a single sideband signal with tone modulation, the carrier is $\cos\omega_c t$ and the message signal is $m(t) = A_m\cos\omega_m t$. (a) Obtain time-domain expressions for the upper sideband and the lower sideband waveforms and sketch them. (b) Obtain expressions for the spectra of the upper and lower sideband and sketch them.

Solution

(a) $m_h(t)$ is obtained by shifting the phase of $m(t)$ by $-90°$.

$$m_h(t) = A_m\cos(\omega_m t - 90°) = A_m\sin\omega_m t$$

$$\phi_{SSB}(t) = m(t)\cos\omega_c t \mp m_h(t)\sin\omega_c t = A_m\cos\omega_m t\cos\omega_c t \mp A_m\sin\omega_m t\sin\omega_c t$$

Rewriting the above to better reveal the sidebands and to facilitate sketching,

$$\phi_{SSB}(t) = \frac{A_m}{2}\{[\cos(\omega_c - \omega_m)t + \cos(\omega_c + \omega_m)t] \mp [\cos(\omega_c - \omega_m)t - \cos(\omega_c + \omega_m)t]\}$$

Replacing $\mp$ with $-$ for the upper sideband and with $+$ for the lower sideband,

$$\phi_{USB}(t) = \frac{A_m}{2}[2\cos(\omega_c + \omega_m)t] = A_m\cos(\omega_c + \omega_m)t$$

$$\phi_{LSB}(t) = \frac{A_m}{2}[2\cos(\omega_c - \omega_m)t] = A_m\cos(\omega_c - \omega_m)t$$

The sketches for $\phi_{USB}(t)$ and $\phi_{LSB}(t)$ are shown in Figures 3.21(a) and (b). Note that tone-modulated SSB waveforms are constant-amplitude sinusoids of different frequencies, unlike the time-varying waveform amplitudes of tone-modulated AM and DSB-SC.

(b) Taking the Fourier transforms of $\phi_{USB}(t)$ and $\phi_{LSB}(t)$ obtained above,

$$\Phi_{USB}(\omega) = A_m\pi[\delta(\omega - \omega_c - \omega_m) + \delta(\omega + \omega_c + \omega_m)]$$

$$\Phi_{LSB}(\omega) = A_m\pi[\delta(\omega - \omega_c + \omega_m) + \delta(\omega + \omega_c - \omega_m)]$$

The sketches for these spectra are shown in Figures 3.21(c) and (d).

PRACTICE PROBLEM 3.11

For a single sideband signal with multi-tone modulation, the message signal is given by $m(t) = 5\cos\omega_m t + 3\sin 2\omega_m t$, and the carrier is $4\cos\omega_c t$. Obtain time-domain expressions for the upper sideband and the lower sideband and sketch their spectra.

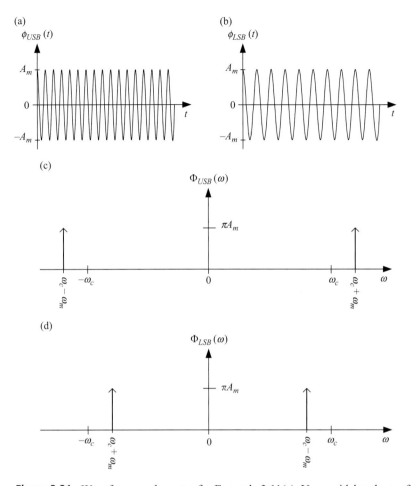

Figure 3.21. Waveforms and spectra for Example 3.11(a). Upper sideband waveform; (b) lower sideband waveform; (c) upper sideband spectrum; (d) lower sideband spectrum.

Answer:

$$\phi_{USB}(t) = 20\cos(\omega_c + \omega_m)t + 8\cos(\omega_c + 2\omega_m)t.$$

$$\phi_{LSB}(t) = 20\cos(\omega_c - \omega_m)t + 8\cos(\omega_c - 2\omega_m)t.$$

Their spectra will be similar to those of Figures 3.21(c) and (d) but, instead of two impulses, $\Phi_{USB}(\omega)$ will have four impulses located at $\pm(\omega_c + \omega_m)$ and $\pm(\omega_c + 2\omega_m)$, while $\Phi_{LSB}(\omega)$ will have four impulses located at $\pm(\omega_c - \omega_m)$ and $\pm(\omega_c - 2\omega_m)$.

3.5.2 Generation of SSB AM

The generation of SSB signals as illustrated in Figure 3.20 is more conceptual than practical. The ideal filters are unrealizable because they entail infinite attenuation of unwanted sidebands

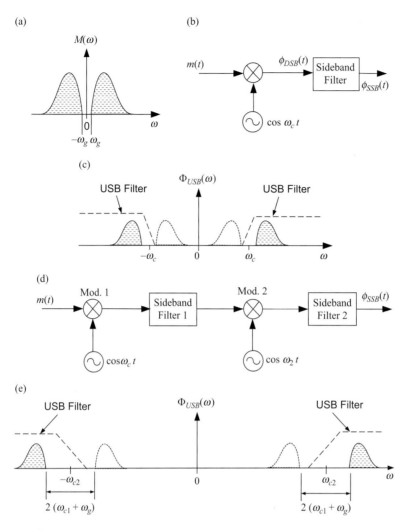

Figure 3.22. Frequency discrimination method of generating SSB. (a) Message signal with dc gap-band; (b) block diagram for one-step generation; (c) USB spectrum from one-step generation; (d) block diagram for two-step generation; (e) USB spectrum from two-step generation.

at the cut-off frequencies. If the message signal spectrum has a dc gap-band like $M(\omega)$ of Figure 3.22(a), the *frequency-discrimination method* can be used. The *phase-discrimination method* exploits the time-domain expressions for SSB signals derived above. It can be used whether or not the message signal has a dc gap-band.

Frequency-discrimination method of generating SSB AM

The frequency-discrimination or *selective filtering method* is illustrated in Figure 3.22. It can be used with message signals such as audio signals, which have a dc gap-band. The gap-band permits the use of gradual cut-off filters. LSB or USB is generated, depending on whether the SSB filter following the DSB modulator is a USB filter or an LSB filter. The illustration in Figure 3.22 is for USB generation. Note that for a dc gap-band of $\omega_g = 2\pi f_g$ in the message signal, the DSB signal to be filtered has a gap-band of $2\omega_g = 4\pi f_g$ at the carrier frequency f_c. For voice signals, f_g is about 300 Hz. To minimize interference from the unwanted sideband, it should be attenuated by at least 40 dB over the corresponding DSB gap-band of $2f_g = 600$ Hz. To ensure sufficient suppression of the unwanted sideband, $2f_g$ should be no less than 1% of f_c. If this criterion is met SSB is generated in a one-step modulation process as depicted in Figures 3.22(b) and (c). When the carrier frequency is too high relative to $2f_g$ for that criterion to be met, the two-step modulation process shown in Figure 3.22(d) and (e) is employed. In the two-step process, an intermediate carrier frequency ω_{c1}, which is small relative to the desired carrier frequency ω_{c2}, is first used to produce an intermediate SSB signal. Good suppression of unwanted sideband is possible in this first step because ω_{c1} is not too high relative to $2\omega_g$. Nevertheless, ω_{c1} must exceed the bandwidth of the message signal bandwidth Ω_m, to prevent overlap of the lower sidebands near the frequency origin in the first modulation step. In the second step the DSB signal has a gap-band of $2(\omega_{c1} + \omega_g)$ at ω_{c2}. A good choice of ω_{c1} requires that $2(\omega_{c1} + \omega_g) \geq 0.01\omega_{c2}$ and that $\omega_{c1} - \Omega_m \geq \omega_g$.

EXAMPLE 3.12

In the two-step frequency discrimination method of generating SSB the desired carrier frequency is 1 MHz and the message signal has a bandwidth B of 5 kHz and a dc gap-band f_g of 300 Hz. The intermediate carrier frequency f_{c1} is to be chosen to the nearest kHz, and it must exceed B by at least f_g. Determine the width of the gap-band at f_{c2}.

Solution
To allow a sufficient gap-band at f_{c2},

$$2\left(f_{c1} + f_g\right) \geq 0.01 f_{c2} \Rightarrow f_{c1} \geq \left(0.005 f_{c2} - f_g\right) = (0.005 \times 10^6 - 300) \text{ Hz} = 4.7 \text{ k Hz}$$

To prevent interference of lower sidebands near the frequency origin,

$$f_{c1} \geq \left(B + f_g\right) = (5 + 0.3) \text{ kHz} = 5.3 \text{ kHz}$$

Choosing the higher of the two values, $f_{c1} \geq 5.3$ kHz.
Rounding (always upward) to the nearest kHz gives $f_{c1} = 6$ kHz.
Thus the gap-band at f_{c2} is $2\left(f_{c1} + f_g\right) = 12.6$ kHz.

PRACTICE PROBLEM 3.12

Repeat Example 3.12, but for $f_{c2} = 900$ kHz, $B = 3$ kHz, and $f_g = 350$ Hz.

Answer: The gap-band at f_{c2} is 10.7 kHz.

Phase-discrimination method of generating SSB AM

This method is illustrated in the block diagram of Figure 3.23. It is based on a direct implementation of the time-domain expression for SSB signals. The upper or in-phase path produces a DSB signal $\phi_{DSB}(t) = m(t)\cos\omega_c t$, while the lower or quadrature path produces $\phi_h(t) = m_h(t)\sin\omega_c t$. The summer combines $\phi_{DSB}(t)$ and $\phi_h(t)$ to yield the upper or lower sideband as desired. Hilbert transformation (or phase shifting by $-\frac{\pi}{2}$) of the sinusoidal carrier $\cos\omega_c t$ to produce $\sin\omega_c t$ is relatively easy. Although a good approximation can be achieved, it is difficult to shift the phase of a typical message signal by $-\frac{\pi}{2}$ over its entire spectrum. The requirements on the phase shifter for the message signal may be relaxed by splitting the $-\frac{\pi}{2}$ phase shift with α and β phase shift networks, such that $\alpha - \beta = -90°$. This method is illustrated in Figure 3.23(b).

(a)

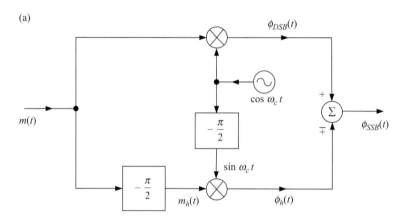

(b)

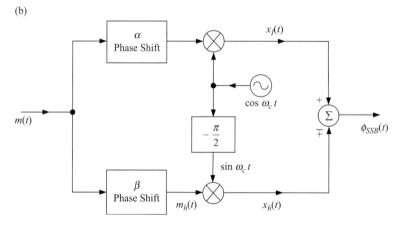

Figure 3.23. Phase-discrimination method of generating SSB AM. (a) Basic block diagram; (b) block diagram employing two networks for signal phase shift.

EXAMPLE 3.13

The carrier used in the phase-discrimination method of generating the SSB of Figure 3.23(a) is $2 \cos \omega_c t$. Find expressions for $\phi_{USB}(t)$ and $\phi_{LSB}(t)$ if $m(t) = A_1 \cos \omega_1 t + A_2 \sin \omega_2 t$.

Solution

$$m_h(t) = A_1 \cos (\omega_1 t - 90°) + A_2 \sin (\omega_2 t - 90°) = A_1 \sin \omega_1 t - A_2 \cos \omega_2 t$$

$$\phi_{SSB}(t) = 2m(t) \cos \omega_c t \mp 2m_h(t) \sin \omega_c t$$
$$= 2[A_1 \cos \omega_1 t + A_2 \sin \omega_2 t] \cos \omega_c t \mp 2[A_1 \sin \omega_1 t - A_2 \cos \omega_2 t] \sin \omega_c t$$

Thus

$$\phi_{USB}(t) = 2[(A_1 \cos \omega_1 t + A_2 \sin \omega_2 t) \cos \omega_c t - (A_1 \sin \omega_1 t - A_2 \cos \omega_2 t) \sin \omega_c t]$$
$$= 2A_1(\cos \omega_c t \cos \omega_1 t - \sin \omega_c t \sin \omega_1 t) + 2A_2(\cos \omega_c t \sin \omega_2 t + \sin \omega_c t \cos \omega_2 t)$$
$$\phi_{USB}(t) = 2A_1 \cos (\omega_c + \omega_1)t + 2A_2 \sin (\omega_c + \omega_2)t$$

$$\phi_{LSB}(t) = 2[(A_1 \cos \omega_1 t + A_2 \sin \omega_2 t) \cos \omega_c t + (A_1 \sin \omega_1 t - A_2 \cos \omega_2 t) \sin \omega_c t]$$
$$= 2A_1(\cos \omega_c t \cos \omega_1 t + \sin \omega_c t \sin \omega_1 t) + 2A_2(\cos \omega_c t \sin \omega_2 t - \sin \omega_c t \cos \omega_2 t)$$
$$\phi_{LSB}(t) = 2A_1 \cos (\omega_c - \omega_1)t + 2A_2 \sin (\omega_c - \omega_2)t$$

PRACTICE PROBLEM 3.13

Repeat Example 3.13, but with the carrier changed to $3 \sin \omega_c t$ and the message signal changed to $m(t) = 4 \sin \omega_1 t - 2 \cos \omega_2 t$.

Answer: $\phi_{USB}(t) = 12 \cos (\omega_c + \omega_1)t - 6 \sin (\omega_c + \omega_2)t$

$$\phi_{USB}(t) = 12 \cos (\omega_c + \omega_1)t - 6 \sin (\omega_c + \omega_2)t$$

3.5.3 Demodulation of SSB AM

Synchronous demodulation is possible with any type of modulation, including SSB AM. Whenever envelope demodulation is possible as in AM, it is preferred because it is much cheaper. Envelope demodulation of SSB is possible if a carrier is added to the SSB at the receiver prior to demodulation. Both synchronous demodulation of SSB and envelope demodulation of SSB plus carrier are discussed below.

Synchronous demodulation of SSB AM

As illustrated in the block diagram of Figure 3.24, synchronous demodulation of SSB is conceptually simple. It consists of multiplying the SSB signal with a local carrier $2 \cos \omega_c t$, which is synchronous with the carrier in the received signal, and passing the product through a low-pass filter. The product of the SSB signal and the carrier is

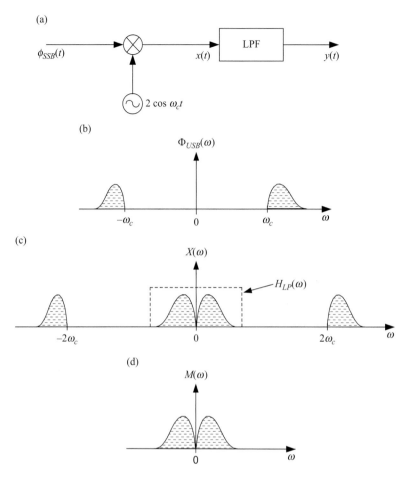

Figure 3.24. Synchronous demodulation of SSB AM. (a) Demodulator block diagram; (b) USB signal spectrum; (c) spectrum of $x(t)$; (d) spectrum of demodulated output signal.

$$x(t) = 2\phi_{SSB}(t)\cos\omega_c t = [m(t)\cos\omega_c t \mp m_h(t)\sin\omega_c t]2\cos\omega_c t$$
$$= m(t) + [m(t)\cos 2\omega_c t \mp m_h(t)\sin 2\omega_c t]$$

After passing $x(t)$ through the low-pass filter, the demodulator output is $y(t) = m(t)$. A frequency-domain illustration of the demodulation operation is shown in Figure 3.24(b) to (d) for the upper sideband case.

Frequency and phase errors in synchronous demodulation of SSB AM

The examples below illustrate the effect of phase and frequency errors in synchronous demodulation of SSB signals. Following the examples, the effects of the errors in SSB and DSB-SC AM will be compared.

EXAMPLE 3.14

For the SSB synchronous demodulator of Figure 3.24 the local carrier contains a frequency error $\Delta\omega$, and is given by $2\cos(\omega_c + \Delta\omega)t$. Obtain the demodulator output if the SSB signal demodulated is an upper sideband version.

Solution

The product of the USB signal and the local carrier is

$$x(t) = \phi_{USB}(t)[2\cos(\omega_c + \Delta\omega)t] = [m(t)\cos\omega_c t - m_h(t)\sin\omega_c t][2\cos(\omega_c + \Delta\omega)t]$$
$$= m(t)[\cos\Delta\omega t + \cos(2\omega_c + \Delta\omega)t] + m_h(t)[\sin\Delta\omega t - \sin(2\omega_c + \Delta\omega)t]$$

After passing $x(t)$ through the low-pass filter, the demodulated output is

$$y(t) = m(t)\cos\Delta\omega t + m_h(t)\sin\Delta\omega t \tag{3.61}$$

PRACTICE PROBLEM 3.14

Repeat Example 3.14 for the case in which the synchronous demodulator with frequency error is used to demodulate a lower sideband version of SSB AM.

Answer: $y(t) = m(t)\cos\Delta\omega t - m_h(t)\sin\Delta\omega t$

The answers to the last example and practice problem show that the mathematical expression for the SSB demodulator output is similar to that for the LSB version of SSB if a USB signal is demodulated, and vice versa. However, the frequency error $\Delta\omega$ is too small to be a valid carrier frequency, so the demodulator outputs are baseband signals and not SSB signals. Figure 3.25(a) shows the message signal spectrum which the demodulator is trying to recover. Consistent with the results of the last example and practice problem, part (b) shows the demodulated signal spectrum if the LSB version is demodulated when the frequency error is $\Delta\omega$, or if USB version is demodulated when the frequency error is $-\Delta\omega$. Clearly, the demodulated signal of Figure 3.25(b) is the message signal with a small frequency up-shift of $\Delta\omega$.

For the sake of simplicity, the message signal of Figure 3.25(a) has no dc gap-band, but typical audio message signals used in SSB systems have a dc gap-band of about 300 Hz. The effect of a small frequency error would be to widen or reduce that gap-band by an amount equal to the frequency error. This causes a small frequency up-shift or down-shift in the demodulated message signal, depending on the algebraic sign of the frequency error. The small frequency shifts translate to small changes in the pitch of the demodulated signal. For most audio signals, up to 5 Hz of frequency error is tolerable, but most commercial communication systems limit the maximum error to 2 Hz. Such small changes in pitch are hardly perceptible to the human ear, and receiver local carriers can easily match transmitter carriers to within a fraction of a Hertz. Thus frequency errors do not pose a serious problem in synchronous demodulation of SSB.

To facilitate a comparison of the effects of frequency error in synchronous demodulation of SSB and DSB-SC, spectra for the DSB-SC demodulator output of Example 3.10 are shown in

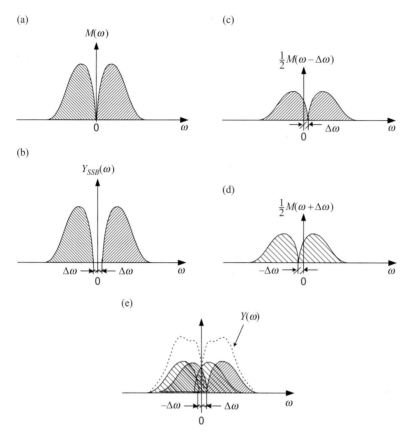

Figure 3.25. Spectral illustration of effect of local carrier frequency error in synchronous detection of SSB and DSB. (a) Message signal spectrum; (b) detector output for SSB (from LSB when frequency error is positive or from USB when frequency error is negative); (c) right-shifted part of DSB detector output due to frequency error; (d) left-shifted part of DSB detector output due to frequency error; (e) interference and distortion in DSB detector output due to small frequency error in the local carrier.

Figure 3.25(c) to (e). The demodulator output is $y(t) = m(t)\cos\Delta\omega$; its spectrum is $Y(\omega) = \frac{1}{2}[M(\omega + \Delta\omega) + M(\omega - \Delta\omega)]$. Note that $\frac{1}{2}M(\omega - \Delta\omega)$ is centered at $\omega = \Delta\omega$ while $\frac{1}{2}M(\omega + \Delta\omega)$ is centered at $\omega = -\Delta\omega$. They are respectively shown in parts (c) and (d) of the figure. Both of them are shown in part (e) of the figure where they overlap and interfere. Their sum $Y(\omega)$ shown in dotted outline is a distorted version of the message signal. Thus frequency errors in receiver local carriers cause more serious problems in synchronous demodulation of DSB-SC than in SSB.

EXAMPLE 3.15
Repeat Example 3.14 for the synchronous demodulation of an SSB upper sideband signal, but for a local carrier of $2\cos(\omega_c t + \alpha)$ which has a phase error of α.

Solution

$$x(t) = \phi_{USB}(t)[2\cos(\omega_c t + a)t] = [m(t)\cos\omega_c t - m_h(t)\sin\omega_c t][2\cos(\omega_c t + a)]$$
$$= m(t)[\cos a + \cos(2\omega_c t + a)] + m_h(t)[\sin a - \sin(2\omega_c t + a)]$$

After low-pass filtering, the demodulator output is

$$y(t) = m(t)\cos a + m_h(t)\sin a \tag{3.62}$$

PRACTICE PROBLEM 3.15

Repeat Example 3.15 for the case in which the synchronous demodulator with phase error of a is used to demodulate a lower sideband version of SSB AM.

Answer: $y(t) = m(t)\cos a - m_h(t)\sin a$

For a better interpretation of the effect of phase error in the local carrier of the SSB synchronous demodulator, consider the frequency domain expression for the demodulator output of Example 3.15.

$$Y(\omega) = M(\omega)\cos a + M_h(\omega)\sin a = [M(\omega)\cos a - j\operatorname{sgn}(\omega)M(\omega)\sin a]$$

i.e. $$Y(\omega) = \begin{cases} M(\omega)[\cos a - j\sin a] = M(\omega)e^{-ja}, & \omega > 0 \\ M(\omega)[\cos a + j\sin a] = M(\omega)e^{ja}, & \omega > 0 \end{cases} \tag{3.63}$$

The last equation shows that a phase error of a in the local carrier of an SSB modulator causes a phase shift of a in the demodulated output. This is not a serious problem in communication of audio signals because the human ear is hardly sensitive to phase distortion. However, phase distortion is a serious problem in video signals.

In Example 3.9, the demodulator output for a DSB-SC synchronous demodulator with a phase error of a in the local carrier was found to be $y(t) = m(t)\cos a$. This represents an attenuation of the demodulated message signal by a factor of $\cos a$. This effect can be a serious problem when a is large or time-varying. When $\cos a$ assumes the extreme value of zero, the demodulated output will be zero. Thus a phase error in the local carrier causes a more serious problem in synchronous detection of DSB-SC than in SSB.

Envelope detection of SSB plus carrier

Suppose a carrier $A_c \cos \omega_c t$ is added to an SSB signal, the resulting SSB plus carrier is

$$\phi_{SSB+C}(t) = A_c \cos \omega_c t + m(t)\cos\omega_c t \mp m_h(t)\sin\omega_c t$$
$$= [A_c + m(t)]\cos\omega_c t \mp m_h(t)\sin\omega_c t = E(t)\cos[\omega_c t + \theta(t)]$$

where $E(t)$ is the envelope and $\theta(t)$ is the phase of $\phi_{SSB+C}(t)$. The phase is not relevant to envelope detection, so it is ignored in the analysis below. The envelope is given by

$$E(t) = \left\{ [A_c + m(t)]^2 + m_h^2(t) \right\}^{1/2} = [A_c^2 + 2A_c m(t) + m^2(t) + m_h^2(t)]^{1/2}$$

Since $m(t)$ and $m_h(t)$ are equal in amplitude, $A_c \gg |m(t)| \Rightarrow A_c \gg |m_h(t)|$, so if $A_c \gg |m_h(t)|$, the third and fourth terms of the last expression can be ignored, in which case

$$E(t) \approx \left[A_c^2 + 2A_c m(t)\right]^{1/2} = A_c \left[1 + \frac{2m(t)}{A_c}\right]^{1/2}$$

Employing binomial series expansion,

$$E(t) \approx A_c + m(t) - \frac{1}{2}\frac{m^2(t)}{A_c} + \cdots \tag{3.64}$$

For $A_c \gg |m(t)|$, the third- and higher-order terms of the binomial expansion are diminishingly small and can be ignored. Thus the envelope is

$$E(t) \approx A_c + m(t) \tag{3.65}$$

The above analysis shows that envelope detection of SSB plus carrier requires that the carrier be much larger in amplitude than the message signal. Recall that envelope detection of AM requires only that the carrier exceed the message signal in amplitude. Thus, power efficiency is even poorer in SSB plus carrier than in AM. To avoid waste of power in the transmitted signal, a local carrier may be generated and inserted into the SSB signal at the receiver, prior to envelope detection. However, the above derivation assumes a synchronous carrier. Consequently, phase and frequency errors in the local carrier have similar minor effects as in synchronous demodulation of SSB. Clearly, the simpler and cheaper envelope detection of SSB plus carrier offers a great advantage.

EXAMPLE 3.16

For an SSB plus carrier signal, the carrier is $A_c \cos \omega_c t$ and the message signal is $2 \cos \omega_m t$. Find the minimum carrier amplitude if a minimum of 26 dB attenuation of second-harmonic distortion relative to the desired message signal amplitude is to be achieved in the envelope detector output.

Solution

26 dB $= -20 \log_{10} \alpha \Rightarrow \alpha = 10^{-26/20} = 10^{-1.3} = \frac{1}{20}$, where $\alpha =$ attenuation factor.

From Eq. (3.64) the desired message signal amplitude is $|m(t)| = |2 \cos \omega_m t| = 2$,

while the amplitude of the second harmonic is $\left|-\frac{m^2(t)}{2A_c}\right| = \left|\frac{4 \cos^2 \omega_m t}{2A_c}\right| = \frac{2}{A_c}$.

The given requirement $\Rightarrow \frac{2/A_c}{2} = \frac{1}{A_c} \le \frac{1}{20} \Rightarrow A_c \ge 20$.

PRACTICE PROBLEM 3.16

Repeat Example 3.16 if the carrier is $A_c \cos \omega_c t$, the message signal is $2 \sin \omega_m t$, and a minimum of 20 dB attenuation of second-harmonic distortion relative to the desired message signal amplitude is to be achieved in the envelope detector output.

Answer: $A_c \ge 10$.

EXAMPLE 3.17

For AM with tone modulation and for SSB plus carrier with tone modulation, the message signal is $2 \cos \omega_m t$ and the carrier is $A_c \cos \omega_c t$. For the AM modulator $A_c = 5$ but for the SSB modulator $A_c = 25$. What is the power efficiency for each of the modulated signals?

Solution

For AM,

$$\mu = \frac{A_m}{A_c} = \frac{2}{5} = 0.4$$

$$\eta_{AM} = \frac{\mu^2}{2 + \mu^2} = 0.074 \ \text{or} \ 7.4\%$$

For SSB

$$P_m = \ \text{power in message signal} = \frac{A_m^2}{2} = \frac{2^2}{2} = 2$$

$$P_c = \ \text{power in carrier} = \frac{A_c^2}{2} = \frac{25^2}{2} = 312.5$$

$$\phi_{SSB+C}(t) = A_c \cos \omega_c t + [m(t) \cos \omega_c t \mp m_h(t) \sin \omega_c t]$$

Power in $m(t) \cos \omega_c t = $ power in $m_h(t) \sin \omega_c t = \frac{P_m}{2}$

$P_s = $ useful power $= $ power in $m(t) \cos \omega_c t \mp m_h(t) \sin \omega_c t = 2\left[\frac{1}{2}P_m\right] = P_m = 2$

$$\therefore \ \eta_{SSB} = \frac{P_s}{P_c + P_s} = \frac{P_m}{P_c + P_m} = \frac{2}{312.5 + 2} = 0.00636 \ \text{or} \ 0.636\%$$

PRACTICE PROBLEM 3.17

The carrier for an AM modulator is $2 \cos \omega_c t$, while the carrier for an SSB plus carrier modulator is $10 \cos \omega_c t$. The message signal for both modulators is $\sin \omega_m t$. Find the power efficiency for the modulated AM and SSB signals.

Answer: $\eta_{AM} = 0.111 \ \text{or} \ 11.1\%$

$\eta_{SSB} = 0.0099 \ \text{or} \ 0.99\%$

3.6   VESTIGIAL SIDEBAND (VSB) AM

Vestigial sideband systems transmit only one sideband and a small fraction (vestige) of the undesired sideband. Typically, VSB bandwidth is about 25 % larger than SSB bandwidth. VSB

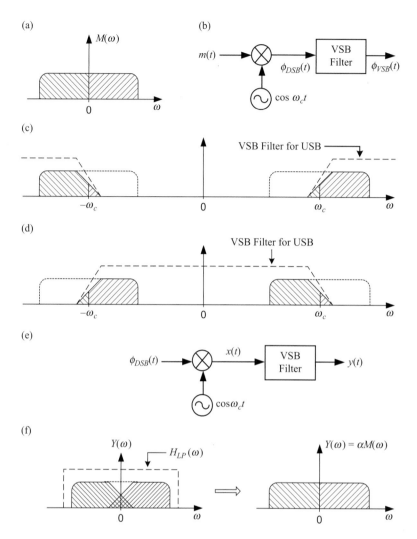

Figure 3.26. Generation and demodulation of VSB. (a) Message signal; (b) block diagram for VSB generation; (c) USB version of VSB; (d) LSB version of VSB; (e) synchronous demodulator for VSB; (f) demodulated output signal.

is particularly suited to message signals which have large bandwidth but no dc gap-band, such as baseband digital signals and video signals. For such signals SSB is not an option; VSB provides the needed bandwidth economy.

Vestigial sideband modulation is illustrated in Figure 3.26. As shown in the block diagram of the modulator of part (b), VSB is obtained by passing a DSB-SC signal through a VSB filter. Unlike SSB sharp cut-off filters, VSB filters are gradual cut-off filters, which are easier to realize. An upper sideband VSB filter and the resulting VSB spectrum are shown in Figure 3.26(c), while the lower sideband filter and the resulting spectrum are shown in part (d). Note that the inclusion of a vestige of the unwanted sideband is counterbalanced by the attenuation of a corresponding part of the desired sideband.

Consider a message signal $m(t) = \cos \omega_m t$, which results in impulses at $\omega_c \pm \omega_m$ in the DSB-SC spectrum, located within the roll-off band of the VSB filter. Let the VSB filter have unit gain in its passband, and let its gains at the location of the impulses be $H(\omega_c - \omega_m) = a$ and $H(\omega_c + \omega_m) = b$. The filter response must be such that $H(\omega_c - \omega_m) + H(\omega_c + \omega_m) = a + b = 1$ if the filter gain of the impulse in the vestigial sideband is to compensate for the attenuation of the impulse in the desired sideband. This is strictly true only if $H(\omega_c)$ consists of a magnitude response and has a zero phase, in which case $H(\omega_c) = |H(\omega_c)|e^{j0} = |H(\omega_c)|$. Indeed, the VSB filters of Figure 3.26 consist of magnitude responses only. The above condition should hold true for all frequency components at $\omega_c \pm v$ within the bandwidth of the DSB signal. Thus, for a message bandwidth of B Hz, the requirement on the VSB filter is that

$$\boxed{|H(\omega_c - v)| + |H(\omega_c + v)| = 1, \quad |v| \leq 2\pi B} \tag{3.66}$$

If the VSB filter is a zero-phase filter, then the above requirement is equivalent to

$$H(\omega_c - v) + H(\omega_c + v) = 1, \quad |v| \leq 2\pi B \tag{3.67}$$

Note that a bandpass filter will suffice as a VSB filter, provided that it has a gradual cut-off in the neighborhood of ω_c which satisfies the above criterion.

EXAMPLE 3.18

The modulator of Figure 3.26(b) is used to produce an upper VSB signal. The frequency response of the VSB filter $H_U(\omega)$ is shown in Figure 3.27(a) for positive frequencies. The carrier is $\cos \omega_c t$ and the message signal is $m(t) = \cos \omega_m t$, where $\omega_c = 10^6$ rad/s and $\omega_m = 2 \times 10^4$ rad/s. Derive expressions for the VSB signal $\phi_{VSB}(t)$ consisting of the undesired and desired sidebonds and the in-phase and quadrature components.

Solution

$$\phi_{DSB}(t) = \cos \omega_m t \cos \omega_c t = \frac{1}{2}[\cos (\omega_c - \omega_m)t + \cos (\omega_c + \omega_m)t]$$

$$\omega_c - \omega_m = 10^6 - 2 \times 10^4 = 0.98 \times 10^6; \quad \omega_c + \omega_m = 10^6 + 2 \times 10^4 = 1.02 \times 10^6$$

From the VSB filter, $H_u(\omega_c - \omega_m) = 0.3; \quad H_u(\omega_c + \omega_m) = 0.7$
Thus, the output of the VSB filter is

$$\phi_{VSB}(t) = \frac{1}{2}[0.3 \cos (\omega_c - \omega_m)t + 0.7 \cos (\omega_c + \omega_m)t]$$

The above expression shows that the gain of the undesired vestigial sideband is 0.3, while the gain of the desired (upper) sideband is 0.7. Continuing,

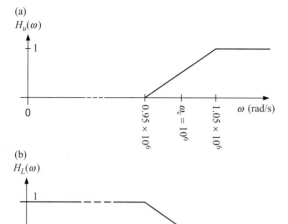

Figure 3.27. VSB shaping filters (a) for Example 3.18 and (b) for Practice problem 3.18.

$$\phi_{VSB}(t) = \frac{1}{2}[0.3(\cos \omega_c t \cos \omega_m t + \sin \omega_c t \sin \omega_m t)]$$

$$+ \frac{1}{2}[0.7(\cos \omega_c t \cos \omega_m t - \sin \omega_c t \sin \omega_m t)]$$

$$\therefore \quad \phi_{VSB}(t) = \frac{1}{2}[\cos \omega_m t \cos \omega_c t - 0.4 \sin \omega_m t \sin \omega_c t] \tag{3.68}$$

PRACTICE PROBLEM 3.18

Repeat Example 3.18 with the carrier and message signal unchanged, but the VSB filter changed to the lower sideband version whose response is shown in Figure 3.27 (b).

Answer:

$$\phi_{VSB}(t) = \frac{1}{2}[\cos \omega_m t \cos \omega_c t + 0.4 \sin \omega_m t \sin \omega_c t] \tag{3.69}$$

Time domain representation of VSB signals

The similarity between the time-domain equations for $\phi_{VSB}(t)$ derived in the last example and practice problem and the expression $\phi_{SSB}(t) = m(t) \cos \omega_c t \mp m_h(t) \sin \omega_c t$ formerly found for SSB is obvious. That similarity is not a coincidence. SSB and VSB belong to a class of bandpass signals whose spectra are not symmetrical about a center frequency. In general, such signals have expressions of the form

$$\boxed{\phi_{BP}(t) \quad = \quad m_I(t) \cos \omega_c t \mp m_Q(t) \sin \omega_c t} \tag{3.70}$$

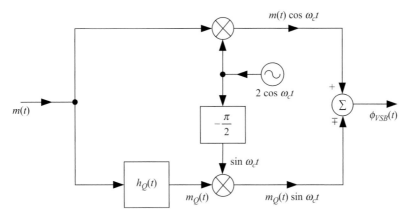

Figure 3.28. Phase discrimination method of generating VSB AM.

where $m_I(t)$ is the in-phase component of the baseband (message) signal, and $m_Q(t)$ is its quadrature component. Combining the results of the last example and practice problem in one equation gives

$$\phi_{VSB}(t) = \frac{1}{2}\left[\cos\omega_m t \cos\omega_c t \mp 0.4\sin\omega_m t \sin\omega_c t\right] \tag{3.71}$$

The constant $\frac{1}{2}$ in the last equation arises from the selective filtering method used for VSB generation in the last example and practice problem, but that method is not the only one that can be used. The $\frac{1}{2}$ can be replaced by 1, since it can easily be cancelled, if necessary, with an amplifier gain of 2. In the last two equations, the $-$ applies to the USB, while the $+$ applies to LSB. For SSB, $m_I(t) = m(t)$ and $m_Q(t) = m_h(t)$. In the last VSB equation, $m_I(t) = m(t) = \cos\omega_m t$, while $m_Q(t) = 0.4\sin\omega_m t$. Usually $m_Q(t)$ depends on the cut-off characteristics of the VSB filter. Thus the general time-domain equation for VSB is

$$\boxed{\phi_{VSB}(t) = m(t)\cos\omega_c t \mp m_Q(t)\sin\omega_c t} \tag{3.72}$$

Based on the above time-domain expression, a phase-discrimination method of generating VSB signals may be implemented as shown in Figure 3.28. Note that $h_Q(t)$ which processes $m(t)$ to yield $m_Q(t)$ has low-pass characteristics. It is a version of the VSB filter $H(\omega)$ shifted down in frequency by ω_c.

EXAMPLE 3.19
Multi-tone modulated VSB

The modulator of Figure 3.26(b) is used to produce a VSB signal. The frequency response of the VSB filter $H_U(\omega)$ is shown in Figure 3.27(a) for positive frequencies. The carrier is $\cos\omega_c t$, and the message signal is $m(t) = 4\cos\omega_m t + 2\cos 2\omega_m t$, where $\omega_c = 10^6$ rad/s and $\omega_m = 2 \times 10^4$ rad/s. Derive a time-domain expression for the VSB signal in terms of the in-phase and quadrature components. Identify $m_I(t)$ and $m_Q(t)$.

Solution

$$\phi_{DSB}(t) = m(t)\cos\omega_c t = [4\cos\omega_m t + 2\cos 2\omega_m t]\cos\omega_c t$$
$$= 2\cos(\omega_c - \omega_m)t + 2\cos(\omega_c + \omega_m)t + \cos(\omega_c - 2\omega_m)t + \cos(\omega_c + 2\omega_m)t$$

$$\omega_c - \omega_m = 10^6 - 2\times 10^4 = 0.98\times 10^6; \quad \omega_c + \omega_m = 10^6 + 2\times 10^4 = 1.02\times 10^6$$

$$\omega_c - 2\omega_m = 10^6 - 2(2\times 10^4) = 0.96\times 10^6; \quad \omega_c + 2\omega_m = 10^6 + 2(2\times 10^4) = 1.04\times 10^6$$

From the VSB filter,

$$H_u(\omega_c - \omega_m) = 0.3; H_u(\omega_c + \omega_m) = 0.7; H_u(\omega_c - 2\omega_m) = 0.1; H_u(\omega_c + 2\omega_m) = 0.9$$

Note that $H_U(\omega)$ of Figure 3.27 is a magnitude function, hence $H_U(\omega)$ is the filter gain constant at the frequency ω.

$$H_u(\omega)\phi_{DSB}(t) = (0.3)2\cos(\omega_c - \omega_m)t + (0.7)2\cos(\omega_c + \omega_m)t$$
$$+ (0.1)\cos(\omega_c - 2\omega_m)t + (0.9)\cos(\omega_c + 2\omega_m)t$$

i.e. $\phi_{VSB}(t) = 0.6[\cos\omega_m t\cos\omega_c t + \sin\omega_m t\sin\omega_c t] + 1.4[\cos\omega_m t\cos\omega_c t - \sin\omega_m t\sin\omega_c t]$
$$+ 0.1[\cos 2\omega_m t\cos\omega_c t + \sin 2\omega_m t\sin\omega_c t] + 0.9[\cos 2\omega_m t\cos\omega_c t - \sin 2\omega_m t\sin\omega_c t]$$
$$= 2\cos\omega_m t\cos\omega_c t - 0.8\sin\omega_m t\sin\omega_c t + \cos 2\omega_m t\cos\omega_c t - 0.8\sin 2\omega_m t\sin\omega_c t$$

Thus $\phi_{VSB}(t) = [2\cos\omega_m t + \cos 2\omega_m t]\cos\omega_c t - 0.8[\sin\omega_m t + \sin 2\omega_m t]\sin\omega_c t$

$$m_I(t) = 2\cos\omega_m t + \cos 2\omega_m t; \quad m_Q(t) = 0.8[\sin\omega_m t + \sin 2\omega_m t]$$

PRACTICE PROBLEM 3.19

Practice problem 3.19: Multi-tone modulated VSB

The modulator of Figure 3.26(b) utilizes the filter response of Figure 3.27(b) to produce the lower sideband version of VSB. The frequency response of the VSB filter $H_u(\omega)$ is shown in Figure 3.27(a) for positive frequencies. The carrier is $\cos\omega_c t$ and the message signal is $m(t) = 6\cos\omega_1 t + 4\cos 2\omega_2 t$, where $\omega_c = 10^6$ rad/s, $\omega_1 = 10^4$ rad/s and $\omega_2 = 3\times 10^4$ rad/s. Derive a time-domain expression for the VSB signal in terms of the in-phase and quadrature components, and identify $m_I(t)$ and $m_Q(t)$.

Answer: $\phi_{VSB}(t) = [3\cos\omega_1 t + 2\cos\omega_2 t]\cos\omega_c t + [0.6\sin\omega_1 t + 1.2\sin 2\omega_2 t]\sin\omega_c t$

$$m_I(t) = 3\cos\omega_1 t + 2\cos\omega_2 t; \quad m_Q(t) = 0.6\sin\omega_1 t + 1.2\sin 2\omega_2 t$$

Synchronous demodulation of VSB

Synchronous demodulation of VSB is illustrated in the block diagram of Figure 3.26(e). The signal $x(t)$ in the block diagram is given by

$$x(t) = \phi_{VSB}(t)[2\cos\omega_c t] = [m(t)\cos\omega_c t + m_Q(t)\sin\omega_c t]2\cos\omega_c t$$
$$= 2m(t)\cos^2\omega_c t + 2m_Q(t)\sin\omega_c t\cos\omega_c t$$

i.e. $x(t) = m(t) + m(t)\cos 2\omega_c t + m_Q(t)\sin 2\omega_c t$

After passing $x(t)$ through the low-pass filter, the demodulator output is $y(t) = m(t)$.

Envelope detection of VSB plus carrier

It can be shown that envelope detection is possible for a VSB plus carrier. Because of the similarity between the time-domain expressions for SSB and VSB, the derivation is very similar to that already discussed for envelope detection of SSB plus carrier. Hence the derivation is not repeated here. Because VSB bandwidth is intermediate between that of AM and SSB, its carrier amplitude requirement is also intermediate between those of the two. Consequently, VSB plus carrier is more power-efficient than SSB plus carrier, but less power-efficient than AM.

Technical note – evolution from monochrome to high-definition TV

The evolution from monochrome to high-definition television (HDTV) is an excellent example of the relentless quest for excellence and the rapid pace of innovation in communications technology. Patented in 1925 by Philo Farnsworth, electronic (monochrome) TV proved superior to earlier mechanical TV endeavors. It was developed, refined, and heavily commercialized in the subsequent years. In 1953 the National Television System Committee (NTSC) adopted as the US standard the compatible (with monochrome TV) color television system developed by RCA/NBC, over the non-compatible color television system developed by CBS two years earlier. The Japanese demonstrated a version of HDTV (the MUSE) in 1981. It was not compatible with the earlier color television system, and it did not see wide adoption. In 1996, the Advanced Television Systems Committee (ATSC) adopted the US HDTV standard, which has many desirable features, and which is compatible with the earlier color television.

Stages in evolution of television

Monochrome (black and white) TV was eventually standardized to an aspect ratio of 4:3 and 525i scanning. The scanning from top to bottom produces 30 frames (images) per second, and each frame is scanned with 525 lines. Each frame consists of two fields. The field of odd numbered lines is followed by the field of the even numbered half of the total 525 lines. The lines in the two fields interlace spatially to give the complete frame of 525 lines. The "i" in 525i refers to this interlacing. The vertical resolution is 525 lines. The horizontal resolution is based on 75% of the width of the screen, which is just the reciprocal of the aspect ratio, so the horizontal resolution is also 525. Consequently, the screen resolution is $525 \times 525 = 275,625$ picture elements (pixels). At 30 frames per second, this amounts to 8.27×10^6 data pulses per second, which corresponds to a bandwidth of 4.135 MHz, or approximately 4.2 MHz. TV employs VSB. To provide 4.2 MHz for one sideband, 1.25 MHz for the vestigial sideband, and 0.5 MHz for the FM audio signal, the Federal Communications Commission (FCC) allows a 6 MHz bandwidth for TV signals.

The compatibility of color TV with monochrome TV means that color TV broadcasts can be received as black and white images with monochrome TV sets, and as color images with color TV sets. Color TV produces color images from red, green, and blue images, each of which entails 525×525 pixels per frame. One would expect it to have three times the bandwidth of monochrome TV, but, as shown in the illustration below, color TV and the later HDTV have the same 6 MHz bandwidth as monochrome TV. How is this possible? First let us see how the bandwidth compression is achieved in color TV.

Let the red, green, and blue image signals be $m_r(t)$, $m_g(t)$, and $m_b(t)$, respectively. This set of signals can be linearly combined or transformed into three other linearly independent signals as follows

$$m_Y(t) = 0.3m_r(t) + 0.59m_g(t) + 0.11m_b(t)$$

$$m_I(t) = 0.6m_r(t) - 0.28m_g(t) + 0.32m_b(t)$$

$$m_Q(t) = 0.21m_r(t) - 0.52m_g(t) + 0.31m_b(t)$$

The above transform is ingenious. It permits compatibility of color and monochrome TV because $m_Y(t)$ is the luminance signal which provides the black and white image in a monochrome TV, while $m_I(t)$ and $m_Q(t)$ are the chrominance or color signals. In color TV, $m_Y(t)$, $m_I(t)$, and $m_Q(t)$ combine to give the color image. The luminance signal is assigned the same 4.2 MHz bandwidth as in monochrome TV. The bandwidth of chrominance signals can be reduced without loss in image perception because the human eye perceives color to a lower resolution relative to luminance. Hence, only the lower 1.6 MHZ and 0.6 MHz portions of $m_I(t)$ and $m_Q(t)$ spectra, respectively, are utilized.

The chrominance signals $m_I(t)$ and $m_Q(t)$ are quadrature amplitude modulated onto a color subcarrier of frequency $f_{cc} = 3.583$ MHz (3.6 MHz for simplicity), with $m_I(t)$ as the in-phase channel input, and $m_Q(t)$ as the quadrature channel input. The sum of the two QAM channel outputs ranges in frequency from 2 MHz to 5.2 MHz. It is passed through a bandpass filter with low and high cut-off frequencies of 2 MHz and 5.2 MHz, respectively. Thus the bandpass filter output contains the modulated $m_Q(t)$ signal, and a lower VSB modulation of $m_I(t)$. This bandpass output is added (multiplexed) onto the baseband luminance signal. The audio signal is frequency modulated onto an audio carrier of frequency $f_a = 4.5$ MHz, and also multiplexed onto the baseband luminance signal. The resulting composite video signal is used to VSB modulate the TV carrier of frequency f_c to yield the TV signal whose spectrum is illustrated below.

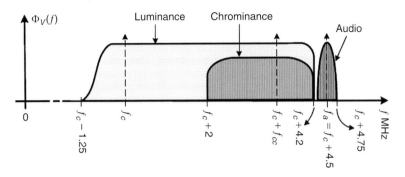

Spectrum of upper VSB color TV signal

An inspection of the spectrum of the upper VSB TV signal shows that the chrominance and luminance spectra occupy overlapping frequency bands. Why would they not interfere with each other? A detailed analysis will show that due to the periodicity of scanning, these signals have discrete spectra with periodic gaps. The color subcarrier is chosen so that the spectra of the chrominance signals lie in the gaps of the luminance signal spectrum.

The superiority of HDTV stems from its much higher resolution and the inherent advantages of digital systems over analog systems. As a digital system, HDTV is much less affected by noise. Because of its higher resolution, the HDTV display can be much larger while retaining high image quality. Its aspect ratio of 16:9 gives a wider field of view. With a small converter box, HDTV can be received at lower image quality with analog TV receivers. HDTV resolution can be 720p, 1080i, or 1080p. The "p" in the resolution specifications refer to the fact that the scanning is progressive (sequential) rather than interlaced. Progressive scanning avoids the image flicker problem of analog TV. The resolution of 1080p has 1920 $\times$ 1080 pixels per frame. This is over seven times the resolution of analog color TV, but HDTV has the same bandwidth of 6 MHz. The question again arises: how is this possible? The answer lies in data compression. Digital signals can be compressed much more efficiently than analog signals. You may be familiar with the compression of a still JPEG picture of many megabytes on the computer into a few tens of kilobytes suitable for attachment to emails. In a similar way, video data can be compressed with an algorithm suitable for motion pictures such as MPEG-2. The compression ratio can be quite high. MPEG-2 can achieve a data compression ratio of about 55 to 1!

3.7 ACQUISITION OF CARRIERS FOR SYNCHRONOUS DETECTION

Problems caused by phase and frequency errors in synchronous detection of suppressed-carrier systems underscore the need for acquisition at the receiver of carriers which match the transmitted carrier in both frequency and phase. Identical quartz crystal oscillators can be used

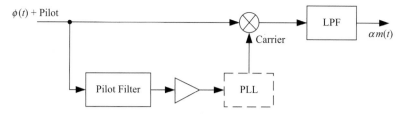

Figure 3.29. Use of pilot carrier in synchronous detection.

to provide very stable and closely matching frequencies at the transmitter and the receiver, but frequency matching is more difficult at high carrier frequencies which entail extremely small crystal dimensions. This method minimizes frequency errors but not phase errors. It may be employed in synchronous demodulation of SSB (especially for audio message signals), in which the effects of frequency and phase errors are less severe than in DSB-SC.

To minimize or eliminate local carrier phase and frequency errors at the receiver it is necessary to do one of the following: transmit a pilot carrier along with the modulated signal, acquire the needed carrier at the receiver from the received signal, or synchronize the local carrier with the received carrier. The *phase-lock loop* (PLL) is a versatile technique for carrier synchronization. It will be discussed in the next chapter because it is widely employed in demodulation of frequency-modulated signals. Three other methods of obtaining a synchronous carrier at the receiver are discussed below.

Pilot carrier

In this method a carrier of very low amplitude relative to the sidebands is transmitted along with the modulated signal. Because of its low amplitude level, the pilot carrier has negligible adverse effect on the power efficiency of the modulated signal. As illustrated in Figure 3.29, the pilot carrier is obtained from the received signal with a narrow bandpass filter centered at the carrier frequency. The carrier so acquired is then amplified and may then be used directly for synchronous detection. In most cases, the recovered carrier is first passed through a PLL in which it is used to synchronize a local oscillator. The output of the PLL is then used for synchronous detection. The PLL box is shown in dotted outline to indicate that it is sometimes omitted.

Signal squaring

The signal-squaring method of carrier acquisition is illustrated in Figure 3.30. Suppose the input $\phi(t)$ is a DSB-SC signal, the squared signal is given by

$$x(t) = [m(t)\cos\omega_c t]^2 = \frac{1}{2}m^2(t) + \frac{1}{2}m^2(t)\cos 2\omega_c t$$

The input to the narrow bandpass filter is $x(t)$; its output is $y_1(t) = \frac{1}{2}m^2(t)\cos 2\omega_c t$.

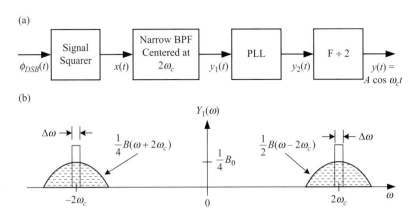

Figure 3.30. Carrier acquisition in DSB-SC AM by signal squaring. (a) Block diagram; (b) spectrum of squared signal.

Let $m^2(t) \Leftrightarrow B(\omega)$, then

$$\frac{1}{2}m^2(t)\cos 2\omega_c t \;\Leftrightarrow\; \frac{1}{4}[B(\omega - 2\omega_c) + B(\omega + 2\omega_c)]$$

The spectrum given on the right-hand side of the above equation is shown in Figure 3.30(b). If the bandwidth $\Delta\omega$ of the narrow bandpass filter is sufficiently narrow, then $B(\omega - 2\omega_c)$ and $B(\omega + 2\omega_c)$ are each approximately constant in amplitude over $\Delta\omega$. Let $B_o/4$ denote that constant amplitude. Then the frequency-domain filter output is given by

$$Y_1(\omega) \approx \frac{B_o\Delta\omega}{4}[\delta(\omega - 2\omega_c) + \delta(\omega + 2\omega_c)]$$

But $\frac{B_o\Delta\omega}{4} = \frac{2\pi\Delta f B_o}{4} = \pi\left[\frac{\Delta f B_o}{2}\right]$. Letting $\beta = \frac{\Delta f B_o}{2}$, where β is a constant, $Y_1(\omega)$ and $y_1(t)$ are respectively,

$$Y_1(\omega) \approx \pi\beta[\delta(\omega - 2\omega_c) + \delta(\omega + 2\omega_c)] \tag{3.73}$$

and

$$y_1(t) \approx \beta\cos 2\omega_c t \tag{3.74}$$

The last two equations show that the output of the narrow bandpass filter is approximately a sinusoid of twice the carrier frequency. With $y_1(t)$ as its input, the PLL produces a pure sinusoid of frequency $2\omega_c$. After passing this sinusoid through the 2:1 frequency divider, the output is the desired synchronous carrier $y(t) = A_c\cos\omega_c t$.

Note that the sign of the input signal is lost in the signal squarer. Consequently, there is a sign ambiguity (or equivalently a phase ambiguity of π radians) in the recovered carrier. A phase error of π radians in the recovered carrier results in a reversal of the polarity of the demodulated signal. This is of little consequence for analog baseband signals because the human ear is insensitive to phase. However, for bipolar digital baseband signals, phase ambiguity cannot be tolerated. Consequently, the signal-squaring method is not directly applicable to such cases.

The Costas loop

The Costas loop is a version of the phase-lock loop that is particularly convenient for local carrier synchronization and demodulation of DSB-SC signals. A block diagram of Costas loop is shown in Figure 3.31. The upper channel is the in-phase channel (*I*-channel), while the lower channel is the quadrature channel (*Q*-channel). The input to each channel is the same DSB-SC signal, but their carriers are in phase quadrature. The phase error between the carrier generated by the local oscillator and the received carrier is $\alpha_e = \alpha_i - \alpha_o$, where α_i is the phase of the received carrier, and α_o is the phase of the local carrier. From the block diagram, it is easy to see that the *I*-channel and the *Q*-channel outputs are respectively $y_I(t) = m(t) \cos \alpha_e$ and $y_Q(t) = m(t) \sin \alpha_e$. Their product is given by $z_1(t) = m^2(t) \cos \alpha_e \sin \alpha_e = \frac{1}{2} m^2(t) \sin 2\alpha_e$. Recall that in the signal-squaring method, an input of $\frac{1}{2} m^2(t) \cos 2\omega_c t$ into the narrowband filter gave an output of $\beta \cos 2\omega_c t$. Similarly the input of $z_1(t) = \frac{1}{2} m^2(t) \sin 2\alpha_e$ to the narrow low-pass filter here will give an output of $z_2(t) = \gamma \sin 2\alpha_e$, where γ is a constant. Thus the input to the *voltage-controlled oscillator* (VCO) is $z_2(t) = \gamma \sin 2\alpha_e$.

> A *voltage-controlled oscillator* (VCO) is an oscillator whose frequency is controlled by an external input voltage.

Suppose the VCO output signal, which is the local carrier, exactly matches the received carrier, then $\alpha_e = 0$, $z_2(t) = 0$, $y_Q(t) = 0$, and the demodulated signal is $y_I(t) = m(t)$. When the received and the local carriers are not exactly matched, α_e, $z_2(t)$, and $y_Q(t)$ have non-zero values, and $y_I(t)$ is only an approximation to $m(t)$. In that case, $z_2(t) = \gamma \sin 2\alpha_e$ is used to change the frequency of the VCO output so that its total phase more closely matches that of the received carrier.

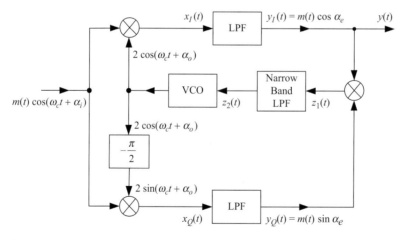

Figure 3.31. Costas phase-locked loop for generation of a synchronous carrier and demodulation of DSB-SC AM.

EXAMPLE 3.20

Derive an expression for the output signal $y(t)$ of the signal-squaring circuit of Figure 3.30 if the input signal to the circuit is a single sideband signal. Determine whether the signal-squaring method is suitable for synchronous carrier recovery in SSB AM.

Solution

The SSB signal can be expressed as

$$\phi_{SSB}(t) = \frac{1}{2}[m(t)\cos\omega_c t \mp m_h(t)\sin\omega_c t] = E(t)\cos[\omega_c t + \psi(t)]$$

where $E(t) = \frac{1}{2}\left[m^2(t) + m_h^2(t)\right]^{1/2}$ is the envelope, and $\psi(t) = \tan^{-1}\left[\frac{\mp m_h(t)}{m(t)}\right]$ is the phase. Thus

$$x(t) = \phi_{SSB}^2(t) = E^2(t)\cos^2[\omega_c t + \psi(t)]$$
$$= \frac{1}{2}E^2(t)\{1 + \cos[2\omega_c t + 2\psi(t)]\} \tag{3.75}$$

The output of the narrow bandpass filter is $y_1(t) = \frac{1}{2}E^2(t)\cos[2\omega_c t + 2\psi(t)]$. In the discussion of the signal-squaring method, it was shown that $y_1(t) = \frac{1}{2}m^2(t)\cos 2\omega_c t$ can be expressed as $y_1(t) \approx \beta\cos 2\omega_c t$. Thus here, $y_1(t) = \frac{1}{2}E^2(t)\cos[2\omega_c t + 2\psi(t)]$ can be expressed as $y_1(t) \approx \beta_1\cos[2\omega_c t + 2\psi(t)]$. After passing this $y_1(t)$ through the 2:1 frequency divider, the signal-squarer output will be

$$y(t) = A\cos[\omega_c t + \psi(t)] \tag{3.76}$$

Because of the time-varying phase $\psi(t)$ in the above output, it cannot be phase synchronous with the transmitted carrier $\cos\omega_c t$. Consequently the signal-squaring method is not suitable for synchronous carrier acquisition in SSB AM.

PRACTICE PROBLEM 3.20

Derive an expression for $z_2(t)$, the input signal to the VCO in the Costas loop of Figure 3.31 if the input signal to the circuit is a single sideband signal. Determine whether the Costas loop is suitable for synchronous carrier generation at the receiver in SSB AM.

Answer: $z_2(t) = \gamma\sin[2\alpha_e + 2\theta(t)]$. Because of the time-varying phase $\theta(t)$, the Costas loop is not suitable for synchronous carrier generation at the receiver in SSB AM.

The last example and practice problem show that the signal-squaring method and the Costas loop are not suitable for carrier recovery in SSB. Because of the similarity of the time-domain expressions for VSB and SSB, reworking the example and practice problem for VSB will lead to results similar to those obtained for SSB. Thus the signal-squaring method and the Costas loop cannot be used for synchronous carrier recovery in SSB or VSB.

3.8 FREQUENCY-DIVISION MULTIPLEXING

Signal multiplexing is a technique which permits the combination of a number signals, in a non-interfering way, to form a composite signal suitable for transmission over a single medium. After transmission, the composite signal can be successfully separated or *demultiplexed* into the original signals. Signals do not interfere with each other if they are orthogonal (non-overlapping in time, frequency, or phase). In frequency-division multiplexing (FDM), signal orthogonality is achieved by ensuring that their spectra do not overlap in frequency.

Frequency-division multiplexing is illustrated in the block diagrams of Figure 3.32. In the multiplexer of Figure 3.32(a), n message signals are frequency-division multiplexed to form a composite signal. Each of the message signals is first band-limited with a low-pass filter prior to modulation. Each modulator translates the corresponding signal to its assigned frequency band. The carrier frequencies increase successively by an amount equal to or greater than the bandwidth allowed for each signal. Consequently, the multiplexed signals do not overlap in frequency. In Figure 3.32(b) demultiplexing is accomplished by separating the signals with bandpass filters. Conventional AM, DSB-SC, SSB, VSB, frequency modulation, or phase modulation can be frequency-division multiplexed. SSB is popular in FDM because of its greater bandwidth economy.

The FDM signal can be transmitted via cable, radio, optical fiber, etc. In the case of radio transmission, the FDM signal is used to modulate a radio frequency carrier at the transmitter. At the receiver, demodulation is first performed to recover the FDM signal from the radio frequency signal. Following demodulation, demultiplexing proceeds as in Figure 3.32(b).

FDM in telephony

Telephony is an important area of application of FDM. Typically, voice signals occupy a frequency range of 300 Hz to 3.1 kHz. In the North American FDM telephony, each voice

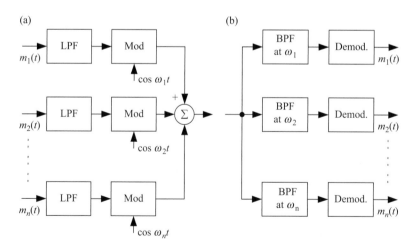

Figure 3.32. Frequency-division multiplexing: (a) multiplexer; (b) demultiplexer.

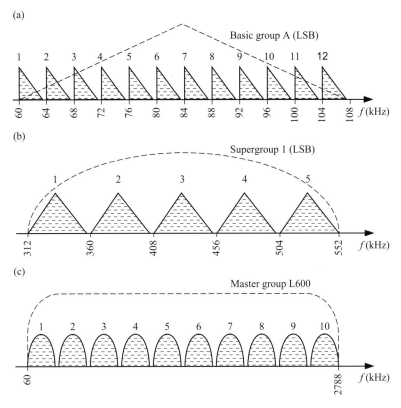

Figure 3.33. Some levels of multiplexing in North American FDM Telephony. (a) Basic group comprising 12 channels; (b) supergroup comprising five basic groups; (c) master group comprising 10 supergroups.

channel is band-limited to 3.1 kHz and assigned a bandwidth of 4 kHz. Thus, a *guard-band* of 900 Hz exists between the multiplexed voice channels. Different levels of multiplexing in the North American FDM telephone system are shown in Figure 3.33. The levels of multiplexing shown employ the lower sideband (LSB) of SSB AM, but upper sideband versions are also used. The first level of multiplexing produces a *basic group* consisting of 12 LSB voice channels and occupying a frequency band of 60 to 108 kHz. The second level is a *supergroup* consisting of five basic groups having a total of 60 voice channels, and occupying a frequency band from 312 to 552 kHz. The third level multiplexes 10 supergroups to produce a *master group* of 600 voice channels. A mastergroup occupies a frequency band from 60 to 2788 kHz.

3.9 APPLICATION: THE SUPERHETERODYNE AM RECEIVER

The superheterodyne AM receiver is an ingenious system capable of receiving any of the 10 kHz bandwidth, 105 AM channels within the commercial AM band of 550 to 1600 kHz.

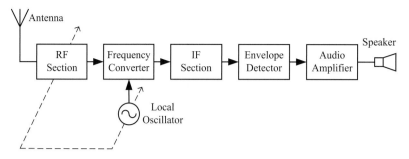

Figure 3.34. The superheterodyne AM receiver.

It employs many electronics principles and many of the techniques discussed in this chapter. A block diagram of the superheterodyne receiver is shown in Figure 3.34. It consists of a radio frequency (RF) tuner, a frequency converter, a local oscillator, an intermediate frequency (IF) section, an envelope detector, and an audio amplifier.

The RF tuner selects the desired channel by amplifying only that channel among the numerous AM channels reaching the antenna. It consists of tunable filters and amplifiers. Tuning is typically accomplished with RLC circuits. The resonant frequency responses of these circuits have low gain and poor frequency selectivity. Most of the required gain and adjacent channel suppression are provided by the IF section.

Consider an RF section which is tuned to an AM channel of carrier frequency f_c. As explained in section 3.3, the output of the frequency converter (mixer) will be an AM signal with a carrier frequency that is the sum or difference of the carrier frequency and the local oscillator frequency, f_{LO}. The difference frequency, rather than the sum, is chosen as the intermediate frequency f_{IF}. The IF section consists of high-gain bandpass amplifier(s) and sharp cut-off filter(s) of 10 kHz bandwidth centered at the intermediate frequency $f_{IF} = f_{LO} - f_c = 455$ kHz. It provides sufficient adjacent channel suppression. The audio amplifier provides the remaining gain to the demodulated signal.

The RF tuner is ganged to the local oscillator in such a way that both are tuned in synchronism, with the local oscillator frequency always exceeding the carrier frequency by the intermediate frequency. As indicated earlier, frequency conversion is also known as heterodyning. Heterodyning in which the local oscillator frequency is greater than the carrier frequency of the input signal is known as *superheterodyning*. To see that superheterodyning is preferable to heterodyning, compare the tuning ratios $f_{LO,max}/f_{LO,min}$ for superheterodyning and for heterodyning.

$$\text{For superheterodyning,} \quad \frac{f_{LO,\max}}{f_{LO,\min}} = \frac{f_{c,\max} + f_{IF}}{f_{c,\min} + f_{IF}} = \frac{(1600 + 455)\text{ kHz}}{(550 + 455)\text{ kHz}} \approx 2 \qquad (3.77)$$

$$\text{For heterodyning,} \quad \frac{f_{LO,\max}}{f_{LO,\min}} = \frac{f_{c,\max} - f_{IF}}{f_{c,\min} - f_{IF}} = \frac{(1600 - 455)\text{ kHz}}{(550 - 455)\text{ kHz}} \approx 12 \qquad (3.78)$$

Consider achieving the tuning range via ganged variable capacitors in the RF tuner and the local oscillator. A large range of capacitance is difficult to achieve in a variable capacitor, hence the preference for superheterodyning, which entails a much smaller tuning range.

The superheterodyne receiver overcomes another important problem – the *image channel* problem. f_{IF} is the difference between f_{LO} and f_c, irrespective of whether the carrier frequency or the local oscillator frequency is the higher of the two. For superheterodyning, the desired carrier frequency is $f_c = f_{LO} - f_{IF}$, but another carrier frequency (the *image frequency*) given by $f'_c = f_{LO} + f_{IF} = f_c + 2f_{IF}$ will also result in a difference frequency of $f'_c - f_{LO} = f_{IF}$. Consider a receiver tuned to a channel of carrier frequency f_c. If an image channel with a carrier frequency of f'_c were to appear at the input to the frequency converter, it would also result in an AM signal of carrier frequency f_{IF} at the outputs of both the frequency converter and the IF section. Such an image channel would then interfere with the desired channel. Fortunately, the image channel is sufficiently suppressed by the RF filter because the desired 10 kHz bandwidth channel centered at f_c is separated from f'_c by a whopping $f'_c - f_c = 2f_{IF} = 910$ kHz. Thus the RF section provides sufficient image channel suppression, while, as indicated earlier, the IF section provides sufficient adjacent channel suppression. As we shall see in the next chapter, the superheterodyne technique is equally applicable to receivers for frequency-modulated and phase-modulated signals.

Summary

1. Modulation facilitates efficient signal transmission through communication media. It permits simultaneous transmission of multiple channels through the same communication medium.

2. Amplitude modulation (AM) is a type of modulation in which the amplitude of the carrier varies linearly with the message signal. It is also the name of a specific type of modulation within this type. It is the type of modulation which is employed in commercial AM broadcasting.

3. Commercial AM carrier frequencies are spaced 10 kHz apart from 550 to 1600 kHz. Each AM channel has a bandwidth of 10 kHz.

4. If the message signal is $m(t)$ and the carrier is $A_c \cos \omega_c t$, then the AM waveform and its spectrum are given respectively by

$$\phi_{AM}(t) = A_c \cos \omega_c t + m(t) \cos \omega_c t$$

$$\Phi_{AM}(\omega) = \frac{1}{2}[M(\omega + \omega_c) + M(\omega - \omega_c)] + \pi A_c[\delta(\omega + \omega_c) + \delta(\omega - \omega_c)]$$

5. The modulation index of the AM signal is $\mu = \frac{m_p}{A_c}$, where m_p is the peak value of the message signal $m(t)$. For envelope detection to be possible, $\mu \geq 1$.

6. The power in $m(t) \cos \omega_c t$ (the message-bearing part or the sidebands) of the AM signal is $P_s = \frac{1}{2}P_m$, where P_m is the power in $m(t)$.

7. For AM the power efficiency is $\eta = \frac{\text{sideband power}}{\text{total power}} = \frac{P_s}{P_c + P_s} = \frac{P_m}{A_c^2 + P_m}$.

8. For tone-modulated AM, the power efficiency is $\eta = \frac{\mu^2}{2 + \mu^2}$.

9. A modulator employing a diode as the non-linear element can operate as a nonlinear modulator if $(A_c + m_p) < V_{on}$, and as a switching modulator if $A_c \gg V_{on}$.

10. Envelope detection is preferred to synchronous detection in AM because it is simpler and cheaper. The RC time constant for the envelope detector should satisfy $B \ll \frac{1}{RC} \ll f_c$. Its optimum value is given by $RC = \sqrt{\frac{1}{Bf_c}}$.

11. DSB-SC is given in the time and frequency domains respectively by

$$\phi_{DSB}(t) = m(t) \cos \omega_c t; \quad \Phi_{DSB}(\omega) = \frac{1}{2}[M(\omega - \omega_c) + M(\omega + \omega_c)].$$

12. The nonlinear, series-bridge, and shunt-bridge DSB-SC modulators are balanced modulators. The ring modulator is a double balanced modulator.

13. QAM transmits two channels in the bandwidth of one DSB-SC channel by using carriers which are in phase quadrature but of the same frequency. The QAM signal is given by $\phi_{QAM}(t) = m_1(t) \cos \omega_c t + m_2(t) \sin \omega_c t$.

   Phase and frequency errors in the receiver local carrier pose serious problems in synchronous detection of QAM.

14. SSB has many advantages. It uses only half the bandwidth of AM or DSB-SC. Its synchronous detection is not much affected by phase and frequency errors of the receiver local carrier. SSB plus carrier can be envelope detected.

15. Hilbert transformation of a signal causes a phase shift of $-j\text{sgn}(\omega)$ in its spectrum. Hilbert transformation in time and frequency domains can be expressed as $m(t) * \frac{1}{\pi t} = m_h(t) \Leftrightarrow M_h(\omega) = -jM(\omega)\text{sgn}(\omega)$.

16. The time-domain equation for SSB is $\phi_{SSB}(t) = m(t)\cos\omega_c t \mp m_h(t)\sin\omega_c t$, where the minus applies to USB and the plus applies to the LSB.

17. VSB is used for message signals which have no dc gap-band, such as digital baseband and video signals. It has advantages similar to, but to a lesser degree than, SSB.

18. The time-domain equation for VSB is $\phi_{VSB}(t) = m_I(t)\cos\omega_c t \mp m_Q(t)\sin\omega_c t$, where the minus applies to USB and the plus applies to the LSB.

19. Some methods for obtaining a synchronous carrier at the receiver include using pilot carriers, signal squaring, and the Costas loop.

Review questions

3.1 Which of the following is not a purpose of modulation?
(a) Transmission of multiple channels through one communication medium.
(b) Increasing signal strength.　(c) Translating signal spectrum to higher frequencies.
(d) Achieving efficient transmission.　(e) Achieving feasible antenna dimension.

3.2 Which of the following is a good example of a message signal?
(a) A radio frequency signal.　(b) A pilot carrier.　(c) A baseband signal.
(d) $A_m\delta(t)$.　(e) $A_m\Pi\left(\frac{t}{\tau}\right)$.

3.3 Which of the following types of modulation is least power-efficient?
(a) QAM.　(b) SSB.　(c) AM.　(d) DSB.　(e) VSB.

3.4 Which of the following types of modulation is most bandwidth-efficient?
(a) SSB.　(b) AM.　(c) DSB.　(d) VSB.　(e) VSB + carrier.

3.5 Which of the following types of modulation is easiest to demodulate?
(a) QAM.　(b) SSB.　(c) AM.　(d) DSB.　(e) VSB.

3.6 Which of the following is a double balanced modulator?
(a) Switching AM modulator.　(b) QAM modulator.　(c) Series-bridge modulator.
(d) Shunt-bridge modulator.　(e) Ring modulator.

3.7 Which of the following is not accomplished in the superheterodyne AM receiver?
(a) Coherent detection.　(b) Radio frequency tuning.　(c) Frequency conversion.
(d) Adjacent channel suppression.　(e) Image channel suppression.

3.8 If $\omega_m = 10\Omega_g$, which of the signals whose waveform or spectrum are given below is not a suitable message signal for the frequency-discrimination method of generating SSB?
(a) $A_m\cos\omega_m t$　(b) $A_m\Pi\left(\frac{\omega}{2\omega_m}\right)$.　(c) $2A_m\cos\omega_m t + A_m\cos 2\omega_m t$.
(d) $A_m\left[\Pi\left(\frac{\omega-\omega_m-\Omega_g}{2\omega_m}\right) + \Pi\left(\frac{\omega+\omega_m+\Omega_g}{2\omega_m}\right)\right]$.

3.9 Which of the following is most vulnerable to phase and frequency errors in synchronous demodulator local carriers?

(a) QAM.　　(b) SSB.　　(c) AM.　　(d) DSB.　　(e) VSB.

3.10 Which of the following is not relevant to synchronous detection?

(a) Signal squaring.　　(b) Pilot carrier.　　(c) Frequency-division multiplexing.

(d) Local oscillator.　　(e) The Costas loop.

Answers: 3.1 (b), 3.2 (c), 3.3 (c), 3.4 (a), 3.5 (c), 3.6 (e), 3.7 (a), 3.8 (b), 3.9 (a), 3.10 (c).

Problems

Section 3.1 Introduction

3.1 A message signal has a bandwidth of 30 kHz. Efficient radiation from an antenna requires that the antenna length be at least a tenth of the transmitted signal wavelength. Determine the minimum antenna length for efficient radiation if: (a) it is to be broadcast without the benefit of modulation, (b) it is broadcast after it is amplitude-modulated with a sinusoidal carrier having a frequency that is 100 times the bandwidth of the message signal.

3.2 Conventional AM broadcast channel carrier frequencies range from 550 to 1600 kHz. Efficient radiation from an antenna requires that the antenna length be at least a tenth of the transmitted signal wavelength. Find the minimum height of the radio mast required for supporting the radio antenna at the extremes of the commercial AM broadcast frequencies.

Section 3.2 Amplitude modulation

3.3 For an amplitude-modulated signal, the carrier is $10 \cos \omega_c t$ and the message signal is $A_m \cos \omega_c t$. Find the modulation index and sketch the modulated waveform if $A_m = 4$, $A_m = 10$, and $A_m = 20$.

3.4 An amplitude-modulated signal is given by $\phi_{AM}(t) = 4[\alpha + 2 \sin \omega_m t] \cos \omega_c t$, where $\omega_m \ll \omega_c$. Find the power efficiency of the AM signal if (a) $\alpha = 10$, (b) $\alpha = 5$.

3.5 For an amplitude-modulated signal, the carrier is $4 \sin \omega_c t$ and the message signal is the triangular waveform of Figure 3.3 with a peak value of M_p. Find the modulation index and the power efficiency if (a) $m_p = 2$, (b) $m_p = 4$

3.6 Figure 3.35(a) shows the waveform of an amplitude-modulated signal as displayed on an oscilloscope. Both the carrier and the message signal are sinusoidal. Find the modulation index and the power efficiency of the amplitude-modulated signal.

3.7 Figure 3.35(b) shows the waveform of an amplitude-modulated signal as displayed on an oscilloscope screen. The carrier is sinusoidal, but the message signal is a periodic triangular waveform. Find the modulation index and the power efficiency of the amplitude-modulated signal.

(a)
$\phi_{AM}(t)$

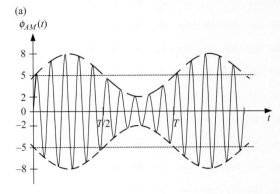

Figure 3.35. Waveforms for Problems 3.6 and
3.7.

(b)
$\phi_{AM}(t)$

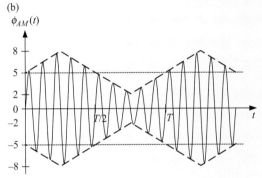

3.8 An AM signal is given by $\phi_{AM}(t) = 4[5 + 3\cos(2\pi \times 10^4 t)]\cos(\pi \times 10^5 t)$ volts. Find the total power dissipated by this AM signal across an equivalent load resistance of 10 Ω.

3.9 Consider a case of multi-tone AM in which the message signal is given by $m(t) = 4\cos\omega_m t + 3\sin 2\omega_m t$. The carrier is $10\cos\omega_c t$, where $\omega_c \gg \omega_m$. Does this AM signal satisfy the envelope-detection criterion? Find the power efficiency of this AM signal.

3.10 For an AM signal with multi-tone modulation, the modulating signal is given by $m(t) = 4\cos\omega_m t + 2\cos 2\omega_m t$ and the carrier is given by $c(t) = 5\cos\omega_c t$, where $\omega_c \gg \omega_m$. Does the AM signal satisfy the envelope-detection criterion? Find an expression for and sketch $\Phi_{AM}(\omega)$, the spectrum of the modulated signal.

3.11 In the nonlinear AM modulator shown in Figure 3.36, $R = 1\ \Omega$. The total diode input voltage v_D is smaller than the diode turn-on voltage, and the diode current is given by $i_D = \beta(4v_D + v_D^2)$. The carrier is $A_c\cos\omega_c t$. The message signal which contains a dc component is $\alpha + m(t)$. The message signal bandwidth is B Hz, where $\omega_c \gg 2\pi B$. Derive an expression for the output signal $y(t)$, if the bandpass filter is centered at ω_c.

3.12 In Figure 3.36, the carrier $A_c\cos\omega_c t$ has an amplitude of 20 V and the diode turn-on voltage is $V_{on} = 0.7$ V. The message signal has a dc component and is given by $m(t) - \lambda$, where λ is a constant. The message signal bandwidth is B Hz, and $\omega_c \gg 2\pi B$. Derive an expression for $y(t)$, the output of the AM modulator.

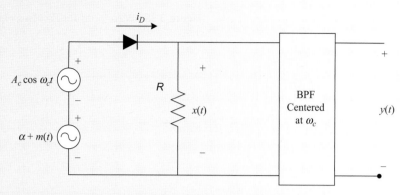

Figure 3.36. AM generator for Problems 3.11 and 3.12.

3.13 The input signal into an envelope detector is an AM signal of carrier frequency 500 kHz. The envelope detector employs a smoothing capacitor of 20 nF. The modulating signal has a bandwidth of 5 kHz. Specify an appropriate value for the resistance in parallel with the smoothing capacitor for a good tracking of the AM envelope. If the AM signal is tone-modulated, what is the maximum value for the resistance if (a) $\mu = 0.5$, (b) $\mu = 0.95$.

3.14 Consider an envelope detector that is used to directly demodulate commercial AM broadcast signals, with carrier frequencies ranging from 550 kHz to 1600 kHz, without the benefit of a heterodyning receiver. The resistance in parallel with the smoothing capacitor has a value of 5 kΩ, and a variable capacitor is used for the smoothing capacitor. Assuming that the capacitor value chosen for each broadcast frequency is the geometric mean of its possible minimum and maximum values, find minimum and maximum values that the variable capacitor must be able to attain in order to permit the demodulation of every possible channel in the entire AM band.

3.15 The parallel combination of the smoothing filter and resistance in the envelope detector performs a smoothing or low-pass filtering function. Thus, the envelope detector and the rectifier detector may appear to be essentially the same circuit. For the AM signal, the carrier frequency is 1 MHz, and the message signal bandwidth is 10 kHz. Determine whether this is true in regard to the required low-pass filter bandwidths by comparing the bandwidth for the low-pass filter entailed in the rectifier detector, and the bandwidth of the low-pass filter entailed in the envelope detector, if the RC time constant is chosen for a good tracking of the AM envelope.

Section 3.3 Double sideband-suppressed carrier (DSB-SC) modulation

3.16 For a DSB-SC carrier signal, the carrier is $A_c \cos \omega_c t$, and the carrier frequency is much greater than the message signal bandwidth. (a) Sketch the modulated signal waveform if (i) the message signal is $0.5A_c \cos \omega_m t$ and (ii) the message signal is the triangular waveform of Figure 3.3 with $M_p = 0.5A_c$. (b) Sketch the modulated signal if (i) the message signal is $2A_c \cos \omega_m t$ and (ii) the message signal is the triangular waveform of Figure 3.3 with $m_p = 2A_c$.

3.17 Consider DSB-SC modulation with a carrier of $A_c \cos \omega_c t$, and for which $\omega_c = 10\omega_m$. Obtain the time-domain expression and the frequency-domain expression, and sketch the spectra for the following modulating signals:

(a) $m(t) = 3 \cos \omega_m t + \cos 2\omega_m t$; (b) $\cos \omega_m t \cos^2 \omega_m t$.

3.18 For a DSB-SC signal, the carrier is given by $A_c \cos 1000t$. Find the power in the modulated signal if the modulating signal is (a) $m(t) = 4 \cos 100t - \sin 200t$, (b) $m(t) = 2[\sin 200t(1 + \sin 100t)]$.

3.19 The input carrier signal for a DSB-SC modulator is $A_c \cos \omega_c t$, and the input message signal is the rectangular signal $m(t) = \Pi\left(\frac{t}{\tau}\right)$, where $\omega_c \gg \frac{2\pi}{\tau}$. (a) Obtain an expression for $\Phi_{DSB}(\omega)$, the spectrum of the modulated signal, and sketch it. (b) Find the energy in the modulated signal.

3.20 The input carrier signal for a DSB-SC modulator is $2 \cos \omega_c t$, and the input message signal is the rectangular signal $m(t) = \Delta\left(\frac{t}{\tau}\right)$, where $\omega_c \gg \frac{2\pi}{\tau}$. (a) Obtain an expression for $\Phi_{DSB}(\omega)$, the spectrum of the modulated signal, and sketch it. (b) Find the energy in the modulated signal.

3.21 Consider a DSB-SC modulator whose input carrier is $2 \cos \omega_c t$. Find an expression for $\Phi_{DSB}(\omega)$, the spectrum of the modulated signal if the message signal is (a) $m(t) = \Pi\left(\frac{t+5}{2}\right) + \Pi\left(\frac{t-5}{2}\right)$, (b) $m(t) = \sin c\left(\frac{\omega t}{2}\right)$. In each case, the carrier frequency is much greater than the message signal bandwidth.

3.22 Shown in Figure 3.37(a) is $\Phi_{DSB}(\omega)$, the spectrum of a DSB-SC modulated signal. (a) Find the energy spectral density $\Psi_\phi(\omega)$ and the autocorrelation function $R_\phi(\tau)$ of the modulated signal. (b) If the carrier is given by $A \cos 100t$, find an expression for the modulating signal, $m(t)$.

3.23 Shown in Figure 3.37(b) is $M(\omega)$, the spectrum of an input message signal for a DSB-SC modulator. The input carrier signal is $\cos \omega_c t$, and $\omega_c \gg \omega_o$. Find: (a) the time domain expression for the modulated signal $\phi_{DSB}(t)$, (b) the energy E_ϕ in the modulated signal.

3.24 You are to design a DSB-SC modulator to give a modulated signal with a carrier frequency of 150 kHz. However, you only have available to you the message signal $m(t)$ of bandwidth 10 kHz, a sinusoidal oscillator whose output is $A_c \cos 1000\pi t$, a box of diodes, and a universal filter which can be configured as a low-pass, bandpass, or high-pass filter with unity gain in its passband. Sketch the circuit you would use in your design. Specify the type of filter in your design, and its passband. What is the expression for the output DSB-SC signal?

3.25 In Figure 3.38, let $A_c \gg |m(t)|$ and $A_c \gg V_{on}$, where V_{on} is the diode turn-on voltage. Assume that the diodes are identical and that R_f, the forward resistance of each diode when conducting, is negligible relative to R. The bandwidth of $m(t)$ is ω_m and $\omega_c \gg \omega_m$. Derive an expression for the output signal $y(t)$.

(a)

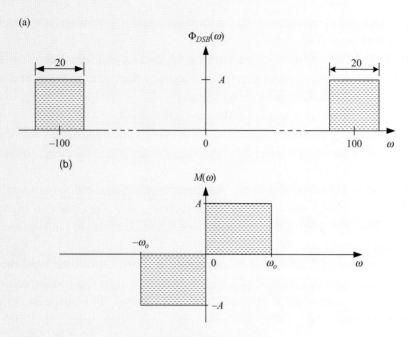

(b)

Figure 3.37. Spectra for Problems 3.22 and 3.23.

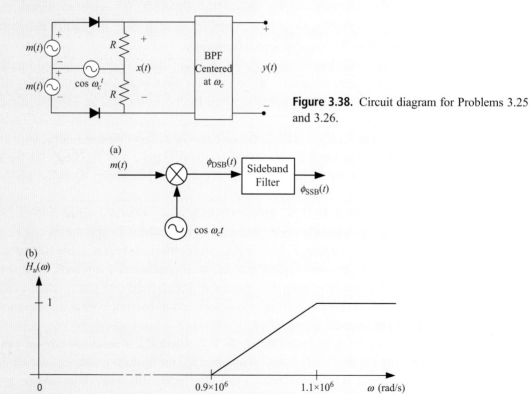

Figure 3.38. Circuit diagram for Problems 3.25 and 3.26.

Figure 3.39. For Problems 3.45 and 3.46. (a) VSB modulator; (b) upper VSB filter.

3.26 In Figure 3.38, $[|A_c| + |m(t)|] < V_{on}$, where V_{on} is the diode turn-on voltage. Suppose $R = 1\,\Omega$ and the diode current is given by $i_D = 0.1[5v_D + v_D^2]$, where $v_D = A_c \cos \omega_c t + m(t)$ is the diode input voltage. (a) Derive an expression for the signal $x(t)$. (b) What is the output signal $y(t)$?

3.27 The DSC-SC synchronous demodulator of Figure 3.12(a) is used to demodulate the signal $\phi_{DSB}(t) = m(t) \cos \omega_c t$. Find the output of the demodulator if the local oscillator carrier at the demodulator is (a) $c(t) = e^{j\omega_c t}$, (b) $c(t) = e^{j(\omega_c t + a)}$, and (c) $c(t) = \cos a e^{j(\omega_c t + a)}$. Note that a is a constant. For which of the local oscillator signals is the demodulated output an undistorted replica of the message signal, save for a possible scaling in amplitude?

3.28 The DSC-SC synchronous demodulator of Figure 3.12(a) is used to demodulate the signal $\phi_{DSB}(t) = A_m \cos \omega_m t \cos \omega_c t$, where $A_m \cos \omega_m t$ is the message signal and $\cos \omega_c t$ is the carrier. Find the output of the demodulator if the local oscillator carrier at the demodulator is (a) $c(t) = a \cos \omega_c t + \beta \sin \omega_c t$, (b) $c(t) = \cos a \cos(\omega_c t + a)$, and (c) $c(t) = \cos a \cos(\omega_c t + a) + \sin a \sin(\omega_c t + a)$, where a is a constant. For which of the local oscillator signals is the demodulated output an undistorted replica of the message signal, save for a possible scaling in amplitude?

3.29 The DSC-SC synchronous demodulator of Figure 3.12(a) is used to demodulate the signal $\phi_{DSB}(t) = m(t) \cos 2\pi f_c t$. The message signal $m(t)$ has a bandwidth of $B = 10$ kHz, where $f_c \gg B$. The low-pass filter has a bandwidth of 12 kHz. The local oscillator carrier is $c(t) = 2 \cos 2\pi(f_c + \Delta f)t$, where $\Delta f = 10$ Hz. (a) Obtain an expression for the demodulator output. (b) Sketch the spectrum of the demodulated output if: (i) the message signal spectrum is similar to the spectrum of Figure 3.12(c) which has no dc gap-band, (ii) the message signal spectrum is similar to the message signal spectrum of Figure 3.22(a) and has a dc gap-band of 200 Hz.

3.30 An amplitude-modulated signal $\phi_{AM}(t) = A_c[1 + m(t)] \cos(\omega_c t + \theta)$, where $m(t)$ is the message signal to be demodulated with the DSC-SC synchronous demodulator of Figure 3.12(a). The local carrier at the demodulator has both a frequency error and a phase error and is given by $c(t) = \cos[(\omega_c + \Delta\omega)t + a]$. (a) Obtain an expression for the output of the demodulator. (b) What is the demodulator output if the frequency error is zero? (c) What is the demodulator output if the phase error is zero?

3.31 In the frequency-translator block diagram of Figure 3.18(a), the input DSB-SC signal is $\phi(t) = m(t) \cos(2\pi \times 10^5 t)$, and the message signal has a bandwidth of 10 kHz. The local oscillator output signal is $2 \cos(1.2\pi \times 10^5 t) + 4 \cos(2.8\pi \times 10^5 t)$. (a) Determine the expression for the product $x(t)$ which shows the individual DSB-SC signals present in it. (b) If the output bandpass filter is chosen to select the component of $x(t)$ with the largest amplitude, give the expression for the output signal and specify f_L and f_H, the low and high cut-off frequencies for the minimal bandwidth ideal bandpass filter required that will select that component.

3.32 In the frequency-translator block diagram of Figure 3.18(a), the input signal is the sum of two DSB-SC signals $m_1(t) \cos(4\pi \times 10^4 t) + m_2(t) \cos(10\pi \times 10^4 t)$. Each of the message signals $m_1(t)$ and $m_2(t)$ has a bandwidth of 5 kHz, and the local oscillator carrier is $\cos\left(16\pi \times 10^4 t\right)$. Give the expression for each of the DSB-SC signals present in $x(t)$.

Section 3.4 Quadrature amplitude modulation

3.33 The input message signals to a QAM modulator are $m_1(t) = A_1 \cos\left(8\pi \times 10^3\right)t$ and $m_2(t) = A_2 \cos\left(12\pi \times 10^3\right)t$. The carrier is $2\sin\left(2\pi \times 10^5 t\right)$. (a) Find an expression for the modulated signal $\phi_{QAM}(t)$. (b) If $\phi_{QAM}(t)$ is applied to the input of a QAM demodulator with a local oscillator carrier of $2\sin\left(\left(2\pi \times 10^5 t\right) + a\right)$, find the demodulated output for channel 1.

3.34 A QAM signal given by $\phi_{QAM}(t) = m_1(t)\cos\omega_c t + m_2(t)\sin\omega_c t$ is the input signal to a QAM demodulator. The local oscillator carrier at the demodulator is the sum of two sinusoids given by $c(t) = \cos\left(\omega_c t + a\right) + \sin\left(\omega_c t + a\right)$. Find the demodulated output signal for each of the two channels of the demodulator.

3.35 Consider a QAM demodulator whose local oscillator carrier is given by $c(t) = e^{j\omega_c t}$. The input QAM signal is $\phi_{QAM}(t) = m_1(t)\cos\omega_c t + m_2(t)\sin\omega_c t$. (a) Find the demodulated output signal for each of the two channels of the demodulator. (b) If the local carrier has a phase error of a and is given by $c(t) = e^{j(\omega_c t + a)}$, obtain expressions for the demodulated output of each of the two channels.

Section 3.5 Single sideband (SSB) amplitude modulation

3.36 The message signal input into a modulator is $m(t) = 5\sin\omega_m t$, and the carrier is $10\cos\omega_c t$, where $\omega_c \gg \omega_m$. Give an expression for and sketch the modulated signal, if the type of modulation performed is: (a) conventional amplitude modulation, (b) DSB-SC amplitude modulation, (c) upper sideband version of SSB amplitude modulation, and (d) lower sideband version of SSB amplitude modulation.

3.37 Consider a single sideband system for which the carrier is given by $A_c \sin\omega_c t$, where $\omega_c = 10\omega_m$. Obtain a time-domain expression for the upper sideband version of the SSB signal if: (a) the message signal is $m(t) = A_m \sin\omega_m t$, and (b) the message signal is $m(t) = A_1 \sin\omega_m t + A_2 \cos 3_m t$.

3.38 For a single sideband system, the carrier is given by $2\cos\omega_c t$, the message signal is $10\sin\omega_m t$, and $\omega_c = 10\omega_m$. (a) Obtain time-domain expressions for the upper sideband and the lower sideband versions of the SSB signal and sketch them. (b) Obtain expressions for and sketch the spectra for the upper sideband and the lower sideband.

3.39 For a single sideband modulator performing multi-tone modulation, the message signal is $8\sin\omega_m t + 4\cos 2\omega_m t$, the carrier is $\cos\omega_c t$ and $\omega_c = 10\omega_m$. Determine the time-domain expressions for the upper sideband and the lower sideband versions of the SSB signal.

3.40 In the phase-discrimination method of generating SSB AM shown in Figure 3.24(a), the carrier is given by $A_c \sin\omega_c t$. The message signal contains a dc component and is given by $m(t) = A_1 \sin \omega_1 t + A_2 \sin \omega_2 t$. Obtain expressions for the upper sideband and the lower sideband versions of SSB.

3.41 A single sideband signal given by $\phi_{SSB}(t) = m(t) \cos \omega_c t + m_h(t) \sin \omega_c t$ is demodulated with the synchronous demodulator the local carrier of which has both a phase error and a frequency error as $2 \cos [(\omega_c + \Delta\omega)t + \alpha]$. (a) Determine an expression for the demodulator output signal. (b). What is the output signal if: (i) the frequency error is set to zero, (ii) the phase error is set to zero.

3.42 A carrier of $24 \cos \omega_c t$ is added to a tone-modulated single sideband signal for which the carrier is $\cos \omega_c t$ and the message signal is $2 \cos \omega_m t$, so that the SSB plus carrier may be envelope-detected. Find in dB the suppression of the second-harmonic distortion relative to the desired signal amplitude.

3.43 A single sideband signal is given by $\phi_{SSB}(t) = 2 \cos (\omega_c - \omega_m)t$, where ω_c is the carrier frequency and ω_m is the message signal frequency. A carrier $A_c \cos \omega_c t$ is added to the SSB signal so that it can be envelope-demodulated. (a) Find an expression for the message signal $m(t)$. (b) Find A_c if the minimum attenuation of the second-harmonic distortion relative to the desired message signal amplitude in the demodulator output is to be 30 dB.

3.44 An amplitude modulated signal $\phi_{AM}(t) = 4[1 + m(t)] \cos \omega_c t$ and a single sideband plus carrier signal $\phi_{SSB+C}(t) = A_c \cos \omega_c t + m(t) \cos \omega_c t + m_h(t) \sin \omega_c t$ are to be envelope-demodulated. The message signal is the same for each modulated signal and is given by $m(t) = 2 \cos \omega_m t$. (a) Determine η_{AM}, the power efficiency for the AM signal. (b) Determine η_{SSB}, the power efficiency for the SSB signal (i) if $A_c = 4$ and (ii) if $A_c = 20$.

Section 3.6 Vestigial sideband (VSB) AM

3.45 The VSB modulator of Figure 3.39(a) is used in generating the upper sideband of a VSB signal. The frequency response of the upper VSB shaping filter is shown as $H_u(\omega)$ in Figure 3.39(b). The message signal is $m(t) = 4 \cos \omega_m t$, and the carrier is $\cos \omega_c t$, where $\omega_c = 10^6$ rad/s and $\omega_m = 0.4 \times 10^5$. Find an expression for the VSB signal and sketch its spectrum.

3.46 The VSB modulator of Figure 3.39(a) is used in generating the upper sideband of a VSB signal. The frequency response of the upper VSB shaping filter is shown as $H_u(\omega)$ in Figure 3.39(b). The message signal is $m(t) = 2 \sin \omega_m t + \cos 3\omega_m t$, and the carrier is

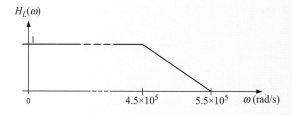

Figure 3.40. Lower VSB shaping filters for Problem 3.47.

$\cos \omega_c t$, where $\omega_c = 10^6$ rad/s and $\omega_m = 0.1 \times 10^5$. Determine an expression for the VSB signal in terms of the in-phase component and the quadrature component.

3.47 The VSB modulator of Figure 3.39(a) is used in generating a VSB signal. The VSB filter used has the frequency response $H_L(\omega)$ shown in Figure 3.40. The message signal is $m(t) = 2 \cos \omega_m t$, and the carrier is $\cos \omega_c t$, where $\omega_c = 5 \times 10^5$ rad/s and $\omega_m = 0.2 \times 10^5$ rad/s. Determine the expression for the VSB signal and sketch its spectrum.

3.48 The VSB modulator of Figure 3.39(a) is used in generating a VSB signal. The VSB filter used has the frequency response $H_L(\omega)$ shown in Figure 3.40. The message signal is $m(t) = 4 \cos \omega_m t + 2 \sin 3\omega_m t$, and the carrier is $\cos \omega_c t$, where $\omega_c = 5 \times 10^5$ rad/s and $\omega_m = 10^4$ rad/s. Determine the expression for the VSB signal in terms of the in-phase component and the quadrature component.

4 | Angle modulation

What we know is not much. What we do not know is immense.

P. S. DE LAPLACE

HISTORICAL PROFILE – Lee De Forest

Lee De Forest (1873–1961) was a prolific US inventor who received over 300 domestic and foreign patents. His inventions are of fundamental importance to radio, radar, telephones, electronics, and the movie industry. By the age of 13 he had invented many gadgets, including a working apparatus for silver-plating. After obtaining his PhD in 1899 from Yale University he worked mainly on radio communication and electronics. He was the first person to use the word "radio" in place of what was previously referred to as "wireless". He founded the De Forest Wireless Telegraph Company in 1902, and the De Forest Radio Telephone Company in 1907. For him the advancement of radio communication was a lifelong enterprise.

In 1907, De Forest patented his most important invention, the *audion* tube. This was a three-terminal gas-filled electronic amplifier tube (later called the triode), the predecessor to the modern transistor. It was later discovered that the triode worked better as a vacuum tube. Since the advent of the triode, signal amplification has been a great enabler for radio, radar, telephony, and electronics. By cascading the triodes (multi-stage amplification as we now know it) he was able to greatly increase signal strengths. In 1912, De Forest recorded in a laboratory notebook that by feeding part of the output of the triode back to the input, he could obtain self-regeneration which emitted a howl. However, he did not fully recognize this discovery as the useful sinusoidal oscillator, and he did not present it as such or apply for a patent for it then. Working independently and using the triode, Edwin H. Armstrong developed the regenerative feedback oscillator, applied for its patent in 1913, and was granted the patent in 1914. De Forest applied for a patent for the regenerative circuit in 1915 and sued for priority over Armstrong's patent. He eventually won in a case that went all the way to the Supreme Court and lasted until 1934, but the radio industry considered the judgment to be in error and always credited Armstrong with the invention.

De Forest's next most important invention was the *phonofilm*, a method of recording the audio signal adjacent to the corresponding images on a movie film, which he patented in 1921. For years, the Hollywood film industry resisted his efforts to sell them on the idea of talking movies, but eventually adopted his phonofilm method after 1927.

Although a prolific inventor, De Forest was a poor businessman. He started and ran 25 companies into bankruptcy. The many lawsuits in which he was involved contributed to many of these business failures. Eventually, he reluctantly sold his patents at very low prices to major communications companies that profited greatly from them. Perhaps there is a lesson here for the engineer. A little knowledge of business principles would not hurt. At the minimum, one could take one or more business-related general education courses.

4.1 INTRODUCTION

Like amplitude modulation, which was discussed in the previous chapter, angle modulation is a form of analog modulation. In amplitude modulation the amplitude of a carrier is varied linearly with the message signal. In angle modulation, the frequency of the carrier is varied to obtain frequency modulation, the phase of the carrier is varied to obtain phase modulation, and a combination of both the frequency and the phase of the carrier is varied to realize generalized angle modulation. This chapter focuses on angle modulation, especially on its more widely used forms of frequency modulation and phase modulation.

Frequency modulation (**FM**) is achieved by varying the frequency of the carrier linearly with the message signal amplitude. *Phase modulation* (**PM**) is achieved by varying the phase of the carrier linearly with the message signal amplitude.

FM and PM are special cases of angle modulation. An important advantage of angle modulation over amplitude modulation is that of better noise-suppression properties. This is why commercial FM radio is less noisy than commercial AM radio. This superior noise performance comes at the expense of larger bandwidth for angle-modulated signals relative to AM signals.

4.2 FREQUENCY, PHASE, AND ANGLE MODULATIONS

The concepts of *instantaneous frequency* and *total instantaneous phase* are pivotal to FM and PM. As the carrier frequency is varied linearly with the message signal amplitude to achieve frequency modulation, the total phase angle varies correspondingly. Similarly, as the phase angle of the carrier is varied linearly with the message signal amplitude to achieve phase modulation, the carrier frequency varies correspondingly.

Consider a message signal $m(t)$ and a sinusoidal carrier $A_c \cos(\omega_c t + \theta_o) = A_c \cos \theta_i(t)$, where $\theta_i(t) = \omega_c t + \theta_o$ is the total instantaneous phase of the sinusoidal carrier of amplitude

A_c and frequency ω_c. The constant phase angle θ_o can be assumed to be zero without any loss of generality. To obtain phase modulation, the total phase is varied proportionally to the message signal $m(t)$ with a phase deviation constant of k_p radians per unit amplitude of $m(t)$. The total instantaneous phase $\theta_i(t)$, the corresponding instantaneous frequency $\omega_i(t)$, and the PM signal $\phi_{PM}(t)$ are given respectively by

$$\theta_i(t) = \omega_c t + k_p m(t), \tag{4.1}$$

$$\omega_i(t) = \frac{d\theta_i(t)}{dt} = \omega_c + k_p m'(t), \tag{4.2}$$

and

$$\boxed{\phi_{PM}(t) = A_c \cos\left(\omega_c t + k_p m(t)\right)} \tag{4.3}$$

To obtain frequency modulation, the carrier frequency is varied proportionally to the message signal $m(t)$ with a frequency deviation constant of k_f Hz per unit amplitude of $m(t)$. The instantaneous frequency $\omega_i(t)$, the instantaneous angle $\theta_i(t)$, and the resulting FM signal are given respectively by

$$\omega_i(t) = \omega_c + k_f m(t), \tag{4.4}$$

$$\theta_i(t) = \int_{-\infty}^{t} \left[\omega_c + k_f m(\lambda)\right] d\lambda = \omega_c t + k_f \int_{-\infty}^{t} m(\lambda) d\lambda, \tag{4.5}$$

And

$$\boxed{\phi_{FM}(t) = A_c \cos\left[\omega_c t + k_f \int_{-\infty}^{t} m(\lambda) d\lambda\right]} \tag{4.6}$$

In each of Eqs. (4.3) and (4.6), the total instantaneous angle of the modulated signal is a specific linear function of the message signal $m(t)$. In general for angle modulation, this angle can be generated as various functions of $m(t)$. The phase angle may be regarded as the output of a system with input signal $m(t)$, impulse response $h(t)$, and output signal $\gamma(t)$. If the system is linear time-invariant, the phase angle may be expressed as

$$\gamma(t) = m(t) * h(t) = \int_{-\infty}^{t} m(\lambda) h(t - \lambda) d\lambda \tag{4.7}$$

With $k = k_p$ for PM and $k = k_f$ for FM, the generalized expression for the angle-modulated signal is

$$\phi_A(t) = A_c \cos\left[\omega_c t + k\gamma(t)\right] = A_c \cos\left[\omega_c t + k\int_{-\infty}^{t} m(\lambda) h(t - \lambda) d\lambda\right] \tag{4.8}$$

As mentioned earlier, FM and PM are special cases of angle modulation. If in the above equation for generalized angle modulation the impulse response is chosen as $h(t) = \delta(t)$ so that $m(t) * h(t) = m(t) * \delta(t) = m(t)$, then the PM expression of Eq. (4.3) is realized. However, if the impulse response is chosen as $h(t) = u(t)$, then $m(t) * h(t) = m(t) * u(t) = \int_{-\infty}^{t} m(\lambda)d\lambda$, and the FM expression of Eq. (4.6) is realized.

Similarities between frequency modulation and phase modulation

Note that if the message signal $m(t)$ is replaced by its derivative $m'(t)$ in Eq. (4.6), a PM signal results. Thus a PM signal can be regarded as an FM signal in which the modulating signal is $m'(t)$ rather than $m(t)$. Consequently, phase modulation can be achieved by directly employing the message signal $m(t)$ as the modulating signal for a phase modulator as in Eq. (4.2), or by first differentiating the message signal and employing its derivative $m'(t)$ as the modulating signal for a frequency modulator. These equivalent methods of generating PM are depicted in Figure 4.1(a).

Two equivalent methods of generation can be similarly deduced for FM signals. If the message signal $m(t)$ is replaced by its integral $\int_{-\infty}^{t} m(\lambda)d\lambda$ in Eq. (4.3), an FM signal results. Thus an FM signal can be regarded as a PM signal in which the modulating signal is $\int_{-\infty}^{t} m(\lambda)d\lambda$ rather than $m(t)$. Consequently, frequency modulation can be achieved by directly employing the message signal $m(t)$ as the modulating signal for a frequency modulator as in Eq. (4.6), or

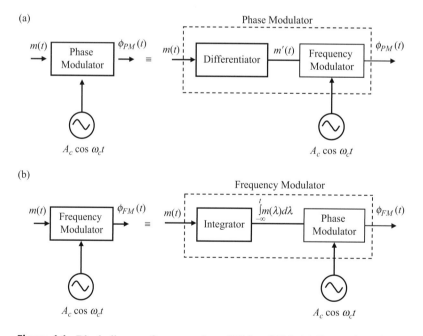

Figure 4.1. Block diagram for generation of PM and FM. (a) Generation of PM using a phase modulator or a frequency modulator; (b) generation of FM using a frequency modulator or a phase modulator.

by first integrating the message signal and employing its integral $\int_{-\infty}^{t} m(\lambda)d\lambda$ as the modulating signal for a phase modulator. These two equivalent methods of generating FM are illustrated in Figure 4.1(b).

FM and PM waveforms

Unlike AM, in which the carrier amplitude is time-varying, Eqs. (4.3), (4.6), and (4.7) show the carrier amplitude is constant in angle-modulated signals. Equation (4.2) indicates that for PM the instantaneous frequency varies linearly with the amplitude of $m'(t)$, while Eq. (4.4) indicates that for FM the instantaneous frequency varies linearly with the amplitude of the $m(t)$. Thus the major distinction between FM and PM waveforms lies in the differences in the variation of their instantaneous frequencies. Their waveforms are examined in the two following examples. Example 4.1 also compares the waveforms for AM, FM, and PM employing the same carrier and message signal.

EXAMPLE 4.1

For a case of tone modulation the carrier is $c(t) = 10 \cos \left(2\pi \times 10^7 t\right)$ and the message signal is $m(t) = 5 \cos \left(2\pi \times 10^4 t\right)$. Sketch the modulated signal waveform if the type of modulation performed is (a) amplitude modulation, (b) frequency modulation with $k_f = 4\pi \times 10^4$, and (c) phase modulation with $k_p = 2$.

Solution

(a) The amplitude-modulated signal is given by

$$\phi_{AM}(t) = [A_c + A_m \cos \omega_m t] \cos \omega_c t = \left[10 + 5 \cos \left(2\pi \times 10^4 t\right)\right] \cos \left(2\pi \times 10^7 t\right)$$
$$= E(t) \cos \left(2\pi \times 10^7 t\right)$$

The envelope $E(t)$ has a minimum value $E_{min} = 10 - 5 = 5$ and a maximum value $E_{max} = 10 + 5 = 15$. The message signal and the carrier are shown in Figure 4.2(a) and (b) respectively, while the AM waveform is shown in Figure 4.2(c).

(b) For the FM signal, the instantaneous carrier frequency is given by

$$f_i = \frac{\omega_i}{2\pi} = \frac{1}{2\pi}\left[\omega_c + k_f m(t)\right] = f_c + \frac{k_f}{2\pi} m(t)$$

$$f_{i,\max} = f_c + \frac{k_f}{2\pi}[m(t)]_{\max} = 10^7 + \frac{4\pi \times 10^4}{2\pi} \times 5 = 10.1 \text{ MHz}$$

$$f_{i,\min} = f_c + \frac{k_f}{2\pi}[m(t)]_{\min} = 10^7 + \frac{4\pi \times 10^4}{2\pi} \times (-5) = 9.9 \text{ MHz}$$

The FM waveform is sketched in Figure 4.2(d).

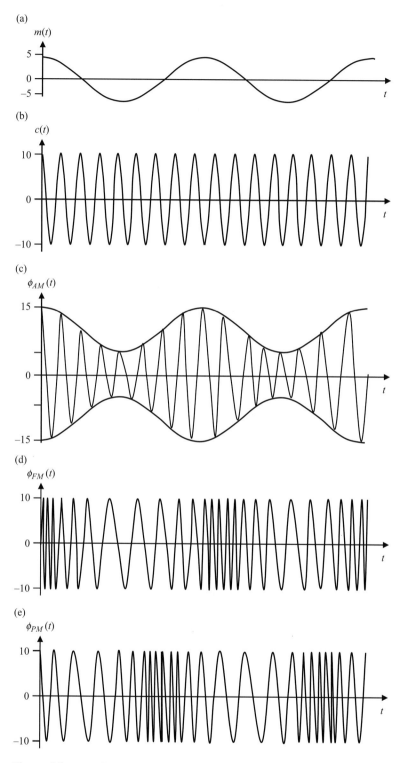

Figure 4.2. Waveforms for Example 4.1. (a) Message signal; (b) the carrier; (c) AM waveform; (d) FM waveform; (e) PM waveform.

(c) For the PM signal, the instantaneous carrier frequency is given by

$$f_i = \frac{\omega_i}{2\pi} = \frac{1}{2\pi} \left[\omega_c + k_p m'(t) \right] = f_c + \frac{k_p}{2\pi} m'(t)$$

$$m'(t) = -5 \times 2\pi \times 10^4 \sin\left(2\pi \times 10^4 t\right) = 10^5 \pi \sin\left(2\pi \times 10^4 t\right)$$

$$f_{i,\max} = f_c + \frac{k_p}{2\pi} [m'(t)]_{\max} = 10^7 + \frac{2}{2\pi} \times 10^5 \pi = 10.1 \text{ MHz}$$

$$f_{i,\min} = f_c + \frac{k_p}{2\pi} [m'(t)]_{\min} = 10^7 + \frac{2}{2\pi} \times \left(-10^5 \pi\right) = 9.9 \text{ MHz}$$

The PM waveform is shown in Figure 4.2(e). Note the similarity between the FM and PM waveforms. Both are constant-amplitude cosine waveforms of varying instantaneous frequency. However, $f_{i,\max}$ and $f_{i,\min}$ occur at the maximum and minimum values of $m(t)$ for FM, and at the maximum and minimum values of $m'(t)$ for PM.

PRACTICE PROBLEM 4.1

For a certain modulator, the carrier is $5 \cos\left(10^8 \pi t\right)$ and the input message signal is $10 \cos\left(4\pi \times 10^4 t\right)$. Find the maximum and minimum instantaneous carrier frequencies and sketch the modulated signal waveform if the type of modulation performed is (a) frequency modulation with $k_f = 10\pi^5$ and (b) phase modulation with $k_p = 2$.

Answer: (a) For the FM signal $f_{i,\max} = 50.5$ MHz, $f_{i,\min} = 49.5$ MHz.
(b) For the PM signal $f_{i,\max} = 50.4$ MHz, $f_{i,\min} = 49.6$ MHz.

The waveforms are similar to the corresponding waveforms of Figures 4.2(d) and (e), except for differences in their amplitudes and their instantaneous frequencies.

EXAMPLE 4.2

Sketch the FM and PM waveforms for the modulating signal shown in Figure 4.3(a) if $k_f = 10^6 \pi$, $k_p = \frac{\pi}{2}$, and the carrier is a sinusoid of amplitude A_c and frequency $f_c = 100$ MHz.

Solution
(a) For the FM signal,

$$f_i = f_c + \frac{k_f}{2\pi} m(t) = 10^8 + \frac{10^6 \pi}{2\pi} m(t)$$

$$f_{i,\max} = 10^8 + \frac{10^6 \pi}{2\pi} [m(t)]_{\max} = 10^8 + \frac{10^6 \pi}{2\pi} (1) = 100.5 \text{ MHz}$$

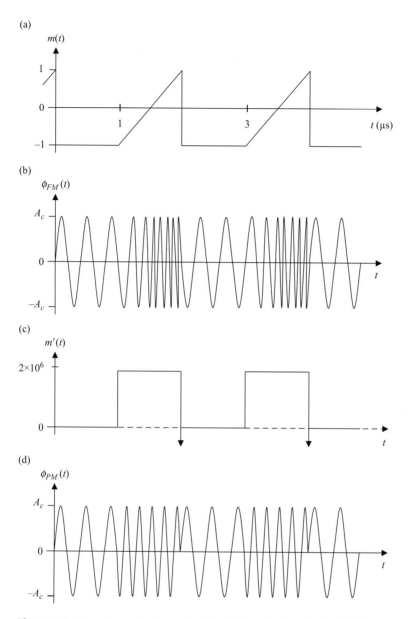

Figure 4.3. Waveforms for Example 4.2. (a) Modulating signal; (b) FM signal waveform; (c) first derivative of modulating signal; (d) PM signal waveform.

$$f_{i,\min} = 10^8 + \frac{10^6\pi}{2\pi}[m(t)]_{\min} = 10^8 + \frac{10^6\pi}{2\pi}(-1) = 99.5 \text{ MHz}$$

The FM waveform is shown in Figure 4.3(b).

(b) For the PM signal, $m'(t)$ is shown in Figure 4.3(c).

$[m'(t)]_{\min} = 0$, while the amplitude of $m(t)$ remains constant at -1.

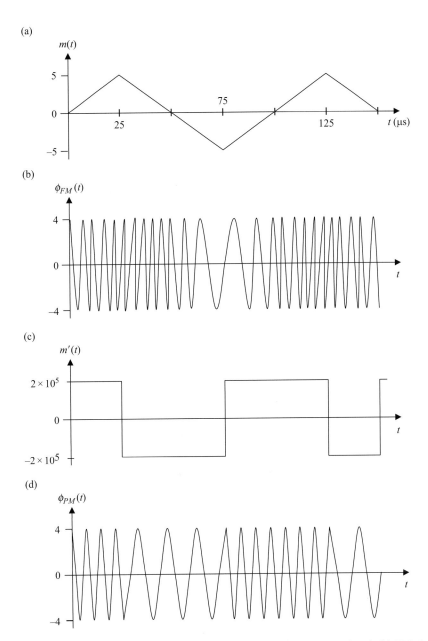

Figure 4.4. Waveforms for Practice problem 4.2. (a) Modulating signal; (b) FM signal waveform; (c) first derivative of modulating signal; (d) PM signal waveform.

$$[m'(t)]_{max} = \text{slope of } m(t) = 2 \times 10^6, \text{ while } m(t) \text{ changes linearly from } -1 \text{ to } 1.$$

$$f_i = f_c + \frac{1}{2\pi} k_p m'(t) = 10^8 + \frac{1}{2\pi}\frac{\pi}{2} m'(t)$$

$$f_{i,\,max} = 10^8 + \frac{1}{2\pi}\frac{\pi}{2}\left[2 \times 10^6\right] = 100.5 \text{ MHz}$$

$$f_{i,\min} = f_c + \frac{1}{2\pi}\frac{\pi}{2}(0) = f_c = 100 \text{ MHz}$$

f_i is 100 MHz in one half cycle of $m'(t)$ and 100.5 MHz in the other. At each jump discontinuity $m(t)$ has a total change of -2, resulting in an impulse of amplitude -2 in $m'(t)$. This impulse results in a change of $-2k_p = -2\left(\frac{\pi}{2}\right) = -\pi$, causing a phase reversal in the carrier at each jump discontinuity of $m(t)$. In general, the presence of an impulse in $m'(t)$ results in a change of the phase of the carrier.

PRACTICE PROBLEM 4.2

The carrier input to a modulator is $4\cos\left(1.8\pi \times 10^8 t\right)$, and the input message signal is the triangular waveform of Figure 4.4(a). Find the maximum and minimum instantaneous carrier frequencies and sketch the modulated signal waveform if the modulation performed is (a) FM with $k_f = 8\pi \times 10^4$, and (b) PM with $k_p = 2\pi$.

Answer: (a) For the FM signal $f_{i,\max} = 90.2$ MHz, $f_{i,\min} = 89.8$ MHz.
(b) For the PM signal $f_{i,\max} = 90.2$ MHz, $f_{i,\min} = 89.8$ MHz.

The FM and PM waveforms are shown in Figure 4.4(b) and (d), respectively.

4.3 BANDWIDTH AND SPECTRUM OF ANGLE-MODULATED SIGNALS

In Eq. (4.8), the expression for generalized angle modulation was given as $\phi_A(t) = A_c \cos\left[\omega_c t + k\gamma(t)\right]$, where $\gamma(t) = m(t) * h(t) = \int_{-\infty}^{t} m(\lambda)h(t-\lambda)d\lambda$, $h(t) = \delta(t)$ for PM, and $h(t) = u(t)$ for FM. Suppose an exponential carrier $A_c e^{j\omega_c t}$ is employed instead of the cosine carrier, the expression for generalized angle modulation becomes

$$\tilde{\phi}_A(t) = A_c e^{j[\omega_c t + k\gamma(t)]} = A_c e^{j\omega_c t} e^{jk\gamma(t)} \tag{4.9}$$

Replacing k by k_p for PM and by k_f for FM, the resulting expressions for the PM and FM signals are respectively

$$\phi_{PM}(t) = \text{Re}\left[\tilde{\phi}_A(t)\right] = \text{Re}\left[A_c e^{j\omega_c t} e^{jk_p\gamma(t)}\right]; \text{ for } \gamma(t) = m(t) * \delta(t) = m(t) \tag{4.10}$$

and

$$\phi_{FM}(t) = \text{Re}\left[\tilde{\phi}_A(t)\right] = \text{Re}\left[A_c e^{j\omega_c t} e^{jk_f\gamma(t)}\right]; \quad \text{for } \gamma(t) = \int_{-\infty}^{t} m(\lambda)d\lambda \tag{4.11}$$

Expanding $e^{jk_f\gamma(t)}$ in a power series, the FM signal can be expressed as

$$\phi_{FM}(t) = \text{Re}\left[A_c e^{j\omega_c t}\sum_{n=0}^{\infty}\frac{j^n k_f^n \gamma^n(t)}{n!}\right]$$

$$= \text{Re}\left[A_c e^{j\omega_c t}\left(1 + jk_f\gamma(t) - \frac{k_f^2\gamma^2(t)}{2!} - \frac{jk_f^3\gamma^3(t)}{3!} + \cdots\right)\right]$$

(4.12)

After taking the real part, the FM signal is

$$\phi_{FM}(t) = A_c\left[\cos\omega_c t - k_f\gamma(t)\sin\omega_c t - \frac{k_f^2\gamma^2(t)}{2!}\cos\omega_c t + \frac{k_f^3\gamma^3(t)}{3!}\sin\omega_c t + \cdots\right] \quad (4.13)$$

The Fourier transform of $\gamma^n(t)$ is the n-fold convolution of the Fourier transform of $\gamma(t)$. Let B Hz be the bandwidth of $m(t)$. Then since $m(t)$ and $\gamma(t) = \int_{-\infty}^{t} m(\lambda)d\lambda$ have the same bandwidth, $\gamma^n(t)$ has a bandwidth of nB Hz. As $n \to \infty$, the bandwidth of $\gamma^n(t)$ becomes infinite. Thus the theoretical bandwidth of FM is infinite! However, the amplitude of the nth term in the power series expansion of Eq. (4.13) becomes negligibly small as n becomes very large. Let N be the highest value of n for which the power series term is significant. Because the instantaneous frequency deviations are symmetrical about the unmodulated carrier frequency, FM is a double sideband system. Thus the effective bandwidth for the FM signal is finite and equal to $2NB$ Hz. A similar derivation for PM also leads to a theoretically infinite bandwidth, but the effective bandwidth is also determined by the number of significant terms in the corresponding power series expansion as $2NB$ Hz. But how is N, and hence the effective bandwidth, determined? The following discussion on narrowband angle modulation will provide some insights.

4.3.1 Narrowband angle modulation

Suppose the frequency deviation constant k_f is so small that $\left|k_f\gamma(t)\right| \ll 1$, then the third- and higher-order terms of Eq. (4.13) are negligibly small, and the FM signal can be approximated by

$$\phi_{FM}(t) \cong A_c\cos\omega_c t - A_c k_f\gamma(t)\sin\omega_c t \quad (4.14)$$

That is

$$\boxed{\phi_{FM}(t) \cong A_c\cos\omega_c t - A_c k_f\left(\int_{-\infty}^{t} m(\lambda)d\lambda\right)\sin\omega_c t} \quad (4.15)$$

For PM, $\gamma(t) = m(t) * \delta(t) = m(t)$, and the condition $\left|k_p\gamma(t)\right| = \left|k_p m(t)\right| \ll 1$ leads to a corresponding approximation for the PM signal as

$$\boxed{\phi_{PM}(t) \cong A_c\cos\omega_c t - A_c k_p m(t)\sin\omega_c t} \quad (4.16)$$

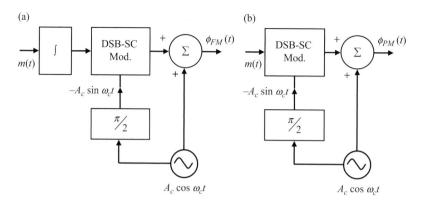

Figure 4.5. Generation of narrowband FM and narrowband PM. (a) Narrowband FM; (b) narrowband PM.

Compare the expressions for FM and PM in the last two equations to the amplitude-modulated signal $\phi_{AM}(t) = A_c \cos \omega_c t + m(t) \sin \omega_c t$. In each of the two expressions, the first term is the carrier, and the second term is a double sideband AM signal. Thus the bandwidth for each is $2B$, where B is the bandwidth of the message signal $m(t)$. Equations (4.15) and (4.16) represent *narrowband FM* (NBFM) and *narrowband PM* (NBPM), respectively. By contrast, FM and PM, which contain many significant higher-order terms in their power series expansion, have much larger bandwidths, and are respectively called *wideband FM* (WBFM) and *wideband PM* (WBPM). The two last equations suggest that NBFM and NBPM can be generated by employing a DSB-SC modulator and carrier insertion. Figure 4.5 shows the block diagrams for generating NBFM and NBPM.

Let the peak values of the message signal $m(t)$ and of its first derivative $m'(t)$ be denoted respectively by m_p and m_p'. The maximum deviation of the instantaneous carrier frequency from the unmodulated carrier frequency is called the *frequency deviation*, denoted by $\Delta \omega$ or Δf. For FM $\Delta \omega = k_f m_p$ ($\Delta f = \frac{1}{2\pi} k_f m_p$); for PM $\Delta \omega = k_p m_p'$ $\left(\Delta f = \frac{1}{2\pi} k_p m_p' \right)$. The ratio of the frequency deviation to the message signal bandwidth is called the *frequency deviation ratio*, denoted by β and given by

$$\beta = \frac{\Delta f}{B} = \frac{\Delta \omega}{2\pi B}$$

β is also referred to as the *modulation index for angle modulation*. The value of β provides a basis for differentiating between narrowband and wideband angle modulation. For narrowband $\beta \ll 1$ while for wideband $\beta \gg 1$. Equivalently, $\Delta f \ll B$ for narrowband, while $\Delta f \gg B$ for wideband. Typically, for NBFM $\beta \le 0.3$ while for WBFM $\beta \ge 5$. NBFM has about the same bandwidth and noise performance as AM. Commercial FM is WBFM, and larger bandwidth is the price paid for its superior noise performance. So why bother with NBFM; what use is it? As we shall subsequently see, it is easier to generate NBFM than WBFM, and the generation of NBFM can serve as an initial stage in the generation of WBFM.

NBFM and NMPM with tone modulation

Let the message signal be $m(t) = A_m \cos \omega_m t$, then $m_p = A_m$, $\omega_m = 2\pi B$, and $\beta = \frac{k_f m_p}{\omega_m} = \frac{k_f A_m}{\omega_m}$.
The $k_f \int_{-\infty}^{t} m(\lambda)d\lambda$ in Eq. (4.15) becomes $k_f \int_{-\infty}^{t} A_m \cos \omega_m \lambda \, d\lambda = \frac{k_f A_m}{\omega_m} \sin \omega_m t$. Thus Eq. (4.15) can be rewritten as

$$\phi_{FM}(t) \cong A_c \cos \omega_c t - A_c \frac{k_f A_m}{\omega_m} \sin \omega_m t \sin \omega_c t \tag{4.17}$$

Or equivalently as

$$\boxed{\phi_{FM}(t) \cong A_c \cos \omega_c t - \beta A_c \sin \omega_m t \sin \omega_c t} \tag{4.18}$$

Rewriting the above equation in order to better reveal the sidebands,

$$\phi_{FM}(t) \cong A_c \cos \omega_c t + \frac{1}{2}\beta A_c \cos(\omega_c + \omega_m)t - \frac{1}{2}\beta A_c \cos(\omega_c - \omega_m)t \tag{4.19}$$

This last equation further shows the similarity between NBFM and AM. The second term in the equation is the upper sideband, while the third term is the lower sideband. From the last equation the spectrum for NBFM with tone modulation can easily be drawn. As shown in the example below, a parallel derivation for NBPM with tone modulation leads to expressions similar to those of Eqs. (4.18) and (4.19).

EXAMPLE 4.3
Starting from Eq. (4.16), derive an expression for narrowband PM with tone modulation, similar to Eq. (4.19), which shows the similarity of narrowband PM to AM, and reveals its upper and lower sidebands.

Solution
As in the FM case, let the message signal be $m(t) = A_m \cos \omega_m t$. Equation (4.16) becomes

$$\phi_{PM}(t) \cong A_c \cos \omega_c t - A_c A_m k_p \cos \omega_m t \sin \omega_c t$$

$$m'(t) = -\omega_m A_m \sin \omega_m t \quad \Rightarrow \quad m'_p = \omega_m A_m$$

The frequency deviation is $\Delta \omega = \omega_m A_m k_p$, so $\beta = \frac{\Delta \omega}{\omega_m} = A_m k_p$.
Thus the PM signal can be rewritten as

$$\boxed{\phi_{PM}(t) \cong A_c \cos \omega_c t - \beta A_c \cos \omega_m t \sin \omega_c t} \tag{4.20}$$

Rewriting the last term of the last equation in terms of sum and difference frequencies,

$$\phi_{PM}(t) \cong A_c \cos \omega_c t - \frac{1}{2}\beta A_c \sin(\omega_c + \omega_m)t - \frac{1}{2}\beta A_c \sin(\omega_c - \omega_m)t \tag{4.21}$$

PRACTICE PROBLEM 4.3

For an FM modulator the input message signal is $m(t) = 8\cos\left(4\pi \times 10^4 t\right)$, $k_f = 5\pi \times 10^2$, and the carrier is $16\cos\left(2\pi \times 10^8 t\right)$. Verify that the modulated signal is NBFM. Obtain an expression for its amplitude spectrum.

Answer: $\beta = 0.1 < 0.3 \Rightarrow$ narrowband frequency modulation.

$$\Phi_{FM}(\omega) = 16\pi[\delta(\omega + \omega_c) + \delta(\omega - \omega_c)] + 0.8\pi[\delta(\omega + \omega_c + \omega_m) + \delta(\omega - \omega_c - \omega_m)]$$
$$- 0.8\pi[\delta(\omega + \omega_c - \omega_m) + \delta(\omega - \omega_c + \omega_m)];$$

where $\omega_c = 2\pi \times 10^8$ rad/s and $\omega_m = 4\pi \times 10^4$ rad/s.

EXAMPLE 4.4

The message signal input to a modulator is $m(t) = 4\cos\left(2\pi \times 10^4 t\right)$ and the carrier is $10\cos\left(10^8 \pi t\right)$. If frequency modulation is performed with $k_f = 10^3 \pi$, verify that the modulated signal is NBFM, obtain an expression for its amplitude spectrum, and sketch it.

Solution

For FM with $k_f = 10^3 \pi$,

$$\beta = k_f \frac{A_m}{\omega_m} = 10^3 \pi \left[\frac{4}{2\pi \times 10^4}\right] = 0.2; \ \beta < 0.3 \Rightarrow \text{narrowband FM.}$$

$$A_c = 10, \ \frac{1}{2}\beta A_c = \frac{1}{2}(0.2)(10) = 1.$$

Thus from Eq. (4.19),

$$\phi_{FM}(t) = 10\cos\omega_c t + \cos(\omega_c + \omega_m)t - \cos(\omega_c - \omega_m)t$$

The corresponding expression for the spectrum is

$$\Phi_{FM}(\omega) = 10\pi[\delta(\omega + \omega_c) + \delta(\omega - \omega_c)] + \pi[\delta(\omega + \omega_c + \omega_m) + \delta(\omega - \omega_c - \omega_m)]$$
$$- \pi[\delta(\omega + \omega_c - \omega_m) + \delta(\omega - \omega_c + \omega_m)];$$

where $\omega_c = 10\pi^8$ rad/s and $\omega_m = 2\pi \times 10^4$ rad/s.

The amplitude spectrum is sketched in Figure 4.6. Note its similarity to an AM spectrum, except for the negative algebraic sign of the lower sideband.

PRACTICE PROBLEM 4.4

For the carrier and message signal of Example 4.4, phase modulation is performed with $k_p = 0.05$. Verify that the modulated signal is narrowband PM and obtain an expression for its spectrum.

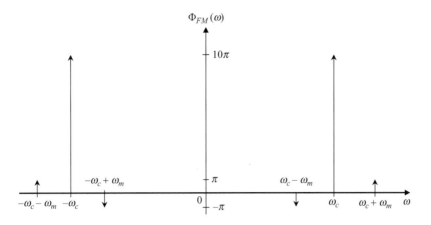

Figure 4.6. Amplitude spectrum for narrowband FM signal of Example 4.4.

Answer: $\beta = 0.2 < 0.3 \Rightarrow$ narrowband phase modulation.

$$\Phi_{PM}(\omega) = 10\pi[\delta(\omega + \omega_c) + \delta(\omega - \omega_c)] - j\pi[\delta(\omega + \omega_c + \omega_m) - \delta(\omega - \omega_c - \omega_m)]$$
$$- j\pi[\delta(\omega + \omega_c - \omega_m) - \delta(\omega - \omega_c + \omega_m)];$$

where $\omega_c = 10\pi^8$ rad/s and $\omega_m = 2\pi \times 10^4$ rad/s.

4.3.2 Wideband angle modulation

Consider again the expression for the FM signal in Eq. (4.11) for an arbitrary value of $k_f \gamma(t)$, and tone modulation with $m(t) = A_m \cos \omega_m t$. As obtained before for this case, $k_f \gamma(t) = \frac{k_f A_m}{\omega_m} \sin \omega_c t = \beta \sin \omega_c t$. Thus, Eq. (4.11) can be written as

$$\phi_{FM}(t) = \text{Re}\left[A_c e^{j\omega_c t} e^{j\beta \sin \omega_m t}\right] \tag{4.22}$$

The $e^{j\beta \sin \omega_m t}$ in the last equation is periodic with period $2\pi / \omega_m$. Thus it can be expanded in an exponential Fourier series as

$$e^{j\beta \sin \omega_m t} = \sum_{n=-\infty}^{\infty} C_n e^{jn\omega_m t}$$

With $\alpha = \omega_m t$, the Fourier coefficients C_n are given by

$$C_n = \frac{\omega_m}{2\pi} \int_{-\pi/\omega_m}^{\pi/\omega_m} e^{j\beta \sin \omega_m t} e^{-jn\omega_m t} dt = \frac{1}{2\pi} \int_{-\pi/\omega_m}^{\pi/\omega_m} e^{j(\beta \sin \alpha - n\alpha)} d\alpha \tag{4.23}$$

The above coefficients C_n are *Bessel functions of the first kind of order n and argument β*, denoted by $J_n(\beta)$. The equation for $\phi_{FM}(t)$ can now be rewritten as

$$\phi_{FM}(t) = \text{Re}\left[A_c \sum_{n=-\infty}^{\infty} J_n(\beta)e^{j(\omega_c t + n\omega_m t)}\right]$$

$$\text{i.e.}\quad \phi_{FM}(t) = A_c \sum_{n=-\infty}^{\infty} J_n(\beta) \cos\left(\omega_c + n\omega_m\right)t \qquad (4.24)$$

Equation (4.24) shows that the FM signal has an infinite number of sidebands centered at $\omega_c \pm n\omega_m$; $n = 1, 2, 3, \ldots$, leading to a theoretically infinite bandwidth. The number of significant coefficients (or significant sidebands) in this Fourier series expansion determines the effective FM bandwidth.

The Bessel function of the first kind is widely tabulated. Values of $J_n(\beta)$ for selected values of n and β are shown in Table 4.1. Graphs of $J_n(\beta)$ for selected values of β are shown in Figure 4.7. The FM spectrum is centered about the carrier frequency, and the sidebands are located symmetrically on either side of the carrier. For the case of tone modulation under

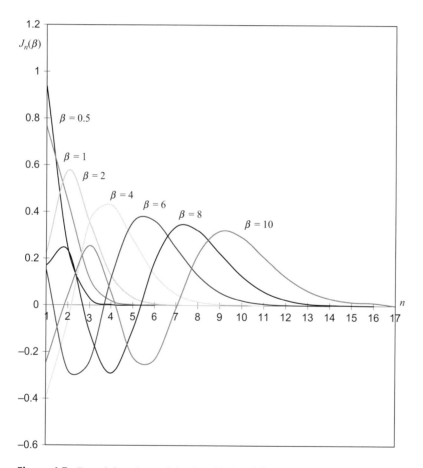

Figure 4.7. Bessel functions of the first kind $J_n(\beta)$ as a function of n for selected values of β.

Table 4.1. **Table of Bessel functions of the first kind $J_n(\beta)$**

n/β	0.1	0.2	0.5	1	2	4	6	8	10
0	0.997	0.990	0.9385	0.7652	0.2239	−0.3971	0.1506	0.1717	−0.2459
1	0.050	0.100	0.2423	0.4401	0.5767	−0.0660	−0.2767	0.2346	0.0435
2	0.003	0.005	0.0307	**0.1149**	0.3530	0.3641	−0.2428	−0.1131	0.2546
3			0.0026	0.0196	**0.1290**	0.4302	0.1150	−0.2910	0.0580
4			0.0002	0.0027	0.0340	0.2812	0.3578	−0.1052	−0.2198
5				0.0020	0.0070	**0.1322**	0.3621	0.1859	−0.2338
6				0.0012	0.0492	0.2457	0.3375	−0.0140	
7				0.0002	0.0154	**0.1296**	0.3204	0.2170	
8					0.0047	0.0565	0.2232	0.3178	
9					0.0034	0.0211	**0.1259**	0.2915	
10						0.0067	0.0602	0.2069	
11						0.0020	0.0246	**0.1231**	
12						0.0006	0.0073	0.0640	
13							0.0029	0.0304	
14							0.001	0.0151	
15								0.0118	

consideration, the message signal frequency f_m is also the message signal bandwidth B. If the maximum number of significant sidebands is $n_{\max} = N$, then the FM bandwidth is $B_{FM} = 2Nf_m = 2NB$. Table 4.1 shows that the sideband amplitudes (Bessel coefficients) $J_n(\beta)$ become vanishingly small as n becomes very large. The unmodulated carrier amplitude is normalized to a value of $A_c = 1$. Thus for the FM signal, $J_0(\beta)$ is the modulated carrier amplitude, and the Bessel coefficients for the various values of n are the sideband amplitudes located at $f_c \pm f_m, f_c \pm 2f_m, f_c \pm 3f_m, \ldots, f_c \pm nf_m$.

Various criteria exist for determining the maximum number of significant sidebands, leading to slightly different bandwidth estimates. The most widely used criterion, which is the one adopted here, is based on determining the number of sidebands containing up to 98% of the total power in the modulated signal. For an FM signal with carrier amplitude A_c, the total power in the modulated signal is the power in the constant-amplitude carrier, given by $P_c = \frac{1}{2}A_c^2$.

The power P_N in the N significant sidebands can be found by summing the power in the modulated carrier and the significant sidebands as

$$P_m = \frac{1}{2} A_c^2 \sum_{n=-N}^{N} J_n^2(\beta) = \left[J_0^2(\beta) + 2 \sum_{n=1}^{N} J_n^2(\beta) \right] P_c \qquad (4.25)$$

The normalized power in an FM signal has a value of 1, and this normalized power is the sum of the power in all the Bessel coefficients. That is

$$J_0^2(\beta) + 2 \sum_{n=1}^{\infty} J_n^2(\beta) = 1 \qquad (4.26)$$

The power in the N significant Bessel coefficients is denoted by $\hat{P}_N$ and given by

$$\hat{P}_m = J_0^2(\beta) + 2 \sum_{n=1}^{N} J_n^2(\beta) \qquad (4.27)$$

Thus N is the minimum value of n such that $\hat{P}_N \geq 0.98$, and $P_N = \frac{1}{2} A_c^2 \hat{P}_N = P_c \hat{P}_N$. For each value of β, the number of significant sidebands can be determined by applying this criterion. In Table 4.1, the last significant Bessel coefficients are shown in bold face for selected values of β between 1 and 10. If β is an integer, $N = \beta + 1$. If β is not an integer, $\beta + 1$ is rounded up to $N = \lceil \beta + 1 \rceil$, the next larger integer value. Thus the bandwidth for FM with tone modulation is

$$\boxed{B_{FM} \simeq 2Nf_m = 2\lceil \beta + 1 \rceil f_m} \qquad (4.28)$$

For the general message signal, $\beta = \Delta f / B$. The FM spectrum will not consist of discrete lines, so $\beta + 1$ should not be rounded to an integer. The FM signal bandwidth is given by

$$\boxed{B_{FM} \simeq 2(\beta + 1)B = 2(\Delta f + B)} \qquad (4.29)$$

The above formula for estimating FM bandwidth is known as *Carson's rule*. It gives the correct bandwidth for WBFM. For narrowband FM with $\beta \ll 1$ or equivalently $\Delta f \ll B$, Carson's rule gives $B_{FM} \simeq 2B$ as previously derived. For *very wideband* FM with $\beta \gg 1$ or equivalently $\Delta f \gg B$, Carson's rule gives $B_{FM} \simeq 2\beta B = 2\Delta f$.

A parallel derivation for phase modulation, starting from Eq. (4.10), leads to an estimate for bandwidth of tone-modulated PM similar to that of Eq. (4.29):

$$\boxed{B_{PM} \simeq 2Nf_m = 2\lceil \beta + 1 \rceil f_m} \qquad (4.30)$$

For the general message signal the corresponding PM bandwidth is

$$\boxed{B_{PM} \simeq 2(\beta + 1)B = 2(\Delta f + B)} \qquad (4.31)$$

In addition to giving the bandwidths, Eqs. (4.28) and (4.30) can also be used for obtaining the discrete spectra for tone-modulated FM or PM.

EXAMPLE 4.5

The input modulating signal to an FM modulator is $A_m \cos 2\pi f_m t$, and the frequency deviation constant is $k_f = 4\pi \times 10^4$. Find β, the FM bandwidth, and sketch the FM amplitude spectrum normalized to an unmodulated carrier of unit amplitude for positive frequencies if (a) $A_m = 2$ and $f_m = 20$ kHz, (b) $A_m = 1$ and $f_m = 10$ kHz, (c) $A_m = 2$ and $f_m = 10$ kHz.

Solution

(a)
$$\beta = \frac{\Delta f}{f_m} = \frac{1}{2\pi} \frac{k_f A_m}{f_m} = \frac{4\pi \times 10^4 \times 2}{2\pi \times 20 \times 10^3} = 2$$

$$B_{FM} = 2(\beta + 1)f_m = 2(2 + 1)20 \text{ kHz} = 120 \text{ kHz}$$

From Table 4.1, the modulated carrier amplitude is $J_0(2) = 0.2239$. The number of significant sidebands is $N = \beta + 1 = 3$. The sidebands are located at $f_c \pm nf_m$ and have amplitudes $J_n(\beta) = J_n(2)$ for $n = 1$, 2, and 3 which are respectively 0.1567, 0.3530, and 0.1290. The amplitude spectrum is shown in Figure 4.8(a).

(b)
$$\beta = \frac{1}{2\pi} \frac{k_f A_m}{f_m} = \frac{4\pi \times 10^4 \times 1}{2\pi \times 10 \times 10^3} = 2$$

$$B_{FM} = 2(\beta + 1)f_m = 2(2 + 1)10 \text{ kHz} = 60 \text{ kHz}$$

Because β is 2 here as in part (a), the number of significant sidebands and their amplitudes and the modulated carrier amplitude are the same as in part (a). However, because f_m is half of the corresponding value in part (a), the bandwidth and the separation between sidebands are half the corresponding values in part (a). The spectrum is shown in Figure 4.8(b).

(c)
$$\beta = \frac{1}{2\pi} \frac{k_f A_m}{f_m} = \frac{4\pi \times 10^4 \times 2}{2\pi \times 10 \times 10^3} = 4$$

$$B_{FM} = 2(\beta + 1)f_m = 2(4 + 1)10 \text{ kHz} = 100 \text{ kHz}$$

From Table 4.1, the modulated carrier amplitude is $J_0(4) = -0.3971$. There are $N = \beta + 1 = 5$ significant sidebands. The sidebands are located at $f_c \pm nf_m$. From Table 4.1, their amplitudes $J_n(\beta) = J_n(4)$ are $-0.0660, 0.3641, 0.4302, 0.2812$, and 0.1322, for $n = 1, 2, 3, 4$, and 5, respectively. The amplitude spectrum is shown in Figure 4.8(c). Although the frequency separation between the sidebands is the same as in part (b), note that because β is larger here, the number of significant sidebands, and hence the bandwidth, is larger here than in part (b).

(a)

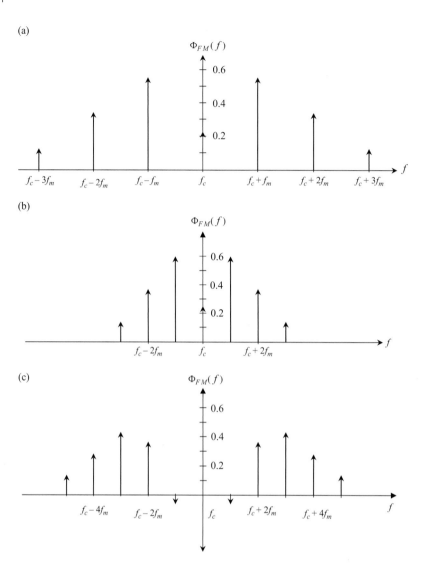

Figure 4.8. Sketch of magnitude spectrum, for FM with tone modulation, for Example 4.5. (a) $\beta = 2, f_m = 20$ kHz; (b) $\beta = 2, f_m = 10$ kHz; (c) $\beta = 4, f_m = 10$ kHz.

PRACTICE PROBLEM 4.5

The message signal input to a modulator is $4\cos\left(2\pi \times 10^4 t\right)$.

(a) If phase modulation is performed with $k_p = 2$, find the PM signal bandwidth.
(b) If frequency modulation is to be performed, find k_f that will make $B_{FM} = B_{PM}$.

Answer: (a) $B_{PM} = 180$ kHz
(b) $k_f = 4\pi \times 10^4$

EXAMPLE 4.6

An FM signal obtained across a 1 Ω resistor is $A_c \cos\left[\omega_c t + k_f \int_{-\infty}^{t} m(\lambda)d\lambda\right]$ volts. The message signal is $m(t) = 2 \sin 2\pi \times 10^4 t$ and A_c is 10. Find (a) the total power in the FM signal, (b) the power in up to the first three sidebands of the FM signal if $k_f = 2\pi \times 10^4$, and (c) the power in up to the first three sidebands of the FM signal if $k_f = 4\pi \times 10^4$.

Solution

(a) The total power is P_c, the power in the sinusoidal carrier of constant amplitude.

$$P_c = \frac{1}{2}A_c^2 = \frac{10^2}{2} = 50 \text{ W}$$

(b)
$$\beta = \frac{1}{2\pi}\frac{k_f A_m}{f_m} = \frac{1}{2\pi}\frac{2\pi \times 10^4 \times 2}{10^4} = 2$$

From Eq. (4.27), the normalized power in three most significant Bessel coefficients is

$$\hat{P}_3 = J_0^2(2) + 2\sum_{n=1}^{3} J_n^2(2)$$
$$= (0.2239)^2 + 2\left[(0.5767)^2 + (0.353)^2 + (0.129)^2\right] = 0.9978$$

Thus the power in up to three significant sidebands is

$$P_3 = \hat{P}_3 P_c = 0.9978(50 \text{ W}) = 49.89 \text{ W}$$

(c)
$$\beta = \frac{1}{2\pi}\frac{k_f A_m}{f_m} = \frac{1}{2\pi}\frac{4\pi \times 10^4 \times 2}{10^4} = 4$$

The normalized power in three most significant Bessel coefficients is

$$\hat{P}_3 = J_0^2(4) + 2\sum_{n=1}^{3} J_n^2(4)$$

$$i.e. \quad \hat{P}_3 = (-0.3971)^2 + 2\left[(-0.066)^2 + (0.3641)^2 + (0.4302)^2\right] = 0.8017$$

$$\therefore \quad P_3 = \hat{P}_3 P_c = 0.8017(50 \text{ W}) = 40.085 \text{ W}$$

This example shows that as β increases, the power is spread into a larger number of sidebands. Consequently the power in a given number of leading sidebands decreases.

PRACTICE PROBLEM 4.6

The output of a phase modulator obtained across a 1 Ω resistor is $A_c \cos\left[\omega_c t + k_p m(t)\right]$ volts, where $m(t) = 2 \cos 2\pi \times 10^4 t$ and A_c is 10. Find the power in the PM signal up to the three most significant sidebands if: (a) $k_p = 1$; (b) $k_p = 3$.

Answer: (a) 49.89 W; (b) 16 W.

EXAMPLE 4.7

(a) The message signal input to a modulator is $10 \cos \left(2\pi \times 10^4 t\right)$. If frequency modulation with frequency deviation constant $k_f = 10^4 \pi$ is performed, find the bandwidth of the resulting FM signal.

(b) If phase modulation is performed using the message signal of part (a), choose a phase deviation constant k_p such that the FM and PM signals will have the same bandwidth.

(c) If the message signal is the multi-tone signal $10 \cos \left(2\pi \times 10^4 t\right) + 5 \cos \left(4\pi \times 10^4 t\right)$, find the modulated signal bandwidth if FM is performed with $k_f = 10^4 \pi$ as in part (a).

(d) Repeat part (c) if PM is performed with $k_p = 0.75$.

Solution

(a)
$$\beta = \frac{1}{2\pi} \frac{k_f A_m}{f_m} = \frac{10^4 \pi \times 10}{2\pi \times 10^4} = 5$$

$$B_{FM} = 2(\beta + 1)f_m = 2(5 + 1)10 \text{ kHz} = 120 \text{ kHz}$$

(b) For the FM and PM bandwidths to be equal, β and Δf will be the same for both.
For FM, $\Delta \omega = k_f m_p = k_f A_m$, but for PM, $\Delta \omega = k_p m'_p$.
Expressing the message signal as $m(t) = A_m \cos \omega_m t$ gives

$$m'(t) = \frac{d}{dt}[A_m \cos \omega_m t] = -\omega_m A_m \sin \omega_m t \Rightarrow m'_p = \omega_m A_m.$$

Thus $k_f A_m = k_p \omega_m A_m \Rightarrow k_p = \frac{k_f}{\omega_m} = \frac{10^4 \pi}{2\pi \times 10^4} = \frac{1}{2}$

(c) Let $m(t) = m_1(t) + m_2(t) = 10 \cos \left(2\pi \times 10^4 t\right) + 5 \cos \left(4\pi \times 10^4 t\right)$
where $m_1(t) = A_1 \cos \omega_1 t = 10 \cos \left(2\pi \times 10^4 t\right) \Rightarrow \omega_1 = 2\pi \times 10^4; \; f_1 = 10 \text{ kHz}$

$$m_2(t) = A_2 \cos \omega_2 t = 5 \cos \left(4\pi \times 10^4 t\right) \Rightarrow \omega_2 = 4\pi \times 10^4; \; f_2 = 20 \text{ kHz}.$$

$f_1/f_2 = \frac{1}{2} = $ rational fraction $\Rightarrow m(t)$ is periodic. Thus peak values of $m_1(t)$ and $m_2(t)$ will align at certain time instants, giving the peak value of $m(t)$ as $m_p = A_1 + A_2 = 10 + 5 = 15$.

$$\Delta f = \frac{1}{2\pi} k_f m_p = \frac{1}{2\pi} \left(10^4 \pi\right) \times 15 = 75 \text{ kHz}$$

The bandwidth of $m(t)$ is the larger of f_1 and f_2, hence, $B = 20$ kHz.

$$\therefore \; B_{FM} = 2(\Delta f + B) = 2(75 + 20) \text{ kHz} = 190 \text{ kHz}$$

(d)
$$m'(t) = \frac{d}{dt}[+A_2 \cos \omega_2 t] = -[\omega_1 A_1 \sin \omega_1 t + \omega_2 A_2 \sin \omega_2 t]$$

$$m'_p = \omega_1 A_1 + \omega_2 A_2 = \left(2\pi \times 10^4 \times 10\right) + \left(4\pi \times 10^4 \times 5\right) = 4\pi \times 10^5.$$

$$\Delta f = \frac{1}{2\pi} k_p m_p' = \frac{1}{2\pi} (0.75) \times 4\pi \times 10^5 = 150 \text{ kHz}.$$

The message signal bandwidth is the same $B = 20$ kHz as in part (c).

$$\therefore \; B_{PM} = 2(\Delta f + B) = 2(150 + 20) \text{ kHz} = 340 \text{ kHz}$$

PRACTICE PROBLEM 4.7

A multi-tone message signal is given by $8\cos\left(2\pi \times 10^4 t\right) + 4\cos\left(10^4 \pi t\right)$.

(a) If FM is performed with $k_f = 10^4 \pi$, determine the bandwidth of the FM signal.
(b) If PM is performed with $k_p = 0.75$, determine the bandwidth of the PM signal.

Answer: (a) $B_{FM} = 140$ kHz
   (a) $B_{PM} = 170$ kHz

EXAMPLE 4.8

(a) If k_f and k_p retain their values in Example 4.7 (a) and (b) respectively, but the amplitude of the message signal is doubled while its frequency remains unchanged, find the bandwidths for the resulting FM and PM signals.
(b) Repeat part (a) if the message signal amplitude is the same as in Example 4.7, but its frequency is doubled.

Solution

(a) For the FM signal, $\beta = \frac{1}{2\pi} \frac{k_f A_m}{f_m} = \frac{10^4 \pi \times 20}{2\pi \times 10^4} = 10$

For the PM signal, $\beta = \frac{k_p m_p'}{\omega_m} = \frac{k_p \omega_m A_m}{\omega_m} = \frac{1}{2} \times 20 = 10$

Thus $B_{FM} = B_{PM} = 2(\beta + 1)f_m = 2(10 + 1)10 \text{ kHz} = 220 \text{ kHz}$

Note that doubling the message signal amplitude doubles β for both FM and PM and increases their bandwidths accordingly.

(b) For the FM signal, $\beta = \frac{1}{2\pi} \frac{k_f A_m}{f_m} = \frac{10^4 \pi \times 10}{2\pi \times 2 \times 10^4} = 2.5$

$$B_{FM} = 2(\beta + 1)f_m = 2(2.5 + 1)20 \text{ kHz} = 140 \text{ kHz}$$

For the PM signal, $\beta = \frac{k_p m_p'}{\omega_m} = \frac{k_p \omega_m A_m}{\omega_m} = \frac{1}{2} \times 10 = 5$

$$B_{PM} = 2(\beta + 1)f_m = 2(5 + 1)20 \text{ kHz} = 240 \text{ kHz}$$

The last two examples show that increasing the message signal bandwidth only slightly increases the FM signal bandwidth but greatly increases the PM signal bandwidth.

PRACTICE PROBLEM 4.8

If $m(t) = 4\cos\left(2\pi \times 10^4 t\right) + 8\cos\left(3\pi \times 10^4 t\right)$, find the modulated signal bandwidth if (a) FM is performed with $k_f = 2\pi \times 10^4$, and (b) PM is performed with $k_p = 0.75$

Answer: (a) $B_{FM} = 270$ kHz

(b) $B_{PM} = 180$ kHz

Most FM systems have a frequency deviation ratio in the range $2 < \beta < 10$. However, it can be shown that Carson's rule somewhat underestimates FM bandwidths for $\beta > 2$. From the point of view of equipment design, an underestimate of FM bandwidth is undesirable. For example, an amplifier or filter with a narrower bandwidth than the actual FM bandwidth would distort the FM signal. The following bandwidth criterion which is more conservative than Carson's rule is often used in equipment design:

$$B_{FM} \simeq 2(\beta + 2)B = 2(\Delta f + 2B) \tag{4.32}$$

EXAMPLE 4.9

For commercial FM, the audio message signal has a spectral range of 30 Hz to 15 kHz, and FCC allows a frequency deviation of 75 kHz. Estimate the transmission bandwidth for commercial FM using Carson's rule and the more conservative rule of Eq. (4.32).

Solution

$$\beta = \frac{\Delta f}{B} = \frac{75 \text{ kHz}}{15 \text{ kHz}} = 5$$

Using Carson's rule: $B_{FM} = 2(\beta + 1)B = 2(5 + 1)15 \text{ kHz} = 180 \text{ kHz}$

Using the conservative rule: $B_{FM} = 2(\beta + 2)B = 2(5 + 2)15 \text{ kHz} = 210 \text{ kHz}$

The bandwidth for commercial FM is 200 kHz. The above example shows that Carson's rule underestimates the actual bandwidth while the conservative rule overestimates it. For equipment design in cases of $\beta > 2$, the conservative estimate is preferable.

PRACTICE PROBLEM 4.9

Repeat Example 4.9 if the message signal is changed to

$$m(t) = 4\cos\left(2\pi \times 10^4 t\right) + 8\cos\left(4\pi \times 10^4 t\right).$$

Answer: Using Carson's Rule: $B_{FM} = 240$ kHz.

Using the conservative rule: $B_{FM} = 280$ kHz.

4.4 GENERATION OF FM SIGNALS

Wideband FM is often simply referred to as FM. Two basic methods are used for generation of FM signals: the *direct method* and *Armstrong's indirect method*. In the direct method, frequency modulation is accomplished with a voltage-controlled oscillator (VCO) which varies the carrier frequency linearly with the message signal amplitude. This method directly achieves sufficient frequency deviation, but frequency instability of the VCO is a setback. In Armstrong's indirect method, the message signal is first used to generate narrowband FM, which has a very small frequency deviation and a small carrier frequency. *Frequency multipliers* and a frequency converter are then used to raise the carrier frequency and the frequency deviation to the desired levels.

The frequency multiplier

The block diagram of a frequency multiplier is shown in Figure 4.9. It consists of a nonlinear device followed by a bandpass filter centered at the desired output frequency. The nonlinear device is defined by its input/output relationship. Suppose the input to the nonlinear device is $\phi(t)$, and $a_1, a_2, a_3, \ldots$ are constants, its output $y(t)$ is given by

$$y(t) = a_1\phi(t) + a_2\phi^2(t) + a_3\phi^3(t) + \cdots + a_n\phi^n(t) \tag{4.33}$$

If the input signal is a narrowband FM signal $\phi_i(t) = A \cos\left[\omega_c t + k_f \int_0^t m(\lambda)d\lambda\right]$ with instantaneous frequency $\omega_i = \omega_c + k_f m(t)$ and frequency deviation is $\Delta\omega = k_f m_p$, the nth term in the output signal of Eq. (4.33) will be an FM signal with n times the carrier frequency and n times the frequency deviation of the input FM signal. Thus, if the output of the nonlinear device is passed through a bandpass filter centered at $n\omega_c$, the filter output will be an FM signal with a carrier frequency $n\omega_c$ given by

$$\phi_o(t) = \phi_{FM}(t) = A_c \cos\left[n\omega_c t + nk_f \int_0^t m(\lambda)d\lambda\right] \tag{4.34}$$

Its instantaneous frequency and frequency deviation are respectively given by

$$n\omega_i = n\omega_c + nk_f m(t) \tag{4.35}$$

$$n\Delta\omega = nk_f m_p \tag{4.36}$$

The frequency multiplier accomplishes an n-fold multiplication of the carrier frequency, the instantaneous frequency, and the frequency deviation. It does so without distorting the information in the modulated signal. Thus, the message signal can be faithfully recovered from a modulated signal that has undergone frequency multiplication.

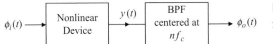

Figure 4.9. Block diagram of a frequency multiplier.

4.4.1 Armstrong's indirect method for generating wideband FM

Armstrong's indirect method of generating WBFM is shown in Figure 4.10. In this method, NBFM is first generated by employing a crystal oscillator in the block diagram of Figure 4.5. The crystal oscillator has the advantage of providing frequency stability. The NBFM is then passed through a cascade of frequency multipliers, a frequency converter, and a second cascade of frequency multipliers.

Why is it necessary to divide the frequency multipliers into two sets, and to interpose a frequency converter between them? To answer the question, consider the following example. Commercial FM with a carrier frequency $f_c = 100$ MHz and a frequency deviation $\Delta f = 75$ kHz is to be generated. The audio message signal has a spectral range of 100 Hz to 15 kHz. The NBFM has a carrier frequency of $f_{c1} = 100$ kHz at which stable crystal oscillators are readily available. The NBFM frequency deviation is $\Delta f_1 = 50$ Hz. This results in a maximum frequency deviation ratio of $\beta_1 = \frac{50 \text{ Hz}}{100 \text{ Hz}} = 0.5$ at the lowest modulating signal frequency. At intermediate and higher frequencies of the modulating signal, $\beta_1 \ll 1$, as required for NBFM. Thus the NBFM carrier frequency and the frequency deviation need to be multiplied by the following factors to yield the desired WBFM carrier frequency and frequency deviation:

$$n = \frac{\Delta f}{\Delta f_1} = \frac{75 \text{ kHz}}{50 \text{ Hz}} = 1500$$

$$n_f = \frac{f_c}{f_{c1}} = \frac{100 \text{ MHz}}{100 \text{ kHz}} = 1000$$

Clearly, the multiplication factors for the carrier frequency and the frequency deviation should be equal. The actual multiplication performed is the first of the above two. It is carried out in two steps of n_1 and n_2 with $n = n_1 \times n_2$. After the first frequency multiplication step, the resulting carrier frequency is $f_{c2} = n_1 f_{c1}$, and the frequency deviation is $\Delta f_2 = n_1 \Delta f_1$. Using the frequency converter, f_{c2} is now reduced to a lower frequency f'_{c2}. To enhance frequency stability, the frequency converter employs a crystal oscillator as the local oscillator. Note that f'_{c2} is the difference between f_{c2} and the frequency of the local oscillator signal f_{LO}, and is given by either $f'_{c2} = f_{LO} - f_{c2}$ or $f'_{c2} = f_{c2} - f_{LO}$. The second frequency multiplication step now follows, resulting in $n_2 f'_{c2} = f_{c3} = f_c$ and $n_2 \Delta f_2 = \Delta f_3$. Note that f_{c3} is equal to the

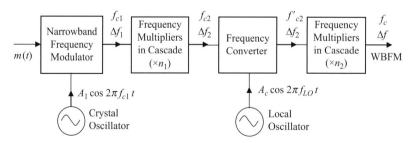

Figure 4.10. Indirect method of generating wideband EM.

desired WBFM carrier frequency f_c. Often Δf_3 differs slightly from Δf, the desired WBFM frequency deviation.

Frequency multipliers with small multiplying factors are preferable to those with large multiplying factors. Consider the nonlinear device characteristic of Eq. (4.33). The coefficients $a_n; n = 1, 2, 3, \ldots$, become very small as n becomes large. Thus large multiplication factors result in weak output signals. High-gain amplifiers can be used to increase the output signal level, but this also amplifies the signal noise. Large frequency multiplication is better achieved with a cascade of many multipliers of small multiplication factors, preferably frequency doublers and triplers. The desired frequency deviation for WBFM may be difficult to achieve with exactness because the overall product of the multiplying factors employed often differs from n, the desired overall multiplication factor.

In the current example, $n = n_1 \times n_2 = 1500$. If $n_1 = 50 = 2 \times 5^2$ and $n_2 = 30 = 2 \times 3 \times 5$, the largest multiplying factor is 5 and an exact realization of $\Delta f = 75$ kHz is achieved. Suppose only frequency doublers and triplers are allowed in the implementation, a reasonable choice will be $n_1 = 48 = 2^4 \times 3$ and $n_2 = 32 = 2^5$. In this case, $n_1 \times n_2 = 1536$, instead of 1500. Consequently, instead of 75 kHz, the frequency deviation realized is $1536 \times \Delta f_1 = 76.8$ kHz. It is still possible to achieve an exact frequency deviation of 75 kHz with this choice of $n_1 = 48$ and $n_2 = 32$. This requires a modification in the NBFM modulator that changes Δf_1 to a new value of $\Delta f_1 = \frac{\Delta f}{1536} = \frac{75000}{1536} = 48.828$ Hz.

EXAMPLE 4.10

In an indirect method for generating WBFM employing the block diagram of Figure 4.10, the audio message signal has a spectral range of 125 Hz to 12 kHz. For the NBFM, the carrier frequency is 200 kHz and the maximum value of frequency deviation ratio is to be $\frac{1}{3}$. For the WBFM, the desired carrier frequency is 108 MHz and the desired frequency deviation is 75 kHz. The maximum multiplying factor for any frequency multiplier in the scheme is to be 5. Complete the design of the scheme by choosing the multiplying factors for the frequency multipliers, and the local oscillator frequency for the frequency converter. Is the desired WBM frequency deviation exactly realized?

Solution

For the NBFM, Δf_1 and the maximum value of β_1 correspond to the minimum message signal frequency. Thus

$$\Delta f_1 = \beta_1 \times 125 \text{ Hz} = \frac{1}{3} \times 125 \text{ Hz} = 41.67 \text{ Hz}$$

$$n = n_1 \times n_2 = \frac{\Delta f}{\Delta f_1} = \frac{75 \text{ kHz}}{41.67 \text{ Hz}} = 1800$$

Let $n_1 = 50 = 2 \times 5^2$ and $n_2 = 36 = 2^2 \times 3^2$.

These specify multiplying factors for frequency multiplier steps 1 and 2.

$$\Delta f_3 = n_1 \times n_2 \times \Delta f_1 = 75 \times 10^3 = \Delta f$$

Thus the desired frequency deviation of 75 kHz is exactly realized.

$$f_{c2} = n_1 f_{c1} = 50 \times 200 \text{ kHz} = 10 \text{ MHz}$$

$$f'_{c2} = \frac{f_c}{n_2} = \frac{108 \text{ MHz}}{36} = 3 \text{ MHz}$$

Thus, the local oscillator frequency for the frequency converter is

$$f_{LO} = f'_{c2} + f_{c2} \text{ or } f_{LO} = f_{c2} - f'_{c2}$$

$$f_{LO} = 13 \text{ MHz or } 7 \text{ MHz}$$

PRACTICE PROBLEM 4.10
In the indirect method for generating WBFM of Figure 4.10, the audio message signal has a spectral range of 100 Hz to 15 kHz. For the NBFM, the carrier frequency is 250 kHz and the maximum value of frequency deviation ratio is 0.25. For the WBFM, the desired carrier frequency is 96 MHz and the desired frequency deviation is 75 kHz. The maximum multiplying factor for any frequency multiplier in the scheme is to be 5. Complete the design of the scheme by choosing the multiplying factors for the frequency multipliers, and the local oscillator frequency for the frequency converter.

Answer: $n_1 = 50 = 2 \times 5^2$ and $n_2 = 60 = 2^2 \times 3 \times 5$.

$$f_{LO} = 14.1 \text{ MHz or } 10.9 \text{ MHz}.$$

EXAMPLE 4.11
Repeat Example 4.10 if only frequency doublers and triplers are allowed in the implementation. If the desired WBM frequency deviation of 75 kHz is not realized, specify a new value to which the NBFM frequency deviation Δf_1 should be changed to achieve a WBFM frequency deviation of 75 kHz with the same values of n_1 and n_2.

Solution
$\Delta f_1 = 41.67$ Hz, and $n = 1800$, as in the previous example.
   Let $n_1 = 48 = 2^4 \times 3$ and $n_2 = 36 = 2^2 \times 3^2$
   These specify multiplying factors for frequency multiplier steps 1 and 2.
   $n_1 \times n_2 = 1728$, so $\Delta f_3 = 1728 \times \Delta f_1 = 72$ kHz.
   Thus the desired frequency deviation of 75 kHz is not achieved.

$$f_{c2} = n_1 f_{c1} = 48 \times 200 \text{ kHz} = 9.6 \text{ MHz}$$

$$f'_{c2} = \frac{f_c}{n_2} = \frac{108 \text{ MHz}}{36} = 3 \text{ MHz}$$

Thus, the local oscillator frequency for the frequency converter is

$$f_{LO} = f'_{c2} + f_{c2} \quad \text{or} \quad f_{LO} = f_{c2} - f'_{c2}$$

To achieve an exact frequency deviation of 75 kHz with $n_1 = 48$ and $n_2 = 36$, the NBFM frequency deviation needs to be changed to

$$\Delta f_1 = \frac{\Delta f}{n_1 \times n_2} = \frac{75000}{1728} = 43.403 \text{ Hz}$$

PRACTICE PROBLEM 4.11

Repeat Practice problem 4.10 if only frequency doublers and triplers are allowed in the implementation. If the desired WBM frequency deviation of 75 kHz is not realized, specify the value to which the frequency deviation β_1 in the NBFM should be changed, to achieve a WBFM frequency deviation of 75 kHz with the same multiplying factors.

Answer: Good choice: $n_1 = 64 = 2^6$, $n_2 = 48 = 2^4 \times 3$, $\Rightarrow n_1 n_2 = 3072$
$f_{LO} = 18$ MHz or 14 MHz; new $\beta_1 = 0.242$.

EXAMPLE 4.12

In a scheme employing the indirect method for generating WBFM, the NBFM has a carrier frequency of $f_{c1} = 250$ kHz and a frequency deviation of $\Delta f_1 = 32.552$ Hz. For the WBFM, the desired carrier frequency is $f_c = 100$ MHz and the desired frequency deviation is $\Delta f = 75$ kHz. Only frequency doublers and triplers are available for frequency multiplication. Instead of implementing the frequency multiplication in two steps separated by a frequency conversion stage as usual, all the frequency multiplication is implemented in one step, followed by a frequency conversion stage. (a) Sketch a block diagram implementation of this scheme. (b) Complete the design of the scheme by specifying the multiplying factors for the cascade of frequency multipliers and the local oscillator frequency for the frequency converter. (c) Comment on the suitability of this scheme compared to the scheme which employs a two-step frequency multiplication.

Solution

(a) The block diagram for the scheme is shown in Figure 4.11. The notation for the frequency and the frequency deviation for the respective stages are shown on the block diagram.

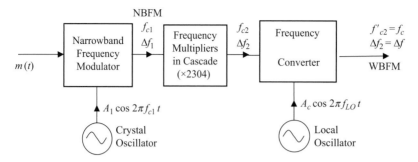

Figure 4.11. Wideband EM generation for Example 4.12.

(a)
$$n = \frac{\Delta f}{\Delta f_1} = \frac{75 \text{ kHz}}{32.552 \text{ Hz}} = 2304.006 \simeq 2304$$

$$n = 2304 = 2^8 \times 3^2$$

The frequency multiplier consists of a cascade of eight doublers and two triplers.

$$f_{c2} = nf_{c1} = 2304 \times 250 \text{ kHz} = 576 \text{ MHz}$$

$f_{c2} > f_c = 100 \text{ MHz} \Rightarrow$ frequency conversion to a lower value of $f'_{c2} = f_c$ is necessary.
Let f_{LO} be the frequency of the local crystal oscillator for the frequency converter.
Either $f'_{c2} = f_{LO} - f_{c2}$ or $f'_{c2} = f_{c2} - f_{LO}$
Thus $f_{LO} = f'_{c2} + f_{c2} = 676 \text{ MHz}$ or $f_{LO} = f_{c2} - f'_{c2} = 476 \text{ MHz}$

(c) Each of the two possible crystal oscillator frequencies obtained in part (b) is very high. It is more difficult to find crystal oscillators operating at such frequencies than the lower crystal oscillator frequencies needed in a two-step multiplication scheme. Therefore, the two-step multiplication scheme is preferable.

PRACTICE PROBLEM 4.12

In a scheme for indirect generation of WBFM, the NBFM has a carrier frequency $f_{c1} = 245 \text{ KHz}$ and a frequency deviation $\Delta f_1 = 28.5 \text{ Hz}$. For the WBFM, the desired carrier frequency is $f_c = 104 \text{ MHz}$ and the desired frequency deviation is $\Delta f = 73.872 \text{ kHz}$. Only frequency doublers and triplers are available for frequency multiplication. Instead of implementing the frequency multiplication in two steps separated by a frequency conversion stage, all the frequency multiplication is implemented in one step, followed by frequency conversion as shown in Figure 4.11. Complete the design of the scheme by specifying the multiplying factors for the cascade of frequency multipliers and the local oscillator frequency for the frequency converter.

Answer: $n = 2592 = 2^5 \times 3^4$; $f_{Lo} = 531.04 \text{ MHz}$ or $f_{Lo} = 739.04 \text{ MHz}$.

4.4.2   The direct method for generating of wideband FM

In the direct method, wideband FM is generated by varying the output signal frequency of a voltage-controlled oscillator (VCO) linearly with the message signal amplitude. Variation of the output signal frequency is usually accomplished by means of a variable reactance in a highly selective resonant circuit. The variable reactance can be a capacitor or an inductor. Most often, it is a variable capacitor, commonly called a *varactor* or a *varicap*, which is essentially a diode operated in reverse bias. The diode depletion region bounded by the n and p regions is in effect a dielectric sandwiched between two conducting surfaces, as in a capacitor. As the reverse bias voltage is increased, the width of the depletion region increases, and the junction capacitance of the diode decreases. As shown in Figure 4.12, a series combination of a dc voltage V_B and the message signal $m(t)$ is applied across the varactor diode C_v. The voltage V_B should be large enough to ensure that the diode is always reverse biased. Message signal variations result in variations in the value of C_v. Let C be the total capacitance due to C_v and C_1. When the carrier is unmodulated the value of C_v is C_{vo}, and is determined only by V_B. Thus, C has an unmodulated value of $C_o = C_1 + C_{vo}$. The total capacitance C varies linearly with the message signal $m(t)$ and is given by

$$C = C_1 + C_v = C_1 + C_{vo} - \alpha m(t)$$

Thus

$$\boxed{C = C_o - \alpha m(t)} \tag{4.37}$$

where α is a constant. The instantaneous carrier frequency ω_i is the resonance frequency of the LC parallel circuit and is given by

$$\omega_i = \frac{1}{\sqrt{LC}} = \frac{1}{\sqrt{L[C_o - \alpha m(t)]}} = \frac{1}{\sqrt{LC_o}}\left[1 - \frac{\alpha m(t)}{C_o}\right]^{-\frac{1}{2}}$$

Recall that the binomial expansion $(1 + x)^n$ can be approximated by

$$(1 + x)^n \simeq 1 + nx \quad \text{for } |x| \ll 1$$

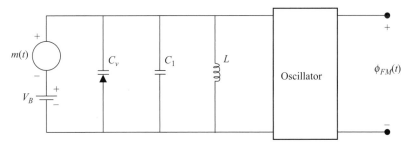

Figure 4.12. Direct generation of wideband FM.

For $\left|\frac{am(t)}{C_o}\right| \ll 1$, the corresponding binomial term in the previous expression for ω_i can be similarly approximated, giving

$$\omega_i \simeq \frac{1}{\sqrt{LC_o}}\left[1 + \frac{am(t)}{2C_o}\right]; \quad \text{for } \left|\frac{am(t)}{C_o}\right| \ll 1 \tag{4.38}$$

The unmodulated carrier frequency corresponds to the unmodulated value of the total parallel capacitance and is given by

$$\omega_c = \frac{1}{\sqrt{LC_o}} \tag{4.39}$$

Let $\omega_{i,\max}$ be the maximum instantaneous carrier frequency attained, and let m_p be the peak value of the message signal $m(t)$. Expressing ω_i and $\omega_{i,\max}$ in terms of ω_c,

$$\omega_i = \omega_c + \frac{am(t)}{2C_o}\omega_c = \omega_c + k_f m(t)$$

$$\omega_{i,\max} = \omega_c + \frac{am_p}{2C_o}\omega_c = \omega_c + k_f m_p$$

The maximum deviation of the capacitance from C_o is $\Delta C = am_p$. Thus

$$\Delta\omega = \frac{am_p}{2C_o}\omega_c = \frac{\Delta C}{2C_o}\omega_c = k_f m_p \tag{4.40}$$

$$\boxed{\therefore \quad \frac{\Delta C}{2C_o} = \frac{\Delta\omega}{\omega_c} = \frac{\Delta f}{f_c}} \tag{4.41}$$

Because $\Delta f/f_c$ is usually a very small fraction, $\Delta C = am_p$ is a small fraction of C_o as required for the approximation in Eq. (4.38) and for linear variation of ω_i with $m(t)$. The direct method usually produces sufficient frequency deviation for WBFM. However, unlike the stable crystal oscillators employed in the indirect method, the voltage-controlled oscillators employed in the direct method lack sufficient frequency stability. Thus the direct method suffers from a setback of frequency instability. This problem is usually remedied by employing negative feedback to stabilize the VCO output frequency.

EXAMPLE 4.13

The direct method of generating wideband FM shown in Figure 4.12 is used to produce FM with an unmodulated carrier frequency of 25 MHz and a frequency deviation of 18.75 kHz. The capacitance of the varactor diode is given by $C_v = 25V^{-\frac{1}{2}}$ pF, $C_1 = 30$ pF and $L = 1$ μH. Determine: (a) the magnitude of the reverse bias voltage V_B, and (b) the maximum and minimum values attained by the varactor capacitance C_v.

Solution

(a) With no modulating signal applied, C_v has its unmodulated value C_{vo} and the VCO output frequency is the unmodulated carrier frequency $f_c = 25$ MHz. Thus

$$C = C_o = C_1 + C_{vo}$$

$$\omega_c = \frac{1}{\sqrt{LC_o}} = 2\pi f_c$$

$$C_o = \frac{1}{(2\pi f_c)^2 L} = \frac{1}{(2\pi \times 25 \times 10^6)^2 \times 10^{-6}} = 40.528 \text{ pF}$$

$$\therefore \ C_{vo} = C_o - C_1 = (40.528 - 30) \text{ pF} = 10.528 \text{ pF}$$

$$\text{i.e.} \ \ C_{vo} = 25V_B^{-\frac{1}{2}} \text{ pF} = 10.528 \text{ pF}$$

$$\therefore \ \ V_B = 5.639 \text{ V}$$

(b) The deviation in the total capacitance is due to the deviation in the variable capacitor C_v. Thus $\Delta C = \Delta C_v$. From Eq. (4.41),

$$\Delta C = \Delta C_v = 2C_o \frac{\Delta f}{f_c} = 2(40.528) \frac{18.75 \times 10^3}{25 \times 10^6} \text{ pF} = 0.061 \text{ pF}$$

$$C_{v,\max} = C_{vo} + \Delta C_v = (10.528 + 0.061) \text{ pF} = 10.589 \text{ pF}$$

$$C_{v,\min} = C_{vo} - \Delta C_v = (10.528 - 0.061) \text{ pF} = 10.467 \text{ pF}$$

PRACTICE PROBLEM 4.13

FM with a carrier frequency of 10 MHz is produced with the direct method shown in Figure 4.12. $L = 4$ µH, $C_1 = 50$ pF, and the maximum value attained by the total capacitance due to C_1 and C_v is 63.8 pF. Determine (a) C_{vo}, the magnitude of the variable capacitance C_v due to the reverse bias voltage V_B alone, and (b) the maximum frequency deviation.

Answer: $C_{vo} = 13.326$ pF
$\Delta f = 37.425$ kHz

4.5   DEMODULATION OF ANGLE-MODULATED SIGNALS

The information in an angle-modulated signal resides in its instantaneous frequency or, equivalently, its instantaneous phase. Demodulation of FM is usually based on some method

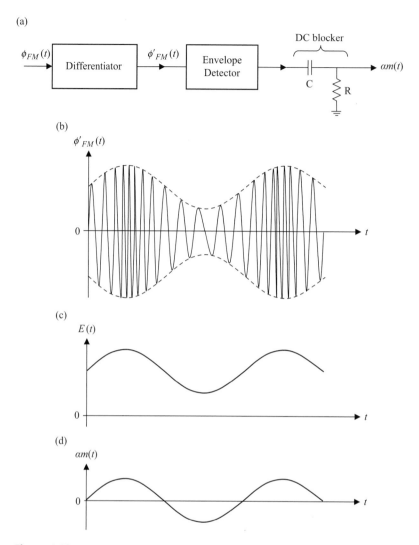

(a)

$\phi_{FM}(t)$ → Differentiator → $\phi'_{FM}(t)$ → Envelope Detector → DC blocker → $am(t)$

(b) $\phi'_{FM}(t)$

(c) $E(t)$

(d) $am(t)$

Figure 4.13. FM demodulator employing an ideal differentiator. (a) Block diagram of demodulator; (b) the derivative of the FM signal; (c) envelope detector output; (d) detected message signal.

of obtaining an output that is proportional to the instantaneous frequency, $\omega_i(t) = \omega_c + k_f m(t)$. In the case of PM, the instantaneous frequency is integrated to yield the instantaneous phase $\theta_i(t) = \omega_c t + k_p m(t)$.

A subclass of demodulators first converts the angle-modulated signal to one whose amplitude is proportional to the instantaneous frequency, thereby yielding a signal that is both amplitude-modulated and angle-modulated. This is usually referred to as FM/PM-to-AM conversion. The resulting signal is then applied to an envelope detector where demodulation proceeds as for AM signals. This approach is discussed in this section. Another approach is the feedback method, which detects the message signal by estimating the phase of the modulated signal with a phase-lock loop. The feedback approach is discussed in the next section.

Ideal differentiator demodulator

FM-to-AM conversion is achieved by passing the FM signal through a system with a frequency response of the form $|H(\omega)| = a_o + a\omega$ over the bandwidth of the FM signal. An ideal differentiator with a frequency response of $H(\omega) = j\omega$ constitutes a very simple implementation of such a network. The ideal differentiator demodulator is shown in Figure 4.13(a). When the differentiator input is the FM signal $\phi_{FM}(t)$, its output is

$$\phi'_{FM}(t) = \frac{d}{dt}\left[A_c\left(\cos\omega_c t + k_f \int_{-\infty}^{t} m(\lambda)d\lambda\right)\right]$$

$$= -A_c[\omega_c + k_f m(t)]\sin\left[\omega_c t + k_f \int_{-\infty}^{t} m(\lambda)d\lambda\right]$$

(4.42)

Note that $\phi'_{FM}(t)$ is both amplitude-modulated and frequency-modulated. After passing $\phi'_{FM}(t)$ through the envelope detector, the output is $E(t) = A_c[\omega_c + k_f m(t)]$. Because the frequency deviation $\Delta\omega = k_f m_p$ is always less than the carrier frequency ω_c, there is no risk of envelope distortion. Next, the envelope detector output is passed through the dc blocking circuit to yield a replica of the message signal, $A_c k_f m(t) = \alpha m(t)$.

EXAMPLE 4.14

Figure 4.14 shows a high-pass filter followed by an envelope detector. The input signal to the circuit is an FM signal given by $\phi_{FM}(t) = A_c \cos\left[\omega_c t + k_f \int_{-\infty}^{t} m(\lambda)d\lambda\right]$. (a) Find the output of the envelope detector if the high-pass filter cut-off frequency is much greater than the carrier frequency for the FM signal. (b) If $R = 100\ \Omega$, and the FM carrier frequency is 100 MHz, choose the value of the capacitance C so that the cut-off frequency of the filter will be ten times the FM carrier frequency.

Solution

(a) The filter frequency response is

$$H(j\omega) = \frac{j\omega RC}{1 + j\omega RC} = \frac{j\omega/\omega_H}{1 + j\omega/\omega_H},$$

where $\omega_H = \frac{1}{RC}$ is the cut-off frequency of the high-pass filter.

For $\omega \ll \omega_H$, the filter response can be approximated as

$$H(j\omega) \simeq \frac{j\omega}{\omega_H} = j\omega RC$$

Multiplication by $j\omega$ in the frequency domain is equivalent to differentiation.
Thus, if $\omega_c \ll \omega_H$, the high-pass filter acts as an ideal differentiator for the FM signal.

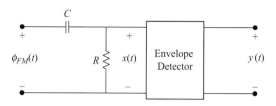

Figure 4.14. Demodulator circuit for Example 4.14.

The high-pass filter output, $x(t)$, is given by

$$x(t) = RC\frac{d}{dt}[\phi_{FM}(t)] = RC\frac{d}{dt}\left[A_c \cos\left(\omega_c t + k_f \int_{-\infty}^{t} m(\lambda)d\lambda\right)\right]$$

$$= -RCA_c\left[\omega_c + k_f m(t)\right]\sin\left[\omega_c t + k_f \int_{-\infty}^{t} m(\lambda)d\lambda\right]$$

The envelope is non-negative, so the algebraic sign does not apply. The envelope detector output is

$$y(t) = A_c\omega_c RC + A_c RC k_f m(t)$$

(b) The high-pass filter cut-off frequency is $f_H = 10f_c = 10^9$ Hz.
Thus, $\omega_H = \frac{1}{RC} = 2\pi f_H = 2\pi \times 10^9$

$$C = \frac{1}{\omega_H R} = \frac{1}{2\pi \times 10^9 \times 100} = 1.59 \text{ pF}$$

PRACTICE PROBLEM 4.14

Repeat part (b) of Example 4.14 if the FM carrier frequency is changed to 8 MHz and the cut-off frequency for the high-pass filter is 12 times the FM carrier frequency.

Answer: $C = 16.58$ pF

Bandpass limiter

For FM demodulation involving envelope detection to work very well, the amplitude of the FM signal should be perfectly constant. Several factors, such as channel noise and channel distortion, introduce undesirable amplitude variations into the FM signal. These in turn distort the demodulated signal. Consequently, it is necessary to rid the FM signal of such undesired amplitude variations prior to demodulation. This is accomplished with a *bandpass limiter*.

The block diagram of a bandpass limiter is shown in Figure 4.15(a). It consists of a *hard limiter* followed by a bandpass filter. A hard limiter is a device with the transfer characteristic shown in Figure 4.15(b), where $x(t)$ is the input and $y(t)$ is the output signal. The output of the hard limiter is 1 when its input signal is positive, but its output is -1 when its input is negative.

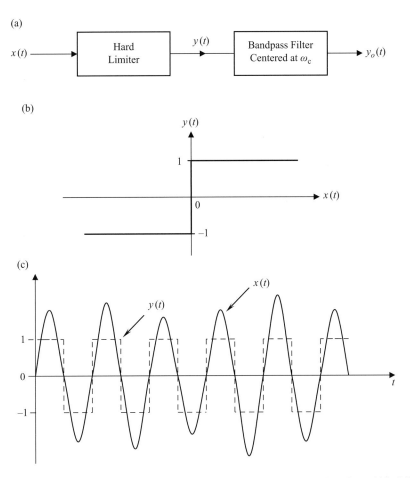

Figure 4.15. Bandpass limiter. (a) Block diagram; (b) transfer function of hard limiter; (c) input and output waveforms of hard limiter.

An input FM signal with a time-varying amplitude $A(t)$ and the output signal are depicted in Figure 4.15(c). The input FM signal is given by

$$x(t) = A(t) \cos \left[\omega_c t + k_f \int_{-\infty}^{t} m(\lambda)d\lambda \right] = A(t) \cos \left[\theta(t) \right] \tag{4.43}$$

where the total angle of the modulated signal is

$$\theta(t) = \omega_c t + k_f \int_{-\infty}^{t} m(\lambda)d\lambda \tag{4.44}$$

In accordance with the transfer function of the limiter, its output can be expressed as

$$y(t) = \begin{cases} 1, & \cos \theta(t) > 0 \\ -1, & \cos \theta(t) < 0 \end{cases} \tag{4.45}$$

Thus the output signal $y(t)$ is a bipolar, square waveform of amplitude 1 with the same zero crossings, and hence the same frequency, as the input FM signal. Commercial FM has a carrier frequency range of 88 MHz to 108 MHz, and a frequency deviation of 75 kHz. Its frequency deviation is less than a thousandth of the carrier frequency, so the FM signal can be considered as approximately sinusoidal. Thus $y(t)$ is approximately a periodic square waveform similar to the periodic square waveform, $p_1(t)$, of Figure 3.17. Hence, $y(t)$ can be expressed as a Fourier series similar to that of $p_1(t)$ in Eq. (3.37) as

$$y(t) = \frac{4}{\pi} \left[\cos \theta(t) - \frac{1}{3} \cos 3\theta(t) + \frac{1}{5} \cos 5\theta(t) - \cdots \right] \tag{4.46}$$

For an integer n, referring to Eq. (4.46), $n\theta(t)$ is given by

$$\cos n\theta(t) = \cos \left[n\omega_c t + nk_f \int_{-\infty}^{t} m(\lambda)d\lambda \right] \tag{4.47}$$

Thus, Eq. (4.47) shows that the hard limiter output contains the desired FM signal of carrier frequency ω_c and its higher-frequency odd harmonics. After passing $y(t)$ through the bandpass filter centered at ω_c the output signal is a constant-amplitude FM signal

$$y_o(t) = \phi_{FM}(t) = \frac{4}{\pi} \cos \left[\omega_c t + k_f \int_{-\infty}^{t} m(\lambda)d\lambda \right] \tag{4.48}$$

The time-delay demodulator

The time-delay differentiator demodulator consists of a time-delay differentiator followed by an envelope detector. The time-delay differentiator implements a discrete time approximation to differentiation. As shown in Figure 4.16, it approximates the first derivative by subtracting a delayed version of its input signal from the input signal, and amplifying the result by a gain constant, which is the reciprocal of the time delay. Let the input to the differentiator be the FM signal $\phi_{FM}(t)$ and let the time delay be τ. Then the time-delayed signal is $\phi_{FM}(t - \tau)$. The time-delay differentiator output, $y(t)$, is given by

$$y(t) = \frac{1}{\tau} [\phi_{FM}(t) - \phi_{FM}(t - \tau)] \tag{4.49}$$

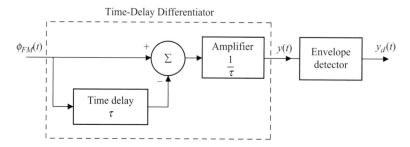

Figure 4.16. Time-delay demodulator.

For $y(t)$ to equal the first derivative of $\phi_{FM}(t)$, τ must be infinitesimally small. That is

$$\phi'_{FM}(t) = \lim_{\tau \to 0} [y(t)] = \lim_{\tau \to 0} \left[\frac{1}{\tau} (\phi_{FM}(t) - \phi_{FM}(t - \tau)) \right] \quad (4.50)$$

If τ is chosen to be sufficiently small, the time-delay differentiator output is a good approximation to the first derivative. The time delay need not be any smaller than is necessary to achieve a reasonably good approximation, otherwise the amplifier gain constant, $\frac{1}{\tau}$, can become exceedingly high. It can be shown that the maximum value of τ for a good approximation is $\frac{T}{4}$, where T is the period of the unmodulated carrier for the FM signal. With such a suitable value of τ,

$$y(t) \simeq \phi'_{FM}(t) \quad (4.51)$$

As previously explained, $\phi'_{FM}(t)$ is both amplitude-modulated and frequency-modulated. Demodulation is completed by passing $\phi'_{FM}(t)$ through an envelope detector.

EXAMPLE 4.15
Determine how small the time delay τ in the time-delay demodulator needs to be in order for the output of the time delay differentiator to be a very good approximation to the first derivative of the input angle-modulated signal.

Solution
Refer to the block diagram of the time-delay demodulator of Figure 4.16. The output of the time-delay differentiator was obtained as

$$y(t) = \frac{1}{\tau} [\phi_{FM}(t) - \phi_{FM}(t - \tau)]$$

Taking Fourier transforms,

$$Y(j\omega) = \frac{1}{\tau} \left[\Phi_{FM}(j\omega) - e^{-j\omega\tau} \Phi_{FM}(j\omega) \right] = \frac{1}{\tau} \Phi_{FM}(j\omega) \left[1 - e^{-j\omega\tau} \right]$$

For $\omega\tau \ll 1$, $e^{-j\omega\tau} \simeq 1 - j\omega\tau \Rightarrow (1 - e^{-j\omega\tau}) \simeq j\omega\tau$
 Thus

$$Y(j\omega) \simeq j\omega \Phi_{FM}(j\omega)$$

Hence, the output, $y(t) \simeq \phi'_{FM}(t)$ for $\omega\tau \ll 1$ or, equivalently, for $\tau \ll \frac{1}{\omega}$.
 For the FM signal, ω is the instantaneous frequency of signal to be demodulated.
 Thus the condition is
 $\tau \ll \frac{1}{\omega_c + \Delta\omega} \simeq \frac{1}{\omega_c} = \frac{1}{2\pi f_c} = \frac{T}{2\pi}$, where T is the period of the unmodulated carrier.
 Thus τ should be $\frac{1}{10} \left[\frac{T}{2\pi} \right]$ or less for a very good approximation, but, as indicated earlier, $\tau_{max} = \frac{T}{4}$ for a reasonably good approximation, which does not require too high an amplifier gain, following the differentiator.

PRACTICE PROBLEM 4.15

For a time-delay demodulator, denote the maximum permissible time delay by τ_{max}, and the time delay required for a very good approximation as τ. Determine τ_{max} and τ if: (a) the signal to be demodulated is commercial FM with a carrier frequency of 100 MHz, and (b) the FM signal to be demodulated is narrowband FM with a carrier frequency of 500 kHz.

Answer: (a) $\tau_{max} = 2.5$ ns, $\tau = 0.159$ ns.

(b) $\tau_{max} = 0.5$ μs, $\tau = 0.0318$ μs.

The slope detector

As shown in Figure 4.17(a), the slope detector is simply a tuned circuit followed by an envelope detector. The amplitude response of the tuned circuit $|H(j\omega)|$ is shown in Figure 4.17(b). It has an approximately linear region over a band of frequencies on either side of the resonant

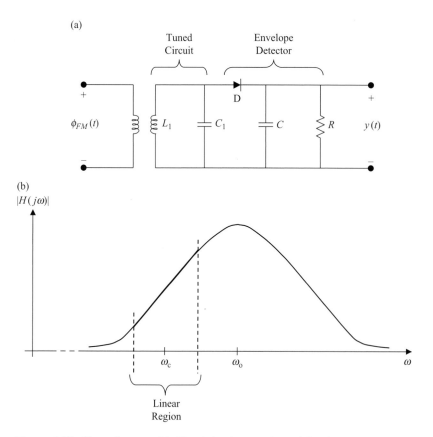

Figure 4.17. Slope detector. (a) Circuit implementation of the slope detector; (b) amplitude response of the slope detector.

frequency. FM-to-AM conversion is achieved by centering the carrier frequency of the FM signal ω_c in the middle of one of these linear regions. This is usually the approximately linear region on the right side, which has a positive slope. Within that region, the frequency response is given by $|H(j\omega)| \simeq a_o + a\omega$.

The slope detector is not a very practical demodulator. Typically, its so-called linear region is not very linear or wide enough for demodulation of wideband FM. In addition, the bias term a_o results in an additional dc term at the envelope detector output.

The balanced discriminator

The balanced discriminator provides a much better performance than the slope detector. Its principle of operation is illustrated in the block diagram of Figure 4.18(a). A circuit implementation is shown in Figure 4.18(b). It consists of two slope detectors so interconnected that the overall output is the difference between the outputs of the two. Let $|H_1(j\omega)|$ be the amplitude response of the upper slope detector circuit and $|H_2(j\omega)|$ the amplitude response of the lower slope detector circuit. The individual amplitude responses and the composite amplitude response are shown in Figure 4.18(c). The composite amplitude response $|H(j\omega)|$ is given by

$$|H(j\omega)| = |H_1(j\omega)| - |H_2(j\omega)| \tag{4.52}$$

The linear regions of the individual amplitude responses can be expressed as $|H_1(j\omega)| \simeq a_o + a_1\omega$ and $|H_2(\omega)| \simeq a_o + a_2\omega$. Within the overlap of these regions, the composite response is

$$|H(j\omega)| = [a_o + a_1\omega] - [a_o + a_2\omega] = (a_1 - a_2)\omega \tag{4.53}$$

Note that there is no bias in the amplitude response of the balanced configuration. The slope detector response is proportional to the instantaneous frequency $\omega_i = \omega_c + k_f m(t)$. However, the response of the balanced configuration is centered at ω_c and proportional to the frequency deviation about the carrier, which is the difference between ω_i and ω_c. In effect, the balanced configuration discriminates between the instantaneous carrier frequency and the unmodulated carrier frequency, hence the name.

The resonance frequencies for the upper and the lower slope circuits are respectively ω_1 and ω_2 or, equivalently, f_1 and f_2. They are displaced from each other so that f_1 is greater than f_c, and f_2 is lower than f_c by the same amount, where f_c is the carrier frequency of the FM signal to be demodulated. Some cancellation occurs between the two nonlinear resonance curves. Consequently, the linear region of the composite response is substantially extended, relative to that of the slope detector, thereby permitting demodulation of wideband FM. A sufficiently wide linear region is obtained when the frequency separation between f_1 and f_2 is about $1.5B$, where B is the 3 dB bandwidth of each resonant circuit. If the bandwidth of each resonant circuit is chosen equal to the bandwidth of the FM signal B_{FM}, then the frequency separation is conveniently $f_1 - f_2 = 1.5B_{FM}$.

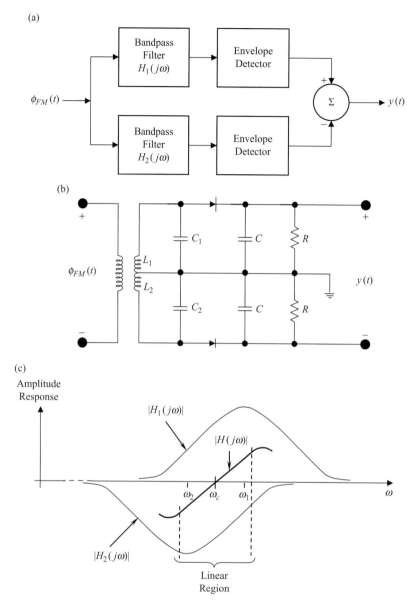

Figure 4.18. Balanced discriminator. (a) Block diagram; (b) circuit diagram; (c) the individual amplitude responses and the composite amplitude response.

EXAMPLE 4.16

Part of a balanced discriminator comprising the upper and the lower slope detectors is shown in Figure 4.19. Each of the inductances L_1 and L_2 has a total (self and mutual) inductance of 20 μH. The FM signal to be demodulated is at the intermediate carrier frequency of 10.7 MHz,

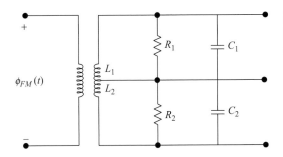

Figure 4.19. Circuit for Example 4.16: the balanced slope detector part of a balanced discriminator.

and has a bandwidth of $B_{FM} = 200$ kHz. The resonance frequencies of the upper and the lower slope detectors are above and below the FM carrier frequency by an equal amount. The bandwidth of each slope detector is B_{FM}, and the separation between their resonance frequencies is $1.5B_{FM}$. Complete the design of the slope detectors by choosing the capacitances C_1 and C_2, and the resistances R_1 and R_2.

Solution

The frequency response for a parallel RLC such as each of the slope detectors is

$$H(j\omega) = \frac{1}{1 + j(\omega C - 1/\omega L)} \tag{4.54}$$

Letting $\omega_o = \frac{1}{\sqrt{LC}}$ be the resonance frequency and $Q = \frac{\omega_o}{BW} = \omega_o CR = \frac{R}{\omega_o L}$ be the quality factor, where BW is the 3 dB bandwidth, the frequency response can be rewritten as

$$H(j\omega) = \frac{1}{1 + jQ(\omega/\omega_o - \omega_o/\omega)} \tag{4.55}$$

For the upper slope circuit, the resonance frequency is

$$f_1 = f_c + \frac{1}{2}(1.5B_{FM}) = \left[10.7 \times 10^6 + \frac{1}{2}(1.5 \times 200 \times 10^3)\right] \text{Hz} = 10.85 \text{ MHz}$$

$$C_1 = \frac{1}{\omega_1^2 L} = \frac{1}{(2\pi \times 10.85 \times 10^6)^2 \times 2 \times 10^{-5}} = 10.76 \text{ pF}$$

$$R_1 = \frac{\omega_1^2 L_1}{BW} = \frac{\omega_1^2 L_1}{2\pi B_{FM}} = \frac{f_1^2 L_1}{B_{FM}} = \frac{(10.85 \times 10^6)^2 \times 2 \times 10^{-5}}{2 \times 10^5} = 11.77 \text{ k}\Omega$$

For the lower slope circuit, the resonance frequency is

$$f_2 = f_c - \frac{1}{2}(1.5B_{FM}) = \left[10.7 \times 10^6 - \frac{1}{2}(1.5 \times 200 \times 10^3)\right] \text{Hz} = 10.55 \text{ MHz}$$

$$C_2 = \frac{1}{\omega_2^2 L_2} = \frac{1}{(2\pi \times 10.55 \times 10^6)^2 \times 2 \times 10^{-5}} = 11.38 \text{ pF}$$

$$R_2 = \frac{\omega_2^2 L_2}{BW} = \frac{f_2^2 L_2}{B_{FM}} = \frac{\left(10.55 \times 10^6\right)^2 \times 2 \times 10^{-5}}{2 \times 10^5} = 11.13 \text{ k}\Omega$$

PRACTICE PROBLEM 4.16

Repeat Example 4.16 for the case in which the FM signal has a carrier frequency of 12 MHz, and a bandwidth of $B_{FM} = 400$ kHz.

Answer: $C_1 = 8.37$ pF, $R_1 = 7.565$ kΩ.
$C_2 = 9.25$ pF, $R_2 = 6.845$ kΩ.

EXAMPLE 4.17

If in Example 4.16 a PM signal given by $\phi_{PM}(t) = 10 \cos \left[10^7 \pi t + 4 \cos \left(2\pi \times 10^4\right)t\right]$ is to be demodulated instead of an FM signal, complete the design of the slope detectors described in that example by choosing the capacitances C_1 and C_2, and the resistances R_1 and R_2. Comment on any additional processing necessary at the demodulator output for this PM signal, relative to the FM signal of Example 4.16.

Solution

From the given PM signal, this is a case of tone modulation.

$$k_p m(t) = 4 \cos \left(2\pi \times 10^4\right)t = 4 \cos \omega_m t,$$

$$f_m = 10 \text{ kHz},$$

$$f_c = 5 \text{ MHz}.$$

$$k_p m'(t) = \frac{d}{dt}\left[4 \cos \omega_m t\right] = -4\omega_m \sin \omega_m t \Rightarrow \Delta\omega = k_p m'_p = 4\omega_m.$$

$$\beta = \frac{\Delta\omega}{\omega_m} = \frac{k_p m'_p}{\omega_m} = \frac{4\omega_m}{\omega_m} = 4$$

$$\therefore \ B_{PM} = 2(\beta + 1)f_m = 2(4 + 1) \times 10 \text{ kHz} = 100 \text{ kHz}$$

For the upper slope circuit, the resonant frequency is

$$f_1 = f_c + \frac{1}{2}(1.5 B_{PM}) = \left[5 \times 10^6 + \frac{1}{2}\left(1.5 \times 10^5\right)\right] \text{ Hz} = 5.075 \text{ MHz}$$

$$C_1 = \frac{1}{\omega_1^2 L_1} = \frac{1}{\left(2\pi \times 5.075 \times 10^6\right)^2 \times 2 \times 10^{-5}} = 49.17 \text{ pF}$$

$$R_1 = \frac{\omega_1^2 L_1}{BW} = \frac{f_1^2 L_1}{B_{FM}} = \frac{(5.075 \times 10^6)^2 \times 2 \times 10^{-5}}{10^5} = 5.15 \text{ k}\Omega$$

For the lower slope circuit, the resonance frequency is

$$f_2 = f_c - \frac{1}{2}(1.5 B_{FM}) = \left[5 \times 10^6 - \frac{1}{2}(1.5 \times 10^5)\right] \text{ Hz} = 4.925 \text{ MHz}$$

$$C_2 = \frac{1}{\omega_2^2 L_2} = \frac{1}{(2\pi \times 4.925 \times 10^6)^2 \times 2 \times 10^{-5}} = 52.22 \text{ pF}$$

$$R_2 = \frac{\omega_2^2 L_2}{BW} = \frac{f_2^2 L_2}{B_{FM}} = \frac{(4.925 \times 10^6)^2 \times 2 \times 10^{-5}}{10^5} = 4.851 \text{ k}\Omega$$

After demodulation of the PM signal and dc removal, the demodulator output will be proportional to $k_p m'(t)$. This will have to be integrated to yield a replica of $m(t)$.

PRACTICE PROBLEM 4.17

The balanced discriminator described in Example 4.16 is used for demodulating the PM signal $\phi_{PM}(t) = 8 \cos\left[2\pi \times 10^7 t + 5 \cos\left(3\pi \times 10^4\right)t\right]$. Complete the design of the slope detectors in the balanced discriminator (Figure 4.19) described in Example 4.16 by choosing the capacitances C_1 and C_2, and the resistances R_1 and R_2.

Answer: $C_1 = 12.33$ pF, $R_1 = 11.41$ kΩ.
$C_2 = 13.013$ pF, $R_2 = 10.81$ kΩ.

4.6   FEEDBACK DEMODULATORS

Feedback demodulators are based on the phase-locked loop (PLL). The PLL employs the principle of negative feedback to continuously estimate the phase of the incoming modulated signal. Recall that the information in an angle-modulated signal resides in its phase. Thus, for PM, the estimated phase provides the demodulated signal, while for FM the instantaneous frequency or the first derivative of the estimated phase provides the demodulated signal. The demodulation techniques discussed in the previous section, which are based on FM/PM-to-AM conversion followed by envelope detection, perform well only in low-noise environments. Feedback demodulators have the advantage of improved performance in noisy environments. Furthermore, they do not involve bulky inductors and can be easily implemented with integrated circuits. Consequently, they are widely used in modern communications systems.

4.6.1 The phase-locked loop

The principle of operation of the PLL is illustrated in the feedback system of Figure 4.20(a). The essential parts are a voltage-controlled oscillator (VCO), a phase detector whose output is a measure of the phase difference between the phase of the incoming angle-modulated signal and the VCO output, and a loop filter. Let the incoming signal $\phi(t)$ and the VCO output signal $v(t)$ be given respectively by

$$\phi(t) = A_c \sin\left[\omega_c t + \psi_i(t)\right] \tag{4.56}$$

and

$$v(t) = A_v \cos\left[\omega_c t + \psi_v(t)\right] \tag{4.57}$$

Although the two signals have the same frequency in the two equations above, in general they can differ in both frequency and phase. The VCO signal tends to track the incoming signal in both frequency and phase, but the tracking may not be exact at a given moment. Because a frequency difference contributes to the total phase difference, the total phase difference alone suffices for purposes of analysis. To see this, suppose the input signal differing in both frequency and phase from the output of the VCO is given by $\phi(t) = A_c \sin\left[(\omega_c + \Delta\omega)t + \psi_{i0}(t)\right] = A_c \sin\left[\omega_c t + \Delta\omega t + \psi_{i0}(t)\right]$. If $\psi_i(t) = \Delta\omega t + \psi_{i0}(t)$, the input signal becomes $\phi(t) = A_c \sin\left[\omega_c t + \psi_i(t)\right]$ as in Eq. (4.56). The multiplier output is

(a)

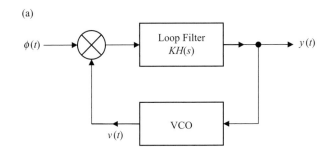

(b)

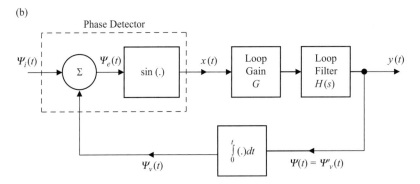

Figure 4.20. The phase-lock loop and its nonlinear model. (a) The phase-locked loop; (b) nonlinear model for the phase-locked loop.

$$x(t) = A_c A_v \sin \left[\omega_c t + \psi_i(t)\right] \cos \left[\omega_c t + \psi_v(t)\right]$$
$$= \frac{1}{2} A_c A_v \sin \left[2\omega_c t + (\psi_i(t) + \psi_v(t))\right] + \frac{1}{2} A_c A_v \sin \left[\psi_i(t) - \psi_v(t)\right] \tag{4.58}$$

The loop filter incorporates a gain constant k and a low-pass filter, which rejects the first term in the above equation. In effect the output $y(t)$ of the loop filter is the convolution of the loop filter impulse response $h(t)$ with the phase error in the last equation. That is

$$y(t) = h(t) * \frac{1}{2} k A_c A_v \sin \left[\psi_i(t) - \psi_v(t)\right]$$
$$= \frac{1}{2} k A_c A_v \int_0^t \sin \left[\psi_i(\lambda) - \psi_v(\lambda)\right] h(t - \lambda) d\lambda \tag{4.59}$$

The phase error between the PLL input signal and the VCO output is given by

$$\psi_e(t) = \psi_i(t) - \psi_v(t) \tag{4.60}$$

Thus the PLL output can be rewritten as

$$y(t) = \frac{1}{2} k A_c A_v \int_0^t \sin \left[\psi_e(\lambda)\right] h(t - \lambda) d\lambda \tag{4.61}$$

Based on the equations above, the PLL can be modeled as shown in Figure 4.20(b). In that model, the PLL input signal is $\psi_i(t)$, the phase angle of the input FM signal. With the unmodulated carrier frequency ω_c known, this phase angle effectively specifies the FM signal as in Eq. (4.56). Similarly, the output of the VCO in this model is its phase angle $\psi_v(t)$, which effectively specifies its waveform as in Eq. (4.57). The phase detector consists effectively of a multiplier and a low-pass filter. When its inputs are the phase angles $\psi_i(t)$ and $\psi_v(t)$, its output is $\sin \left[\psi_i(t) - \psi_v(t)\right]$. Because of the nonlinearity of the sine function, this is a *nonlinear model*. Note that the gain constant k in Eq. (4.59) is given by

$$k = k_a k_e \tag{4.62}$$

where k_e is the gain constant of the phase detector and k_a is an adjustable gain which may be varied to set the overall loop gain to the desired level. For an input signal of amplitude A_c, the overall loop gain is given by

$$G = \frac{1}{2} k A_v \tag{4.63}$$

The PLL output signal $y(t)$, which is the input signal to the VCO, is determined by the phase error $\psi_e(t)$. The phase angle of the VCO output is the integral of the PLL output signal $y(t)$. Thus

$$y(t) = \frac{d\psi_v(t)}{dt} \tag{4.64}$$

The phase angle and the instantaneous frequency of the VCO are given respectively by

$$\psi_v(t) = \int_0^t y(\lambda)d\lambda \tag{4.65}$$

and

$$\omega_v(t) = \omega_c + \frac{d\psi_v(t)}{dt} = \omega_c + y(t) \tag{4.66}$$

The VCO output is set to the unmodulated carrier frequency ω_c. Suppose the incoming signal is an unmodulated carrier of frequency ω_c, and the VCO is perfectly tracking the incoming carrier, then $\psi_e(t) = 0$, $y(t) = 0$, the VCO remains at its set point, and the phase angles of the VCO output and the incoming signal are equal except for the $90°$ phase difference of a sine wave relative to a cosine wave. When the PLL input is an FM signal and the VCO output is tracking it well, $\psi_e(t) \simeq 0$. Hence, $\psi_v(t) \simeq \psi_i(t)$, and the loop is said to be phase-locked. Thus the first derivatives of these phase angles, which are the respective frequency deviations, are approximately equal. That is

$$\frac{d\psi_v(t)}{dt} = y(t) \simeq k_f m(t) = \frac{d\psi_i(t)}{dt} \tag{4.67}$$

Consequently

$$y(t) \simeq k_f m(t) \tag{4.68}$$

Thus the output of the PLL is the demodulated message signal. When the input to the PLL is a PM signal and the loop is phase-locked, $\psi_v(t) \simeq \psi_i(t) = k_p m(t)$, the PLL output is $\frac{d\psi_v(t)}{dt} = y(t) \simeq k_p m'(t)$. Thus for PM the message signal is recovered by integrating the PLL output signal $y(t)$.

When the PLL is tracking well the phase error $\psi_e(t)$ is very small and $\sin[\psi_e(t)] \simeq \psi_e(t)$. Thus the phase detector is now a linear device, and the PLL is a linear time-invariant system. The resulting linear model for the PLL is shown in Figure 4.21. In the following subsections, the second-order PLL is analyzed with the linear model, while the simpler first-order PLL is analyzed with both the linear and the nonlinear models.

First-order phase-locked loop
The order of the PLL is determined by the specification of its loop filter. For a first-order PLL, the loop filter impulse response is $h(t) = \delta(t)$. Alternatively, in the Laplace transform domain,

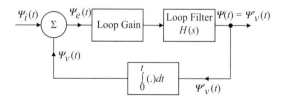

Figure 4.21. Linear model of the phase-locked loop.

the loop filter is given by $H(s) = 1$. The first-order PLL will first be analyzed with the non-linear PLL model. Later on, it will be analyzed with the linear model.

Consider the nonlinear model of Figure 4.20(b) employing a loop filter of impulse response is $h(t) = \delta(t)$. From Eq. (4.59) to Eq. (4.63), the output of the PLL can be written as

$$y(t) = \frac{d\psi_v(t)}{dt} = G \sin \left[\psi_i(t) - \psi_v(t) \right] * \delta(t) = G \sin \left[\psi_i(t) - \psi_v(t) \right] \qquad (4.69)$$

Thus

$$\frac{d\psi_v(t)}{dt} = G \sin \left[\psi_i(t) - \psi_v(t) \right] = G \sin \psi_e(t) \qquad (4.70)$$

If the PLL input signal has a frequency error of $\Delta\omega$ relative to ω_c, then the input signal is

$$\phi(t) = A_c \sin \left[(\omega_c + \Delta\omega)t + \psi_{i0} \right] = A_c \sin \left[\omega_c t + (\Delta\omega t + \psi_{i0}) \right] \qquad (4.71)$$

The phase angle and the phase error of this input signal are respectively given by

$$\psi_i(t) = \Delta\omega t + \psi_{i0} \qquad (4.72)$$

$$\frac{d\psi_e(t)}{dt} = \frac{d\psi_i(t)}{dt} - \frac{d\psi_v(t)}{dt} = \Delta\omega - \frac{d\psi_v(t)}{dt} \qquad (4.73)$$

Replacing $\frac{d\psi_v(t)}{dt}$ by its value in Eq. (4.70),

$$\boxed{\frac{d\psi_e(t)}{dt} = \Delta\omega - G \sin \psi_e(t)} \qquad (4.74)$$

A plot of $\frac{d\psi_e(t)}{dt}$ versus $\psi_e(t)$ for the above equation is shown in Figure 4.22. Such a plot of a function versus its derivative is known as a *phase-plane plot*. This plot is very helpful in understanding how the PLL achieves phase lock. The PLL can only operate on points which lie on its operating curve, the phase-plane plot. Because time increment dt is always positive, $d\psi_e(t)$ is positive whenever $\frac{d\psi_e(t)}{dt}$ is positive. Thus in the positive half plane of the phase-plane plot, changes in the operating point produce increases in $\psi_e(t)$, or a movement of the operating point towards the right. Conversely, in the lower half plane, $\frac{d\psi_e(t)}{dt}$ and hence $d\psi_e(t)$ are negative. Thus changes in the operating point produce decreases in $\psi_e(t)$, or a movement of the operating point towards the left. When the frequency error $\frac{d\psi_e(t)}{dt}$ is zero, the phase error $\psi_e(t)$ stops

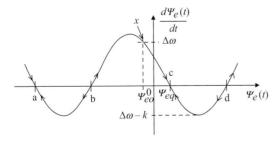

Figure 4.22. Phase-plane plot for a first-order phase-locked loop.

changing. Thus the intercepts of the operating curve with the $\frac{d\psi_e(t)}{dt} = 0$ axis, such as points a, b, c, and d on the plot, are equilibrium points of the system.

Consider a system that is operating at one of these equilibrium points. Due to a small perturbation the operating point is slightly displaced from its point of equilibrium. If the point of operation was initially displaced to the positive half plane, it will move towards the right as previously explained. If it was initially displaced to the lower half plane, it will move towards the left. For points a and c the movements following small perturbations tend to restore operation to the initial equilibrium points as indicated by the direction of the arrows. For points b and d, movements due to small perturbations move the operating point further away from the initial equilibrium points as indicated by the direction of the arrows. Thus, a and c are points of stable equilibrium, while points b and d are points of unstable equilibrium.

Suppose a system is operating at the point x at a certain time instant as shown in Figure 4.22. At this point of operation, the frequency error is $\Delta\omega$, and the phase error is ψ_{eo}. This is not a point of equilibrium, and it is on the upper half plane. Hence, the operating point moves towards the right until it reaches the point of stable equilibrium c. At the point c, the frequency error has been reduced to zero, but the phase error has a non-zero equilibrium value ψ_{eq}. The PLL is said to have achieved phase-lock. Thus at phase-lock, a first-order PLL actually achieves only a frequency lock, while it retains a constant phase error. Equilibrium operating points exist only if the operating curve intersects the $\frac{d\psi_e(t)}{dt} = 0$ axis. Setting $\frac{d\psi_e(t)}{dt} = 0$ in Eq. (4.74) yields $\Delta\omega = G\sin\psi_e(t)$. Thus, for the loop to lock, the frequency error $\Delta\omega$ must be smaller than the overall loop gain G.

EXAMPLE 4.18

A first-order phase-locked loop is operating at phase lock when a frequency change of $\Delta f = 10$ Hz occurs in the input signal. (a) Determine the loop gain G such that the equilibrium phase error will be 10°. (b) For the loop gain so determined, find the equilibrium phase error corresponding to a step frequency change of 30 Hz. (c) Find the maximum frequency error for which that loop gain can achieve a phase-lock.

Solution

(a) $\frac{d\psi_e(t)}{dt} = \Delta\omega - G\sin\psi_e(t)$, where $\psi_e(t)$ is the phase error, and $\frac{d\psi_e(t)}{dt}$ is the frequency error.
At phase lock (or equilibrium), $\frac{d\psi_e(t)}{dt} = 0 = \Delta\omega - G\sin\psi_e(t) = 0$.
Thus, at phase-lock, $G = \frac{\Delta\omega}{\sin\psi_e(t)}$

$$\therefore \quad G = \frac{2\pi\Delta f}{\sin\psi_e(t)} = \frac{2\pi \times 10}{\sin 10^o} = 361.83.$$

(b) At equilibrium, $\frac{d\psi_e(t)}{dt} = 0 = \Delta\omega - G\sin\psi_e(t)$.
thus $\psi_{eq} = \sin^{-1}\left[\frac{\Delta\omega}{G}\right] = \sin^{-1}\left[\frac{2\pi\Delta f}{G}\right]$

$$\Delta f = 30 \text{ Hz} \Rightarrow \psi_{eq} = \sin^{-1}\left[\frac{2\pi\Delta f}{G}\right] = \sin^{-1}\left[\frac{2\pi \times 30}{361.83}\right] = 31.4^{\circ}$$

(c) $\Delta\omega - G\sin\psi_e(t) = 0$ for equilibrium.

For max frequency step, $\Delta\omega_{max} = G$.

$$\Delta f_{max} = \frac{\Delta\omega_{max}}{2\pi} = \frac{G}{2\pi} = \frac{361.83}{2\pi} = 57.59 \text{ Hz}$$

PRACTICE PROBLEM 4.18

A first-order phase-locked loop has a loop gain of $G = 80\pi$. (a) Find the equilibrium phase error corresponding to step frequency changes of 15 Hz and 30 Hz. (b) Repeat part (a) if the loop gain is doubled.

Answer: (a) $\psi_{eq} = 22.02°; \psi_{eq} = 48.59°$

(b) $\psi_{eq} = 10.81°; \psi_{eq} = 22.02°$

Second-order phase-locked loop

For a second-order PLL, the loop filter is given in the time domain by

$$h(t) = \delta(t) + au(t)$$ (4.75)

and its Laplace transform is

$$H(s) = \frac{s+a}{s}$$ (4.76)

For the linear PLL model of Figure 4.21, the transfer function of the output Ψ_v with respect to the input Ψ_i is

$$\frac{\Psi_v}{\Psi_i} = \frac{GH(s)}{s + GH(s)}$$ (4.77)

The error signal is

$$\Psi_e(s) = \Psi_i(s) - \Psi_v(s) = \left[1 - \frac{\Psi_v(s)}{\Psi_i(s)}\right]\Psi_i(s) = \frac{s}{s + GH(s)}\Psi_i(s)$$ (4.78)

Substituting $H(s)$ of Eq. (4.76) into the last equation gives

$$\frac{\Psi_e(s)}{\Psi_i(s)} = \frac{s^2}{s^2 + Gs + Ga}$$ (4.79)

The above equation can be rewritten as

$$\boxed{\frac{\Psi_e(s)}{\Psi_i(s)} = \frac{s^2}{s^2 + 2\zeta\omega_o s + \omega_o^2}}$$

(4.80)

where ω_o is the *undamped natural frequency* and ζ is the *damping factor* for the system, given respectively by

$$\omega_o = \sqrt{Ga}$$

(4.81)

and

$$\zeta = \frac{1}{2}\sqrt{\frac{G}{a}}$$

(4.82)

Let the input signal to the PLL be as given in Eq. (4.71), then as in Eq. (4.72) its phase angle is $\psi_i(t) = \Delta\omega t + \psi_{i0}$. Thus, the Laplace transform of its phase angle is

$$\Psi_i(s) = \frac{\Delta\omega}{s^2} + \frac{\psi_{io}}{s}$$

(4.83)

Substituting $\Psi_i(s)$ into Eq. (4.79),

$$\Psi_e(s) = \frac{s^2}{s^2 + Gs + Ga}\left[\frac{\Delta\omega}{s^2} + \frac{\psi_{io}}{s}\right]$$

(4.84)

The question of interest here is whether the second-order PLL can eventually eliminate the initial frequency and/or phase error in the PLL input signal. To answer this question, we can employ the final-value theorem to find the final value of the phase error, given enough time as

$$\psi_e(\infty) = \lim_{s \to 0}\left[s\Psi_e(s)\right] = 0$$

(4.85)

Recall from Eq. (4.72) that the phase angle of the PLL input signal includes both the initial frequency and phase errors. The last result shows that at phase lock, the second-order PLL eliminates both the frequency error and the phase error. This is an important advantage over the first-order PLL, which can eliminate only the frequency error.

EXAMPLE 4.19

A second-order phase-locked loop is to have a natural frequency of 200 rad/s and a damping factor of $\frac{1}{\sqrt{2}}$. Specify the loop gain G, the loop filter transfer function $H(s)$, and the impulse response $h(t)$.

Solution

The transfer function of the phase error with respect to the input signal phase angle is

$$\frac{\Psi_e(s)}{\Psi_i(s)} = \frac{s^2}{s^2 + Gs + Ga} = \frac{s^2}{s^2 + 2\zeta\omega_o s + \omega_o^2}$$

$$\omega_o = \sqrt{Ga} \Rightarrow \omega_o^2 = (200)^2 = Ga; \quad a = \frac{40{,}000}{G}$$

$$\xi = \frac{1}{2}\sqrt{\frac{G}{a}} \Rightarrow \xi^2 = \frac{1}{2} = \frac{G}{4a}$$

Substituting the value of a,

$$G = 2a = 2 \times \frac{40{,}000}{G} \Rightarrow G^2 = 80{,}000$$

$$\therefore \quad G = \sqrt{80{,}000} = 282.843.$$

$$a = \frac{40{,}000}{G} = 100\sqrt{2}.$$

The second-order PLL loop filter has the form $H(s) = \frac{s+a}{s}$.
  Thus $H(s) = \frac{s+100\sqrt{2}}{s}$.
  Also $H(s) = \frac{s+a}{s} = 1 + \frac{a}{s}$.
  Taking the inverse Laplace transform, $h(t) = \delta(t) + 100\sqrt{2}u(t)$.

PRACTICE PROBLEM 4.19
A second-order PLL has a loop gain of 200 and a loop filter impulse response of $h(t) = \delta(t) + 100u(t) = \delta(t) + au(t)$. (a) Determine the natural frequency and the damping factor for the PLL. (b) Repeat part (a) if the loop gain is doubled. (c) Repeat part (a) if the constant a in the impulse response is doubled, but the loop gain remains 200.

Answer: (a) $\omega_o = 100\sqrt{2}$ rad/s; $\xi = \frac{\sqrt{2}}{2}$.
  (b) $\omega_o = 200$ rad/s; $\xi = 1$.
  (c) $\omega_o = 200$ rad/s; $\xi = \frac{1}{2}$.

Comment: Doubling either G or a increases ω_o by a factor of $\sqrt{2}$. However, doubling G increases ξ by a factor of $\sqrt{2}$, while doubling a reduces ξ by a factor of $\sqrt{2}$.

4.6.2 Frequency-compressive feedback demodulator

The frequency-compressive feedback method is illustrated in the block diagram of Figure 4.23. It is essentially a PLL in which the loop filter is replaced with a bandpass filter in cascade with a frequency discriminator of the type discussed earlier. Consider the signals indicated in the block diagram. As in the PLL the incoming FM signal is given by

$$\phi(t) = A_c \sin\left[\omega_c t + \psi_i(t)\right] \tag{4.86}$$

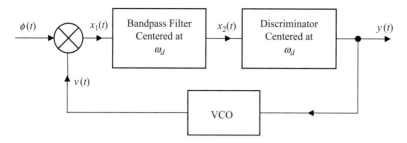

Figure 4.23. Frequency-compressive feedback demodulator.

The center frequency of the discriminator response is ω_d. The VCO center frequency $\omega_c - \omega_d$ is offset from the unmodulated carrier frequency ω_c by ω_d. Thus the VCO output signal and the multiplier output are respectively

$$v(t) = A_v \cos\left[(\omega_c - \omega_d)t + \psi_v(t)\right] \tag{4.87}$$

and

$$x_1(t) = \frac{1}{2}A_c A_v \left\{ \sin\left[(2\omega_c - \omega_d)t + (\psi_i(t) + \psi_v(t))\right] + \sin\left[\omega_d t + \psi_i(t) - \psi_v(t)\right] \right\} \tag{4.88}$$

The bandpass filter centered at ω_d suppresses the term at the relatively high frequency of $2\omega_c - \omega_d$ in the last equation. Together with the signal multiplier, it constitutes a phase detector whose output $x_2(t)$ is given by

$$x_2(t) = \frac{1}{2}A_c A_v \sin\left[\omega_d t + \psi_i(t) - \psi_v(t)\right] \tag{4.89}$$

The phase angle of the VCO output signal $\psi_v(t)$ is the integral of the demodulator output signal $y(t)$, which is also the VCO input signal. The VCO gain constant is k_v. Thus $\psi_v(t)$ and $\psi_e(t)$, the phase angle of $x_2(t)$, are given respectively by

$$\psi_v(t) = k_v \int_0^t y(\lambda)d\lambda \tag{4.90}$$

and

$$\psi_e(t) = \psi_i(t) - \psi_v(t) = \psi_i(t) - k_v \int_0^t y(\lambda)d\lambda \tag{4.91}$$

The discriminator output is proportional to the derivative of the phase angle of its input signal. Letting the gain constant of the discriminator be k_d, its output signal is given by

$$y(t) = k_d \frac{d\psi_e(t)}{dt} = k_d \frac{d}{dt}\left[\psi_i(t) - k_v \int_0^t y(\lambda)d\lambda\right] = k_d \frac{d\psi_i(t)}{dt} - k_d k_v y(t)$$

Rearranging the last equation, and recalling that the derivative of $\psi_i(t)$ is the frequency deviation of the input FM signal, the output yields the demodulated signal as

$$y(t) = \frac{k_d}{1 + k_d k_v} \frac{d\psi_i(t)}{dt} = \frac{k_d}{1 + k_d k_v} \left[k_f m(t) \right] \qquad (4.92)$$

Prior to differentiation of the discriminator, and the discriminator gain of k_d, the signal at the discriminator input is $\frac{1}{1+k_d k_v} \psi_i(t) = \frac{1}{1+k_d k_v} k_f \int_0^t m(\lambda)d\lambda$. On comparing this with the phase angle of the incoming FM signal $\psi_i(t) = k_f \int_0^t m(\lambda)d\lambda$, it is clear that the frequency deviation of the FM signal at the discriminator input has been reduced by a factor of $\frac{1}{1+k_d k_v}$ relative to that of the original FM signal. This is the frequency compression for which this method is named. By using a large enough gain product $k_d k_v$, the compression can be so high as to reduce the incoming wideband FM to a narrowband FM at the discriminator input. An important advantage of this method is the reduced noise bandwidth and improved noise performance of the demodulator. Another obvious advantage is the ease of obtaining discriminators with improved linear responses over a narrower frequency range.

EXAMPLE 4.20

An FM signal is to be demodulated with a frequency-compressive feedback system. The modulating signal is $4\cos(10^4 \pi t)$, and $k_f = 8\pi \times 10^4$. The frequency deviation at the input to the discriminator is to be 10 kHz. The VCO gain constant is to be twice the discriminator gain constant. Specify the values for both gain constants.

Solution

For the input FM signal, $\Delta f = \frac{1}{2\pi} k_f A_m = \frac{1}{2\pi} (8\pi \times 10^4) \times 4 = 160$ kHz

Frequency compression ratio $= \frac{10 \text{ kHz}}{\Delta f} = \frac{10 \text{ kHz}}{160 \text{ kHz}} = \frac{1}{16} = \frac{1}{1+k_v k_d}$.

$$k_v = 2k_d \Rightarrow \frac{1}{16} = \frac{1}{1 + 2k_d^2}$$

$$2k_d^2 = 15 \Rightarrow k_d = 2.739$$

$$k_v = 2k_d = 5.477$$

PRACTICE PROBLEM 4.20

An FM signal is to be demodulated with a frequency-compressive feedback system. The modulating signal for the FM signal is $6\cos(10^4 \pi t)$, and $k_f = 10^5 \pi$. The frequency deviation at the input to the discriminator is to be 15 kHz. The VCO gain constant is to be equal to the discriminator gain constant. Specify the values for both gain constants.

Answer:

$$k_d = k_v = 4.359$$

Other phase-locked loop applications

The ability of the phase-locked loop to generate a local carrier/signal which matches an input carrier/signal in phase and frequency finds important application in many areas. Its application in the Costas loop for the acquisition of synchronous carriers at the receiver for use in synchronous detection was discussed in section 3.7. Two other applications of the PLL are discussed below.

Phase-locked loop implementation of frequency multiplication

In this scheme, harmonics of the input signal are generated. The VCO quiescent frequency is set to the frequency of one of these harmonics. The PLL locks onto the selected harmonic. Thus, if the nth harmonic was selected and the incoming signal is $A_i \cos [\omega_c t + \psi(t)]$, the PLL output signal will be $A_o \cos [n\omega_c t + n\psi(t)]$.

The scheme is illustrated in Figure 4.24. Suppose the input signal is sinusoidal or an angle-modulated signal (which is approximately sinusoidal), the limiter produces a bipolar square waveform as shown in Figure 4.24(b). This waveform contains only odd harmonics of the input signal frequency. If the square waveform is used as the input to the phase detector, the frequency multiplication achieved can only be an odd multiple of the input signal frequency. To permit frequency multiplication by either an even or an odd number, the limiter output should be passed through the pulse-shaping circuit. This circuit produces the waveform of narrow unipolar pulses shown in Figure 4.24(c). This waveform has the Fourier series spectrum of Figure 4.24(d), which contains both even and odd harmonics, thereby permitting the multiplier output frequency to be an even or odd multiple of the input signal frequency. If it is known that the desired output frequency is an odd multiple of the input signal frequency, the pulse-shaping circuit can be omitted.

Phase-locked loop implementation of frequency division or multiplication

This scheme can be used to perform frequency division or frequency multiplication. The scheme is illustrated in Figure 4.25. Suppose the frequency of the input signal $x(t)$ is f_c, and the minimum output frequency desired from the system is $\frac{1}{N} f_c$, the quiescent frequency of the VCO is set to $f_v = \frac{1}{N} f_c$. The VCO output $p(t)$ is a train of narrow pulses such as the one shown in Figure 4.24(c). This pulse train has the spectrum shown in Figure 4.24(d), which consists of harmonics at frequencies $n f_v$, $n = 1, 2, 3, \ldots$ The PLL phase-locks the harmonic component at a frequency of $N f_v = f_c$ onto the input signal. The narrow bandpass filter is centered at a frequency of f_o. Thus, the output signal will have a frequency of $f_o = n f_v = \frac{n}{N} f_c$.

The output frequency for the system is the frequency of the harmonic passed by the narrow-band filter centered at $f_o = n f_v = \frac{n}{N} f_c$. If $\frac{n}{N}$ is a fraction, the system performs frequency division, but if $\frac{n}{N}$ is an integer, the system performs frequency multiplication. Suppose $N = 8$ and frequency division with output frequency $f_o = \frac{1}{4} f_c$ is desired, then n should be 2 so that $f_o = \frac{n}{N} f_c = \frac{2}{8} f_c = \frac{1}{4} f_c$. However, if frequency multiplication with output frequency $f_o = 4 f_c$ is desired, then n should be 32 so that $f_o = \frac{n}{N} f_c = \frac{32}{8} f_c = 4 f_c$. In each case, the pass-band of the narrowband filter should be centered at the desired output frequency f_o.

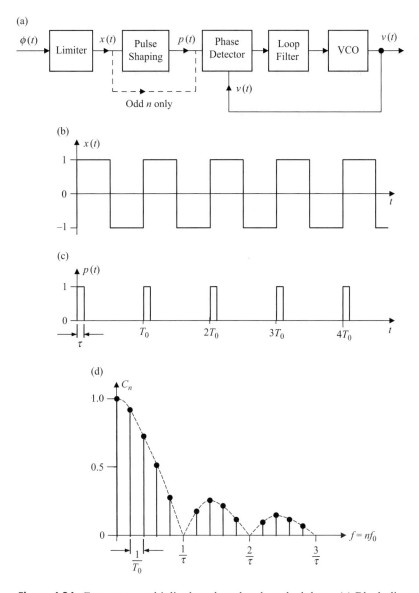

Figure 4.24. Frequency multiplier based on the phase-lock loop. (a) Block diagram; (b) output signal of limiter; (c) narrow pulse train output, $p(t)$, of pulse-shaping circuit; (d) Fourier coefficients of $p(t)$.

4.7 INTERFERENCE IN ANGLE MODULATION

Suppose a receiver is tuned to a channel of carrier frequency ω_c, any signal close enough in frequency to ω_c could interfere with the desired channel. Although the desired channel is a modulated signal $\phi(t)$, and the interfering signal may or may not be a modulated signal, the analysis is considerably simplified by regarding both as unmodulated sinusoidal carriers. Recall

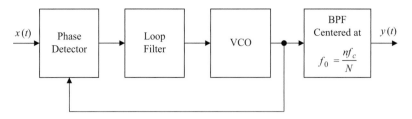

Figure 4.25. Frequency divider/multiplier based on the phase-locked loop.

that commercial FM can be considered as approximately sinusoidal because its frequency deviation is less than 0.1% of its carrier frequency. Let the unmodulated version of $\phi(t)$ and the interfering signal be given respectively by $A_c \cos \omega_c t$ and $A_i \cos (\omega_c + \omega_i)t$, where ω_i is the frequency difference between the frequency of the interfering signal and ω_c. The signal at the receiver input is $x(t) = A_c \cos \omega_c t + A_i \cos (\omega_c + \omega_i)t = [A_c + A_i \cos \omega_i t] \cos \omega_c t - A_i \sin \omega_i t \sin \omega_c t$. Thus

$$x(t) = A(t) \cos [\omega_c t + \psi(t)] \tag{4.93}$$

where $A(t)$ is the envelope of $x(t)$ and $\psi(t)$ is its phase angle given by

$$\psi(t) = \tan^{-1} \left[\frac{A_i \sin \omega_i t}{A_c + A_i \cos \omega_i t} \right] \tag{4.94}$$

In angle modulation the information resides in the phase, so it is the phase (not the envelope) that is relevant to the demodulator output. If the interfering signal amplitude is much smaller than the desired channel carrier amplitude (i.e. $A_i \ll A_c$), the above phase angle can be approximated as

$$\psi(t) \simeq \frac{A_i}{A_c} \sin \omega_i t \tag{4.95}$$

The signal applied to the demodulator, $x(t) = A(t) \cos [\omega_c t + \psi(t)]$, has a phase angle of $\psi(t)$ and an instantaneous frequency of $\omega_c + \psi'(t)$. If the modulated signal $\phi(t)$ is a PM signal, the demodulated output is

$$\boxed{y_d(t) = \psi(t) = \frac{A_i}{A_c} \sin \omega_i t \quad \text{for PM}} \tag{4.96}$$

If $\phi(t)$ is an FM signal, the demodulated output is $\psi'(t) = \frac{\omega_i A_i}{A_c} \cos \omega_i t$. Thus

$$\boxed{y_d(t) = \psi'(t) = \frac{\omega_i A_i}{A_c} \cos \omega_i t \quad \text{for FM}} \tag{4.97}$$

The last two equations show that for both PM and FM the interference amplitude at the demodulator output is inversely proportional to the amplitude of the carrier. Thus, a strong channel suppresses weak interference, resulting in what is known as the *capture effect*.

For a weak interfering signal that is within the audio range of a strong angle-modulated channel, the strong channel effectively suppresses (captures) the weak interference. This is known as the **capture effect**.

In order to compare the effect of interference in angle modulation and amplitude modulation, consider the case in which the desired channel is an AM channel. For AM, the envelope $A(t)$ is the relevant term in Eq. (4.93). This envelope and its approximation for $A_i \ll A_c$ are given respectively by

$$A(t) = \left[(A_c + A_i \cos \omega_i t)^2 + (A_i \sin \omega_i t)^2 \right]^{\frac{1}{2}} \tag{4.98}$$

and

$$A(t) \simeq A_c + A_i \cos \omega_i t; \quad \text{for } A_i \ll A_c \tag{4.99}$$

Thus for an approximately sinusoidal AM signal with interference, the demodulated output is the envelope given by the last equation. The interference amplitude in the AM demodulator output is simply A_i; it is not inversely proportional to A_c as it is in angle modulation. Consequently, angle modulation has much better noise suppression properties than AM. A strong AM channel cannot suppress or capture weak interference, so there is no capture effect in AM. Because of its superior noise performance, the interference amplitude need be only 6 dB below the desired channel amplitude for its effective suppression in angle modulation. However, for good suppression in AM, the amplitude of the interference should be at least 35 dB below that of the desired channel.

Figure 4.26(a) illustrates graphically the interference amplitude at the demodulator output as a function of frequency for PM, FM, and AM. For FM, the interference amplitude increases linearly with frequency. Consequently for FM, interference is much weaker at low modulating signal frequencies. For AM and PM, the interference amplitude is constant with frequency, but it is much stronger in AM than in PM demodulator output.

Preemphasis and deemphasis

Preemphasis is the boosting of the higher-frequency portion of the message signal spectrum transmitter prior to transmission. **Deemphasis** is the attenuation of the demodulated receiver output corresponding to the higher-frequency portion of the spectrum of the message signal plus noise/interference.

Deemphasis reverses the distorting effect of preemphasis on the received message signal. Because the noise/interference is attenuated in the receiver without prior boosting at the transmitter, preemphasis at the transmitter followed by deemphasis at the receiver has the effect of decreasing noise power relative to signal power, thereby increasing the signal-to-noise ratio at the receiver output.

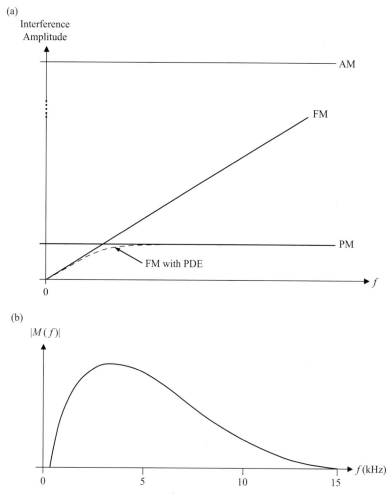

Figure 4.26. Interference amplitude at demodulator output and message signal amplitude as a function of frequency. (a) Interference amplitude in AM, PM, FM, and FM with preemphasis and deemphasis; (b) amplitude spectrum for a typical music signal.

Noise and interference have much in common. With respect to the desired channel, both are unwanted. White noise, the most prevalent channel noise, has a constant power spectral density. Thus, white noise can be viewed as consisting of constant-amplitude sinusoids (or sinusoidal interference signals) of all frequencies. Therefore, the preceding analysis for interference is applicable to channel noise. Figure 4.26(a) depicts noise amplitudes as a function of frequency at the demodulator output for PM, FM, and AM.

Figure 4.26(b) shows an example of the amplitude spectrum of a typical message signal, the audio music signal. Consider the output of an FM demodulator consisting of such an audio signal plus additive noise. The noise is weak at low frequencies where the signal is strong, but it is strong at higher frequencies where the signal is weak. Thus noise would tend to destroy the high-frequency content of the output audio signal. In the case of PM, since the noise amplitude is constant at all frequencies, the high-frequency output signal will fare better than in FM, but its low-frequency component has to cope with more noise than its FM counterpart.

Preemphasis and deemphasis is a scheme which is employed in commercial FM to obtain a better noise performance than would otherwise be possible. In effect, the scheme transmits the strong low-frequency component of the modulating signal as FM, but it transmits the weak high-frequency component of the signal as PM. In this way, commercial FM makes the best of both worlds by employing FM and PM, each in the frequency range of the modulating signal in which it has the superior noise performance. Thus, commercial FM is neither pure FM nor pure PM, but a combination of both.

Preemphasis is implemented prior to modulation in the FM transmitter, and deemphasis is implemented after demodulation in the receiver. Preemphasis is implemented by passing the message signal through a preemphasis filter $H_p(j\omega)$, whose Bode magnitude response is shown in Figure 4.27(a). A circuit implementation of this filter response is shown in Figure 4.27(b). In FM broadcasting, the corner frequency ω_1 usually corresponds to a natural frequency of 2.12 kHz, and the corner frequency ω_2 corresponds to a natural frequency of about 30 kHz, which is well beyond the audio bandwidth of 15 kHz. By inspection of its Bode magnitude plot, the preemphasis filter response is

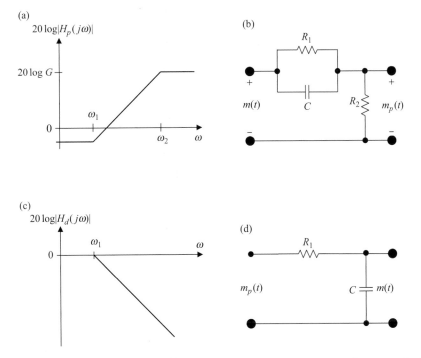

Figure 4.27. Preemphasis and deemphasis in FM. (a) Preemphasis filter magnitude response; (b) preemphasis filter; (c) deemphasis filter magnitude response; (d) deemphasis filter.

$$H_p(j\omega) = G\left(\frac{\omega_1 + j\omega}{\omega_2 + j\omega}\right) \tag{4.100}$$

In the equation above, G is a gain constant. Analysis of the circuit yields the filter response in terms of the circuit components as

$$H_p(j\omega) = \frac{R_2}{R_1 + R_2}\left[\frac{1 + j\omega R_1 C}{1 + \frac{j\omega R_1 R_2 C}{R_1 + R_2}}\right] = \frac{R_2}{R_1 + R_2}\frac{\omega_2}{\omega_1}\left[\frac{\omega_1 + j\omega}{\omega_2 + j\omega}\right] \tag{4.101}$$

where ω_1 and ω_2 are the corner frequencies given by

$$\omega_1 = \frac{1}{R_1 C} \tag{4.102}$$

and

$$\omega_2 = \frac{R_1 + R_2}{R_1 R_2 C} \tag{4.103}$$

In FM broadcasting, the time constant of $R_1 C$ is usually 75 μs. For $\omega \gg \omega_2$, the filter response is $H_p(j\omega) \simeq G$. Thus, G is the high-frequency gain. Comparing Eq. (4.100) with Eq. (4.101), G is given by

$$G = \frac{R_2}{R_1 + R_2}\frac{\omega_2}{\omega_1} \tag{4.104}$$

For very low and very high frequencies, this filter can be approximated as

$$H_p(j\omega) \simeq \begin{cases} \dfrac{R_2}{R_1 + R_2}, & \text{for } \omega \ll \omega_1 \\ \dfrac{R_2}{R_1 + R_2}\dfrac{\omega_2}{\omega_1}, & \text{for } \omega \gg \omega_2 \end{cases} \tag{4.105}$$

Between ω_1 and ω_2, the preemphasis filter can be roughly approximated as

$$H_p(j\omega) \simeq \frac{j\omega}{\omega_1}\frac{R_2}{R_1 + R_2}; \quad \text{for } \omega_1 < \omega < \omega_2 \tag{4.106}$$

Thus, in the audio frequency range of 2.12 kHz to 30 kHz, the preemphasis filter is essentially a differentiator, so that it is the derivative of the message signal in that frequency range that is frequency-modulated. In effect, phase modulation is performed between 2.12 kHz and the upper message signal bandwidth limit, although it is frequency modulation that is performed below 2.12 kHz.

By applying different gains at different frequencies, the preemphasis filter has a distorting effect on the message signal. This distortion needs to be cancelled out with a deemphasis filter after demodulation. The deemphasis filter $H_d(j\omega)$ has the Bode magnitude response shown in

Figure 4.27(c) and the circuit implementation shown in Figure 4.27(d). By inspection of its Bode plot, the deemphasis filter response is

$$\boxed{H_d(j\omega) = \frac{\omega_1}{\omega_1 + j\omega}}$$ (4.107)

Analysis of the filter circuit yields the filter response as

$$H_d(j\omega) = \frac{1}{1 + j\omega R_1 C} = \frac{\omega_1}{\omega_1 + j\omega}$$ (4.108)

where

$$\omega_1 = \frac{1}{R_1 C}$$ (4.109)

Because the corner frequency ω_1 is the same for both the preemphasis and deemphasis filters, R_1 and C are the same for both filters. At low and high frequencies, the deemphasis filter can be approximated as

$$H_d(j\omega) \simeq \begin{cases} 1, & \text{for } \omega \ll \omega_1 \\ \dfrac{\omega_1}{j\omega}, & \text{for } \omega \gg \omega_1 \end{cases}$$ (4.110)

Within the audio signal bandwidth, $H_p(j\omega)H_d(j\omega) \simeq \frac{R_2}{R_1+R_2}$. This approximation is virtually exact below 2.12 kHz. The approximation is good enough but is not as accurate between 2.12 kHz and the audio bandwidth (see the example below). Thus at the receiver output, deemphasis cancels out the distortion due to preemphasis as desired. Deemphasis reduces both the noise and the signal amplitude in the demodulated output for frequencies above ω_1. For the message signal, this reduction compensates for the boost to its high-frequency region due to preemphasis in the transmitter. The noise which contaminated the signal in the channel did not enjoy the benefit of such an earlier boost. In effect, deemphasis restores the message signal to its original form prior to preemphasis, while substantially reducing the noise. Thus the preemphasis and deemphasis scheme results in a significant improvement in noise performance.

EXAMPLE 4.21

An audio signal has non-zero spectral content between 30 Hz and 15 kHz. It is used as the modulating signal for FM with preemphasis and deemphasis. The preemphasis and deemphasis filters and their amplitude responses are as shown in Figure 4.27. For the responses, the corner frequencies are $f_1 = 2.12$ kHz and $f_2 = 30$ kHz. Ideally, deemphasis in the receiver should exactly cancel out the effect of preemphasis in the transmitter. Determine ρ, the ratio of the actual value of the product of the preemphasis and deemphasis filter responses at the corner frequency of 2.12 kHz and at the audio signal bandwidth of 15 kHz to the ideal value of the product of the responses required for exact cancellation.

Solution

The product of the filter preemphasis and deemphasis filter responses is

$$H_p(j\omega)H_d(j\omega) = \left[\frac{R_2}{R_1 + R_2}\frac{\omega_2}{\omega_1}\left(\frac{\omega_1 + j\omega}{\omega_2 + j\omega}\right)\right]\left[\frac{\omega_1}{\omega_1 + j\omega}\right] = \frac{R_2}{R_1 + R_2}\frac{\omega_2}{\omega_2 + j\omega}$$

The ideal value of the product of the responses is obtained for $\omega \ll \omega_1$ as

$$H_p(j\omega)H_d(j\omega) = \frac{R_2}{R_1 + R_2}.$$

At a frequency $\omega = 2\pi f$, the ratio of actual to ideal product of magnitude of responses is,

$$\rho = \left|\frac{\omega_2}{\omega_2 + j\omega}\right| = \frac{\omega_2}{\sqrt{\omega_2^2 + \omega^2}} = \frac{f_2}{\sqrt{f_2^2 + f^2}}.$$

At

$$f = f_1, \rho = \frac{30}{\sqrt{30^2 + 2.12^2}} = 0.9975 \text{ or } \rho\% = 99.75\%$$

At

$$f = 15 \text{ kHz}, \rho = \frac{30}{\sqrt{30^2 + 15^2}} = 0.8944 \text{ or } \rho\% = 89.44\%$$

PRACTICE PROBLEM 4.21

Repeat Example 4.21 if the upper corner frequency of the preemphasis filter is changed to $f_2 = 15$ kHz. The lower corner frequency is still $f_1 = 2.12$ kHz, and the audio signal spectral range is still 30 Hz to 15 kHz. Comment on the results, relative to those of Example 4.21.

Answer: At $f = f_1 = 2.12$ kHz, $\rho\% = 99.02\%$
At $f = f_2 = 15$ kHz, $\rho\% = 70.71\%$

Comment: The results deviate much more from the ideal than those of Example 4.21. Consequently, the distortion introduced by preemphasis is not so well cancelled out by deemphasis in this case. The higher the upper corner frequency f_2 is than the audio bandwidth, the better the cancellation of preemphasis distortion by deemphasis. This is because the higher the frequency f_2, the more linear is the response of the preemphasis and deemphasis filter curves over the message signal spectrum beyond f_1. Typically f_2 is chosen as 30 kHz, which is double the audio bandwidth of 15 kHz.

EXAMPLE 4.22

You are to design RC filter circuits for preemphasis and deemphasis. The preemphasis filter is to have a lower corner frequency of 2.122 kHz and an upper corner frequency of 30 kHz. You have available two 30 kΩ resistors, and these are to be two of the three resistors in your two filters. (a) Specify the other components in your design. (b) Determine the high-frequency gain and the low-frequency gain for your preemphasis filter.

Solution

(a) The preemphasis and deemphasis filter circuits are as shown in Figure 4.27. The two resistors in the preemphasis filter are unequal, thus one of the two equal resistors must be used in each of the two circuits as R_1.

$$\omega_1 = \frac{1}{R_1 C} \Rightarrow C = \frac{1}{\omega_1 R_1} = \frac{1}{2\pi f_1 R_1} = \frac{1}{2\pi \left(2.122 \times 10^3\right)\left(30 \times 10^3\right)} = 2.5 \text{ nF}$$

Thus, $C = 2.5$ nF for the preemphasis circuit and for the deemphasis circuit.
For the preemphasis circuit, $\omega_2 = \frac{R_1 + R_2}{R_1 R_2 C}$

$$\therefore R_2 = \frac{R_1}{\omega_2 C R_1 - 1} = \frac{30 \times 10^3}{2\pi \times 30 \times 10^3 \left(2.5 \times 10^{-9}\right)\left(30 \times 10^3\right) - 1} = 2.284 \text{ k}\Omega$$

(b)
$$H_p(j\omega) \simeq \begin{cases} \dfrac{R_2}{R_1 + R_2}, & \omega \ll \omega_1 \\ \dfrac{R_2}{R_1 + R_2}\dfrac{\omega_2}{\omega_1}, & \omega \gg \omega_2 \end{cases}$$

Thus the low-frequency gain and the high-frequency gain are respectively

$$H_p(j0) = \frac{R_2}{R_1 + R_2} = \frac{2.284 \text{ k}\Omega}{(30 + 2.284) \text{ k}\Omega} = 0.07$$

and

$$H_p(j\infty) = \frac{R_2}{R_1 + R_2}\frac{\omega_2}{\omega_1} = \frac{2.284 \text{ k}\Omega}{(30 + 2.284) \text{ k}\Omega}\left(\frac{2\pi \times 30}{2\pi \times 2.212}\right) = 1$$

PRACTICE PROBLEM 4.22

Design RC filter circuits for preemphasis and deemphasis such that the preemphasis filter has a lower corner frequency of 2.122 kHz and an upper corner frequency of 30 kHz. You are to use a 10 nF capacitor in each of the circuits. (a) Specify the other components in your design. (b) Determine the high-frequency gain and the low-frequency gain for the preemphasis filter.

Answer: (a) $R_1 = 7.5$ kΩ; $R_2 = 570.9$ Ω.
(b) $H_p(j0) = 0.0707$; $H_p(j\infty) = 1$.

4.8   FM BROADCASTING

Commercial FM broadcasting employs carrier frequencies in the range 88 to 108 MHz. The channels are spaced 200 kHz apart, and the frequency deviation is 75 kHz. The modulating signal is typically an audio signal with a bandwidth of 15 kHz, and the frequency deviation ratio is $\beta = 5$.

Commercial FM broadcasting comprises both monophonic and stereophonic FM. Monophonic FM was well established before the advent of stereophonic FM. Following its authorization by the FCC in 1961, stereophonic FM soon took the upper hand. At the time of its authorization, the FCC ruled that stereophonic FM had to be compatible with the previously existing monophonic FM. This meant that stereophonic FM, with its two (left and right) audio channels, must have the same channel bandwidth of 200 kHz and the same frequency deviation of 75 kHz as monophonic FM. Moreover, the new stereophonic FM had to be engineered so that it could be received (in mono form) with the previously existing monophonic FM receivers. Thus stereophonic FM transmitters and receivers incorporate special schemes which implement compatibility with monophonic FM as required. Like AM receivers, commercial FM receivers are superheterodyne receivers. The monophonic superheterodyne FM receiver is discussed below.

The superheterodyne FM receiver

A block diagram of the superheterodyne receiver for monophonic FM is shown in Figure 4.28. Some modifications that adapt it for use with stereophonic FM will be discussed later. It is very similar to the superheterodyne AM receiver, except for a few differences. Its intermediate frequency (IF) is 10.7 MHz instead of 455 kHz as in AM. It uses a frequency discriminator or a phase-lock loop for demodulation. The FM demodulator is preceded by a bandpass limiter, which removes any amplitude variation acquired in the channel. The FM demodulator is followed by a deemphasis filter which corrects for message signal distortion introduced by preemphasis in the transmitter, while reducing the noise power in the demodulated signal.

As in the AM superheterodyne receiver, the RF tuner is ganged to, and hence simultaneously tuned with, the local oscillator. With the inputs from the RF tuner and the local oscillator, the mixer converts the carrier frequency of the selected FM channel to the IF carrier frequency. The local

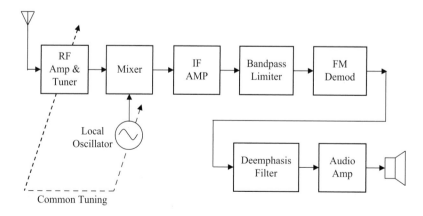

Figure 4.28. The superheterodyne FM receiver.

oscillator frequency is given by $f_{LO} = f_c + f_{IF}$, where f_c is the carrier frequency of the selected channel, and f_{IF} is the intermediate frequency of 10.7 MHz. As in the AM superheterodyne receiver, the RF section provides sufficient image channel suppression, while the IF section provides sufficient adjacent channel suppression. However, adjacent channel interference is not as big a problem in FM as in AM because FM is good at suppressing noise and interference.

4.9   APPLICATION: STEREO PROCESSING IN FM BROADCASTING

FM stereo processing is driven largely by the need for compatibility with monophonic FM receivers. Both the sum and the difference of the left and the right microphone signals are produced and used. The sum of the left and right signals is the monophonic signal required for monophonic receivers. The sum and the difference signals are used to reconstruct the left and right signals at the output of the stereophonic FM receiver.

A block diagram of the stereo signal processing in the FM transmitter is shown in Figure 4.29 (a). For notational simplicity, the left and the right microphone signals $l(t)$ and $r(t)$ are simply denoted by l and r, respectively, and their frequency-domain representations are denoted by L and R. Following their formation, the sum and difference signals $l + r$ and $l - r$ are passed through preemphasis circuits to yield $(l + r)_p$ and $(l - r)_p$, respectively. Like the original microphone signals, each of these combinations is a baseband signal with a 15 kHz bandwidth. The preemphasized difference signal $(l - r)_p$ is DSB-SC modulated with a 38 kHz carrier. The

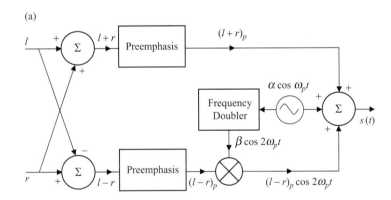

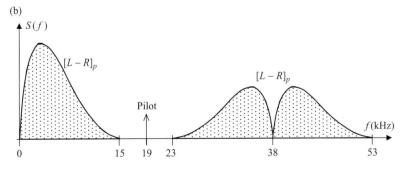

Figure 4.29. Stereo signal processing in FM. (a) Stereo signal processing in FM transmitter; (b) spectrum of composite baseband stereo signal.

38 kHz carrier of amplitude β is obtained by frequency-doubling a 19 kHz pilot carrier of small amplitude α. The $(l+r)_p$ signal, the pilot carrier, and the DSB-SC modulated $(l-r)_p$ signal are then frequency-division multiplexed to form a composite baseband stereo signal $s(t)$ which serves as the modulating signal input to the FM modulator. Denoting the frequency of the pilot carrier by ω_p, the composite baseband stereo signal is given by

$$s(t) = (l+r)_p + \alpha \cos \omega_p t + \beta (l-r)_p \cos 2\omega_p t \qquad (4.111)$$

The positive frequency spectrum $S(f)$ of the composite stereo signal is shown in Figure 4.29(b). Note that the DSB-SC signal $\beta(l-r)_p \cos 2\omega_p t$ is a frequency-translated version of $(l-r)_p$ centered at 38 kHz. Its lower sideband has a lower frequency limit of $(38 - 15)$ kHz $= 23$ kHz, which is 8 kHz above the 15 kHz bandwidth of $(l+r)_p$. It is in the middle of this 8 kHz wide band-gap that the pilot carrier is inserted. As a result, the pilot carrier can easily be recovered at the receiver with a narrowband filter centered at 19 kHz. Had a 38 kHz pilot carrier been used instead, it would not have the benefit of location in the band-gap, and any attempt to recover it with a narrow bandpass filter would incur contamination by the sidebands of the DSB-SC signal $\beta(l-r)_p \cos 2\omega_p t$.

The processing performed in the FM receiver is shown in Figure 4.30. This processing is performed on the composite stereo signal $s(t)$ which is the output of the FM demodulator. If the receiver is monophonic, its 15 kHz low-pass filter following the demodulator selects the $(l+r)_p$ signal. This is then deemphasized to yield the monophonic signal $l+r$. Thus the monophonic FM receiver output is taken from the upper arm of Figure 4.30 prior to the signal summer. In an FM stereo receiver, further processing proceeds as follows. In the lower arm the DSB-SC signal is recovered with a 23–53 kHz bandpass filter, while in the middle arm the pilot carrier is recovered with a narrowband filter centered at 19 kHz. From the recovered pilot the frequency doubler produces a 38 kHz carrier, which is used for synchronous detection of the DSB-SC signal to yield $(l-r)_p$. This $(l-r)_p$ is then deemphasized to yield $l-r$. Adding $l-r$ to $l+r$ with upper signal summer yields the left signal. Subtracting $l-r$ from $l+r$ with lower signal summer yields the right signal.

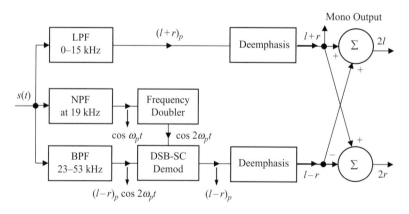

Figure 4.30. Stereo signal processing in an FM receiver.

Summary

1. Denoting the expressions for frequency-modulated (FM), phase-modulated (PM), and angle-modulated signals by $\phi_{FM}(t)$, $\phi_{PM}(t)$, and $\phi_A(t)$: $\phi_A(t) = A_c \cos[\omega_c t + k\gamma(t)] = A_c \cos\left[\omega_c t + k\int_{-\infty}^{t} m(\lambda)h(t-\lambda)d\lambda\right]$. $\phi_{FM}(t) = A_c \cos\left[\omega_c t + k_f\int_{-\infty}^{t} m(\lambda)d\lambda\right]$; if $h(t) = u(t)$ and $k = k_f$ in $\phi_A(t)$. $\phi_{PM}(t) = A_c \cos\left(\omega_c t + k_p m(t)\right)$; if $h(t) = \delta(t)$ and $k = k_p$ in $\phi_A(t)$.

2. For FM the instantaneous frequency and its maximum values are respectively $\omega_i(t) = \omega_c + k_f m(t)$ and $\omega_{i,max}(t) = \omega_c + k_f[m(t)]_{max}$.

   For PM they are respectively $\omega_i(t) = \omega_c + k_p m'(t)$ and $\omega_i(t)_{max} = \omega_c + k_p[m'(t)]_{max}$.

3. The expressions for narrowband FM (NBFM) and narrowband PM (NBPM) are given respectively as:

$$\phi_{FM}(t) \cong A_c \cos\omega_c t - A_c k_f \left(\int_{-\infty}^{t} m(\lambda)d\lambda\right) \sin\omega_c t$$

$$\phi_{PM}(t) \cong A_c \cos\omega_c t - A_c k_p m(t) \sin\omega_c t.$$

   For tone-modulated narrowband FM and narrowband PM the expressions are

$$\phi_{FM}(t) \cong A_c \cos\omega_c t - \beta A_c \sin\omega_m t \sin\omega_c t,$$

$$\phi_{PM}(t) \cong A_c \cos\omega_c t - \beta A_c \cos\omega_m t \sin\omega_c t.$$

4. Tone-modulated FM with message signal $m(t) = A_m \cos\omega_m t$ and carrier $A_c \cos\omega_c t$ is given by $\phi_{FM}(t) = A_c \sum_{n=-\infty}^{\infty} J_n(\beta) \cos(\omega_c + n\omega_m)t$, where β is the frequency deviation ratio and $J_n(\beta)$ are Bessel functions of the first kind of order n and argument β. The coefficients of $J_n(\beta)$ which are significant give the discrete spectrum, determine the power within a given number of sidebands, and determine the bandwidth of the tone-modulated FM signal as $B_{PM} \simeq 2\lceil\beta + 1\rceil f_m$. Similar remarks and expressions hold for tone-modulated PM.

5. FM and PM bandwidths are given by Carson's rule respectively as $B_{FM} \simeq 2(\beta + 1)B = 2(\Delta f + B)$ and $B_{PM} \simeq 2(\beta + 1)B = 2(\Delta f + B)$.

6. An n-fold frequency multiplier increases by n-fold the carrier frequency and the frequency deviation of the input FM/PM signal without distorting the information in the modulated signal. If the multiplier input signal is $\phi_i(t) = A \cos\left[\omega_c t + k_f\int_0^t m(\lambda)d\lambda\right]$, its output signal is $\phi_o(t) = \phi_{FM}(t) = A_c \cos\left[n\omega_c t + nk_f\int_0^t m(\lambda)d\lambda\right]$.

7. In the indirect method of generating wideband FM/PM, frequency multiplication is performed on narrowband FM/PM in two steps. After the first step, the carrier frequency is reduced by frequency conversion. After the second frequency multiplication step, the carrier frequency and frequency deviation are at the desired WBFM/PM levels.

8. In the direct method for generating wideband FM/PM, the frequency of a VCO output signal varies linearly with the message signal about the desired carrier frequency, producing sufficient frequency deviation. Unlike the indirect method which employs crystal oscillators, this method suffers from VCO frequency instability, but it is less noisy than the indirect method, which suffers from noise due to the frequency multipliers.

9. Many FM demodulators effectively differentiate an FM signal to give an amplitude- and frequency-modulated signal $\phi'_{FM}(t) = -A_c \left[\omega_c + k_f m(t) \right] \sin \left[\omega_c t + k_f \int_{-\infty}^{t} m(\lambda) d\lambda \right]$. The message signal is then recovered from $\phi'_{FM}(t)$ through envelope detection. Such FM demodulators include the ideal differentiator demodulator, the time-delay demodulator, the slope detector, and the balanced discriminator. Prior to demodulation, the FM signal is stripped of amplitude variations due to distortion and channel noise with a bandpass limiter. Feedback demodulators employ the phase-locked loop (PLL) to directly detect the message signal from the FM signal. Similar remarks apply to PM demodulators.

10. In a PLL the voltage-controlled oscillator (VCO) signal tracks the incoming signal in both frequency and phase. Because a frequency difference contributes to the total phase difference, the total phase difference alone is sufficient for the system analysis. The input signal and the VCO output are respectively $\phi_i(t) = A_c \sin \left[\omega_c t + \psi_i(t) \right]$ and $v(t) = A_v \cos \left[\omega_c t + \psi_v(t) \right]$. In phase-lock the PLL output is the demodulated message signal $y(t) = \frac{d\psi_v(t)}{dt} \simeq \frac{d\psi_i(t)}{dt} \simeq k_f m(t)$.

11. For a first-order PLL, the loop filter is given by $h(t) = \delta(t)$ or in s-domain $H(s) = 1$, while for a second-order PLL the loop filter is given by $h(t) = \delta(t) + au(t)$ or in s-domain $H(s) = \frac{s+a}{s}$.

12. When the PLL is tracking well the phase error is small, $\sin \left[\psi_e(t) \right] \simeq \psi_e(t)$, the phase detector and hence the PLL is linear, otherwise the PLL is nonlinear. The second-order PLL is analyzed with the linear model, while the first-order PLL can be analyzed with the linear or the nonlinear model.

13. The phase plot $\frac{d\psi_e(t)}{dt} = \Delta\omega - G \sin \psi_e(t)$ is the operating curve for the first-order PLL. At its equilibrium points, the frequency error is $\frac{d\psi_e(t)}{dt} = 0$ but the phase error has a non-zero value of $\psi_e(t) = \sin^{-1} (\Delta\omega/G)$. Thus at phase-lock, a first-order PLL achieves only a frequency lock, while it retains a constant phase error.

14. For the second-order PLL, $\Psi_e(s) = \frac{s^2}{s^2 + Gs + Ga} \left[\frac{\Delta\omega}{s^2} + \frac{\psi_{i0}}{s} \right]$. Given enough time (final value theorem), $\psi_e(\infty) = \lim_{s \to 0} \left[s\Psi_e(s) \right] = 0$. Thus at phase-lock, the second-order PLL eliminates both the frequency error and the phase error since the total phase error contains the frequency error.

15. The interference amplitude at the demodulator output is $y_d(t) = \psi(t) = \frac{A_i}{A_c} \sin \omega_i t$ for PM and $y_d(t) = \psi'(t) = \frac{\omega_i A_i}{A_c} \cos \omega_i t$ for FM. In each case, the interference is inversely proportional to the carrier amplitude, resulting in good interference (and noise) suppression. The interference (and noise) amplitude at the demodulator output is constant in PM but proportional to the message signal frequency in FM.

16. Preemphasis at the transmitter boosts the message signal spectrum at high frequencies while deemphasis at the receiver restores the message signal spectrum to its original form. In effect preemphasis and deemphasis employ FM at low message signal

frequencies where the noise is low and PM at higher message signal frequencies where the noise would have been higher had FM been employed. This results in better noise performance.

17. The preemphasis and deemphasis filter responses are given respectively by $H_p(j\omega) = G\left(\frac{\omega_1 + j\omega}{\omega_2 + j\omega}\right)$ and $H_d(j\omega) = \frac{\omega_1}{\omega_1 + j\omega}$, where $\omega_1 = 2\pi f_1$ and $\omega_2 = 2\pi f_2$ are the lower and upper corner frequencies, and G is a gain constant. Typically f_1 is 2.12 kHz and f_2 is 30 kHz.

18. Commercial FM carrier frequencies range from 88 to 108 MHz. The channels are spaced 200 kHz apart, and the frequency deviation is 75 kHz. It employs the superheterodyne receiver. Stereophonic FM is compatible with monophonic FM.

Review questions

4.1 Tone-modulated narrowband FM with message signal $m(t) = 10\cos 10^4 \pi t$ cannot have
(a) $B_{FM} = 10$ kHz.   (b) $B_{FM} = 11$ kHz.   (c) $\beta < 5$.   (d) Discrete spectrum.
(e) Continuous spectrum.

4.2 You need to generate NBFM of bandwidth B_{FM} and NBPM of bandwidth B_{PM} using the message signal $m(t) = 8\cos 10\pi^4 t$, but you have only two phase modulators. Which of the following is true?
(a) An integrator is needed.   (b) A differentiator is needed.   (c) $B_{FM} > B_{PM}$.
(d) $B_{PM} = 5 \times 10^3 B_{FM}$.   (e) A phase filter is needed.

4.3 Which one of the following is not a permissible formula for determining WBFM bandwidth when the message signal is an audio music signal of bandwidth B Hz?
(a) $B_{FM} = 2(\beta + 1)B$.   (b) $B_{FM} = 2\lceil\beta + 1\rceil B$.   (c) $B_{FM} = 2(\beta B + B)$.
(d) $B_{FM} = 2(\Delta f + B)$.

4.4 The output signal of an FM modulator has a bandwidth of 140 kHz when the message signal is $2\cos 2\pi \times 10^4 t$. When the message signal is $3\cos 4\pi \times 10^4 t$ the FM bandwidth will be:
(a) 220 kHz.   (b) 160 kHz.   (c) 200 kHz.   (d) 120 kHz.   (e) 260 kHz.

4.5 The output signal of a PM modulator has a bandwidth of 80 kHz when the message signal is $2\cos 2\pi \times 10^4 t$. When the message signal is $4\cos 3\pi \times 10^4 t$ the PM bandwidth is:
(a) 120 kHz.   (b) 180 kHz.   (c) 210 kHz.   (d) 8 kHz.   (e) 90 kHz.

4.6 NBFM with $f_{c1} = 50$ kHz and $\Delta f_1 = 40$ Hz is to be used for indirect generation of WBFM with $f_c = 86.4$ MHz and $\Delta f = 92.16$ kHz. Which of the following is the most appropriate choice of multiplying factors n_1 and n_2?
(a) 48 and 48.   (b) 48 and 36.   (c) $\left(2^4 \times 3\right)$ and $\left(2^2 \times 3^2\right)$.
(d) $\left(2^4 \times 3\right)$ and $\left(2^4 \times 3\right)$.   (e) $\left(2^2 \times 3^2\right)$ and $\left(2^4 \times 3\right)$.

4.7 NBFM with $f_{c1} = 100$ kHz and $\Delta f_1 = 50$ Hz is to be used for indirect generation of WBFM with $f_c = 96$ MHz and $\Delta f = 64.8$ kHz. Which of the following is an appropriate choice of the local oscillator frequency for the frequency converter?

(a) 0.6 MHz. (b) 6.27 MHz. (c) 3.6 MHz. (d) 6.67 MHz.
(e) 0.6 MHz.

4.8 Which of the following types of signal processing does FM demodulation not involve?

(a) FM–AM conversion. (b) Hard limiting. (c) Bandpass limiting.
(d) Frequency conversion. (e) Envelope detection.

4.9 Which of the following does a first-order phase-locked loop not have?

(a) Nonlinear model. (b) Linear model. (c) Phase plot. (d) Zero phase error at phase-lock. (e) Zero frequency error at phase-lock.

4.10 Which of the following is not true for interference and noise in FM and PM?

(a) Reduced by preemphasis/deemphasis. (b) Suppressed by large carrier.
(c) Emphasized and deemphasized. (e) Captured by strong channel.
(d) Strong at high signal frequencies in FM.

Answers: 4.1(e), 4.2(a), 4.3(b), 4.4(a), 4.5(c), 4.6(d), 4.7(b), 4.8(d), 4.9(d), 4.10(c),

Problems

Section 4.2 FM, PM, and angle modulation

4.1 The input signals into a modulator are the carrier $c(t) = 4 \sin 10^7 \pi t$ and the message signal $m(t) = 3 + 2 \sin 10^4 \pi t$.

(a) If frequency modulation with $k_f = 10^5 \pi$ is performed, find the maximum and minimum instantaneous frequencies and sketch the modulated waveform.

(b) If phase modulation with $k_p = 20$ is performed, find the maximum and minimum instantaneous frequencies and sketch the modulated waveform.

4.2 The input signals into a modulator are the carrier $c(t) = 5 \cos \left(2\pi \times 10^7 t\right)$ and the message signal shown in Figure 4.31. If frequency modulation is performed with $k_f = 4\pi \times 10^4$, find the maximum and minimum instantaneous frequencies and sketch the modulated waveform.

4.3 Phase modulation is performed using a cosine wave carrier with a frequency of 80 MHz and the message signal shown in Figure 4.32.

(a) Sketch $m'(t)$, the derivative of the message signal.

(b) Find the maximum and minimum instantaneous frequencies and sketch the modulated waveform if the phase deviation constant is 4π.

Section 4.3 Bandwidth and spectrum of angle-modulated signals

4.4 Starting from Eq. (4.10), obtain Eq. (4.20) for narrowband phase modulation if $\left|k_p m(t)\right| \ll 1$, utilizing the appropriate definitions for $\gamma(t)$ and β in phase modulation.

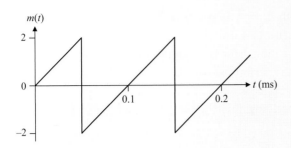

Figure 4.31. Message signal for Problems 4.2, 4.13, and 4.14.

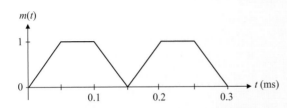

Figure 4.32. Message signal for Problem 4.3.

4.5 Frequency modulation is performed with a message signal given by $m(t) = 2 \cos \omega_m t$, and a carrier given by $c(t) = 5 \cos (10 \omega_m t)$, where $\omega_m = 3\pi \times 10^4$.
  (a) What is the maximum possible value of the frequency deviation constant k_f if the modulated signal is narrowband FM.
  (b) If $k_f = 1.5\pi \times 10^3$, obtain an expression for the spectrum of the narrowband FM signal and sketch its positive frequency amplitude spectrum.

4.6 Repeat Problem 4.5 but for the case of narrowband phase modulation. For part (b) the phase deviation constant is $k_p = 0.1$. Obtain an expression for the PM spectrum, but do not sketch it.

4.7 The modulating signal input into a modulator is $m(t) = 5 \cos (10^4 \pi t)$ and the carrier is $c(t) = 8 \cos (10^8 \pi t)$.
  (a) If frequency modulation with $k_f = 4\pi \times 10^3$ is performed, determine the number of significant sidebands in the modulated signal.
  (b) Find the bandwidth of the FM signal.

4.8 The modulating signal input into a modulator is given by $6 \sin (3\pi \times 10^4 t)$, and the carrier is $10 \cos (4\pi \times 10^6 t)$.
  (a) If frequency modulation with $k_f = 2\pi \times 10^4$ is performed, determine the bandwidth of the FM signal based on the number of significant sidebands.
  (b) If phase modulation with $k_p = \frac{1}{3}$ is performed, determine the bandwidth of the PM signal based on the number of significant sidebands.

4.9 The modulating signal input into a modulator is given by $3 \cos (2\pi \times 10^4 t)$, and the carrier is $A_c \sin \omega_c t$. The output of the modulator is passed through a filter whose frequency response is given by $H(j\omega) = \text{rect} \left[\frac{\omega + \omega_c}{3.4\pi \times 10^4} \right] + \text{rect} \left[\frac{\omega - \omega_c}{3.4\pi \times 10^4} \right]$.
  (a) If frequency modulation with $k_f = 4\pi \times 10^4$ is performed, determine the power in the filter output if (i) $A_c = 1$ and (ii) $A_c = 10$.

(b) If phase modulation with $k_p = \frac{2}{3}$ is performed, determine the power in the filter output if (i) $A_c = 1$ and (ii) $A_c = 10$.

4.10 The message signal input to a modulator is given by $A_m \cos(2\pi \times 10^4 t)$.

(a) If frequency modulation is performed with $k_f = 5\pi \times 10^4$, find the FM bandwidth when (i) $A_m = 2$ and (ii) $A_m = 4$.

(b) If phase modulation is performed with $k_p = 2.25$, find the PM bandwidth when (i) $A_m = 2$ and (ii) $A_m = 4$.

4.11 Repeat Problem 4.10, but with the frequency of the message signal doubled so that the message signal is given by $A_m \cos(4\pi \times 10^4 t)$.

4.12 A multi-tone modulating signal is given by $8 \cos(2\pi \times 10^4 t) + 4 \cos(5\pi \times 10^4 t)$.

(a) If FM is performed with $k_f = 10^4 \pi$, find the FM bandwidth.

(b) If PM is performed with $k_p = 0.5$, find the PM bandwidth.

4.13 Assume that the bandwidth of the message signal of Figure 4.31 is band-limited to the third harmonic of its fundamental frequency, and that frequency modulation with a frequency deviation constant of $k_f = 10^5 \pi$ is performed.

(a) Find the frequency deviation of the FM signal.

(b) Find the phase deviation constant such that if phase modulation is performed, the PM signal will have the same bandwidth as the FM signal.

(c) Find the bandwidth of the FM signal using Carson's rule.

(d) Find the bandwidth of the FM signal using the conservative rule of Eq. (4.32).

4.14 The message signal of Figure 4.31 is passed through a low-pass filter with a bandwidth of 42 kHz before it is used as the input to a modulator.

(a) If FM with $k_f = 12\pi \times 10^4$ is performed, determine the FM bandwidth.

(b) If PM with $k_p = 8\pi$ is performed, determine the PM bandwidth.

(c) Repeat parts (a) and (b) if the bandwidth of the low-pass filter is doubled.

Section 4.4 Generation of FM signals

4.15 Figure 4.33 shows a simple circuit implementation of a frequency multiplier. The input FM signal is $\phi(t) = A_c \sin\left[2\pi \times 10^7 t + \sin\left(2\pi \times 10^4 t\right)\right]$, $x(t)$ is a half wave rectified version of the input signal, the output signal $\phi_n(t)$ is a version of the input signal that has been frequency-multiplied by a factor of n and $L = 5$ µH. Complete the design of the frequency multiplier by choosing the values for C and R so that a frequency multiplication factor of 4 is achieved, and the 3 dB frequencies of the series resonant circuit are one frequency deviation of the output FM signal above and below the resonant frequency.

4.16 Frequency multiplication by a factor of 4 is to be implemented with a cascade of two frequency doublers, each similar to the simple multiplier shown in Figure 4.33. The input FM signal has a carrier frequency of 10 MHz, the message signal is a cosine wave of amplitude 5, frequency 10 kHz, and $k_f = 1$, and $L = 5$ µH for each of the frequency doublers.

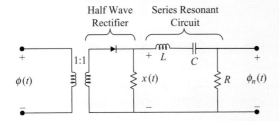

Half Wave   Series Resonant
Rectifier   Circuit

Figure 4.33. Frequency multiplier for Problems 4.15 and 4.16.

(a) Complete the design of the first frequency doubler in the cascade by choosing the values for C and R so that a frequency doubling is achieved, and the 3 dB frequencies of the series resonant circuit are one frequency deviation of the output FM signal above and below the resonant frequency.

(b) Repeat part (a), but for the second frequency doubler in the cascade.

4.17 For a narrowband FM signal, the modulating signal has a frequency range of 200 Hz to 10 kHz, the carrier frequency is 200 kHz, and the maximum value of the frequency deviation ratio is 0.5. Wideband FM with a carrier frequency of 100 MHz and a frequency deviation of 75 kHz is to be produced from the narrowband FM, utilizing the indirect method of generating WBFM shown in Figure 4.10.

(a) If $n_1 = 1.2n_2$, determine the values of Δf_1, Δf_2, f_{c2}, and f'_{c2}, as defined by the notation in Figure 4.10.

(b) Complete the design of the indirect WBFM generator by determining the multiplication factors for the individual multipliers in cascade for n_1 and n_2, and the two possible local oscillator frequencies if the maximum multiplying factor for any frequency multiplier is 5.

4.18 Wideband FM is to be generated from narrowband FM using the indirect method of Figure 4.10. For the narrowband FM, $f_{c1} = 125$ kHz and $\Delta f_1 = 30$ Hz. For the desired wideband FM, $f_c = 80$ MHz and $\Delta f = 48$ kHz.

(a) Suppose the largest multiplication factor allowed for an individual multiplier is 5, and $n_1 = n_2$. Complete the design of the indirect WBFM generator by specifying the constituent multiplying factors for n_1 and n_2, and by specifying the two possible local oscillator frequencies.

(b) Repeat part (a) if only frequency doublers and triplers are allowed, n_1 and n_2 should be as close in value as possible but are not necessarily equal, and the frequency deviation of the WBFM may be higher or lower than, but should be as close as possible to, the desired value of 48 kHz.

(c) Determine a new value of Δf_1 such that with the values of n_1 and n_2 found in part (b), the frequency deviation of the WBFM will be the desired 48 kHz.

4.19 Repeat Problem 4.18 if the message signal used for producing narrowband FM has a frequency range of 50 Hz to 5 kHz. For the narrowband FM, $\beta_{1,\max} = 0.25$ and $f_{c1} = 50$ kHz. For the wideband FM, $\Delta f = 80$ kHz and $f_c = 100$ MHz.

4.20 WBFM is to be generated from NBFM by employing the scheme shown in Figure 4.11 in which all frequency multiplication is accomplished in one step prior to frequency conversion. For the narrowband FM, $f_{c1} = 200$ kHz and $\Delta f_1 = 50$ Hz. For the desired wideband FM, $f_c = 100$ MHz and $\Delta f = 90$ kHz.

(a) Complete the design of this WBFM generator by specifying the multiplying factors for the individual frequency multipliers if the maximum multiplying factor for any of them is 5. Specify the two possible frequencies of the local oscillator.

(b) Repeat part (a) if only frequency doublers and triplers can be used, such that the frequency deviation for the resulting WBFM is as close as possible to 90 kHz. What is the value of this frequency deviation?

(c) What is the disadvantage of this scheme relative to the usual scheme in which frequency multiplication is accomplished in two steps with a frequency conversion step between them?

4.21 The direct method of generating WBFM shown in Figure 4.12 has $L = 5$ μH, $C_1 = 40$ pF, and the maximum value attained by the total capacitance due to C_1 and C_v is 50.8 pF. The FM signal has a carrier frequency of 20 MHz and $\Delta f = 25$ kHz.

(a) Determine C_{vo}, the magnitude of the variable capacitance C_v when no modulating signal is applied.

(b) Determine the maximum and minimum values attained by the variable capacitance C_v.

4.22 Complete the design of the circuit for the direct method of generating WBFM shown in Figure 4.12 by choosing the value of the fixed capacitor C_1 if $L = 1$ μH, $V_B = 6$ V, and the carrier frequency of the output FM signal is to be 20 MHz. The message signal is given by $m(t) = 0.24 \cos 10^4 \pi t$, and the voltage across the varactor diode capacitor is given by $C_v = 20 \text{ pF} / \sqrt{1 + \frac{V_R}{0.75}}$, where V_R is the reverse bias voltage across the varactor diode capacitor. Determine the maximum frequency deviation for the output FM signal.

4.23 The direct method of WBFM generation shown in Figure 4.12 is used for generating an FM signal with a carrier frequency of 40 MHz and a frequency deviation of 25 kHz. It is desired to increase the carrier frequency to 90 MHz and the frequency deviation to 75 kHz. Design a block diagram implementation of a scheme that will achieve this objective by employing one frequency multiplication step and one frequency conversion step for the following cases.

(a) Frequency conversion is preceded by frequency multiplication.

(b) Frequency conversion is followed by frequency multiplication.

Section 4.5 Demodulation of angle-modulated signals

4.24 The circuit of Figure 4.14 is a good approximation to an ideal differentiator demodulator if $f_c \leq 0.1 f_H$, where f_c is the carrier frequency of the FM signal and f_H is the cut-off frequency of the RC high-pass filter. The input FM signal is given by $4 \cos \left[10^8 \pi t + 10^4 \pi (5 \sin 10^4 \pi t) \right]$.

(a) Specify the value of the resistance R such that $f_c = 0.1 f_H$ when $C = 5$ pF.

(b) Give an expression for the output signal of the demodulator.

4.25 The time-delay differentiator of Figure 4.16 is a good approximation to an ideal differentiator if $\tau \ll \frac{1}{\omega} \simeq \frac{1}{\omega_c} = \frac{T}{2\pi}$, but the maximum allowable value of time delay is $\tau_{max} = T/4$.

(a) Find the range of values τ_{max} for commercial FM with a carrier frequency range of 88 MHz to 108 MHz.

(b) A more accurate definition of time delay is $\tau \ll \frac{1}{\omega_i}$, where ω_i is the instantaneous frequency of the FM signal. Find the maximum percentage error in τ_{max} due to use of the approximate expression $\tau \ll \frac{1}{\omega_c}$ instead of the more accurate one, for commercial FM with a carrier frequency of 98 MHz and frequency deviation of 75 kHz.

4.26 An FM signal given by $\phi_{FM}(t) = 8 \cos \left[2\pi \times 10^7 t + 5 \sin \left(3\pi \times 10^4 t \right) \right]$ with a bandwidth of B_{FM} is to be demodulated using a balanced discriminator, the balanced slope detector portion of which is shown in Figure 4.19. The resonance frequencies of the upper and the lower slope detectors are located symmetrically on either side of the FM carrier frequency and are $0.75B_{FM}$ from it. Each of the slope detectors has a bandwidth of B_{FM}. Complete the design of the balanced slope detector by specifying the values of C_1, C_2, R_1, and R_2, if $L_1 = L_2 = 5$ μH.

4.27 The balanced discriminator used for demodulating a PM signal is shown in Figure 4.19. The PM signal $\phi_{PM}(t) = 5 \sin \left[2\pi \times 10^7 t + 4 \cos \left(2\pi \times 10^4 t \right) \right]$ has a bandwidth of B_{PM}. The upper slope detector and lower slope detector each have a bandwidth of $2B_{FM}$, and their resonance frequencies are $1.5B_{FM}$ above and below the PM carrier frequency, respectively. Complete the design of the balanced slope detector by specifying the values of C_1, C_2, R_1, and R_2, if $L_1 = L_2 = 4$ μH.

Section 4.6 Feedback demodulators

4.28 (a) A first-order phase-locked loop has a loop gain of 300. Find the maximum frequency error for which it can achieve phase-lock.

(b) Determine the equilibrium phase error corresponding to a frequency change of 15 Hz.

(c) Determine the frequency error corresponding to an equilibrium phase error of 30°.

4.29 (a) A first-order phase-locked loop has an equilibrium phase error of 15° due to a frequency error of 20 Hz. Determine the loop gain for the phase-locked loop.

(b) With the loop gain found in part (a), determine the equilibrium phase errors corresponding to frequency changes of 30 Hz and 40 Hz.

(c) Repeat part (b) if the loop gain determined in part (a) is doubled.

4.30 A second-order phase-locked loop has a damping factor of 0.8 and a loop gain of 240. Specify its natural frequency ω_o, its loop filter transfer function $H(s)$, and the impulse response of the loop filter $h(t)$.

4.31 (a) A second-order phase-locked loop has a loop gain of 200 and an impulse response of $h(t) = \delta(t) + \beta u(t)$, where $\beta = 200$. Determine the natural frequency and the damping factor for the system.

(b) Repeat part (a), if the constant β is doubled, but the loop gain is unchanged.

(c) Repeat part (a), if the loop gain is doubled, but the constant β is unchanged.

4.32 A frequency-compressive feedback system is to achieve a frequency compression ratio of 10:1. For the input FM signal, the modulating signal is $5 \sin (10^4 \pi t)$, and the frequency deviation constant is $k_f = 4\pi \times 10^4$. The VCO gain constant and the discriminator gain constant are equal. Specify the discriminator gain constant and determine the frequency deviation at the input to the discriminator.

4.33 (a) In Figure 4.24 the input signal is an FM signal with a carrier frequency of $\frac{1}{3}f_c$. Indicate the frequency setting(s) you would make in any relevant block(s) in order to obtain an output FM signal with a carrier frequency of f_c.

    (b) If the output frequency is to be $\frac{2}{3}f_c$, indicate the corresponding frequency setting(s) for the relevant block(s) in the system.

4.34 (a) In Figure 4.25 the input signal is an FM signal with a carrier frequency of $\frac{1}{3}f_c$. Indicate the frequency setting(s) you would make in any relevant block(s) in order to obtain an output FM signal with a carrier frequency of f_c.

    (b) If the output frequency is to be $\frac{8}{9}f_c$, indicate the corresponding frequency setting(s) for any relevant block(s) in the system.

Section 4.7 Interference in angle modulation

4.35 You are to design the preemphasis and deemphasis filters shown in Figure 4.27 to have upper and lower corner frequencies of 2.12 kHz and 30 kHz, respectively. The two capacitors in the two circuits are to be equal, and R_2 is a 2 kΩ resistor. Specify the values of R_1 and C.

4.36 (a) A preemphasis filter has corner frequencies of $f_1 = 2$ kHz and $f_2 = 40$ kHz, and the corresponding deemphasis filter has a corner frequency of $f_1 = 2$ kHz. Determine the ratio of the actual product of the preemphasis and deemphasis filter responses to the ideal value of the products of the responses required for exact cancellation at 2 kHz and 15 kHz.

    (b) Repeat part (a) if f_2 is changed to 15 kHz.

4.37 For the preemphasis filter circuit shown in Figure 4.27, $C = 2.4$ nF, $R_1 = 28$ kΩ, and $R_2 = 2.5$ kΩ. Determine the upper and lower corner frequencies, f_1 and f_2. Determine also the low- and high-frequency gains of the circuit.

4.38 (a) For the preemphasis and deemphasis filters shown in Figure 4.27, $R_1 = 32$ kΩ in each of the filters. Complete the design of the filters by choosing the other components if the upper and lower corner frequencies are $f_1 = 2.12$ kHz and $f_2 = 30$ kHz.

    (a) Determine the high- and the low-frequency gains for the preemphasis filter.

Section 4.8 FM broadcasting

4.39 For commercial FM, the carrier frequencies range from 88 MHz to 108 MHz. In the superheterodyne FM receiver, the local oscillator frequency is $f_{LO} = f_c + f_{IF}$, where f_c is the carrier frequency and f_{IF} is the intermediate frequency.

(a) Determine the range of image frequencies for commercial FM. What roles do the RF tuner and the IF filter play in suppression of image channels and adjacent channel interference?

(b) The RF tuner in the superheterodyne FM receiver consists of a 2 μH inductor in parallel with a variable capacitor and a variable resistance. When tuned to a desired channel, the tuning circuit has a resonance frequency equal to the channel carrier frequency, and a bandwidth equal to the FM channel bandwidth of 200 kHz. Determine the maximum and minimum values for the variable capacitance and the variable resistance.

4.40 In the stereo signal-processing block diagram of Figure 4.33(b), let $\beta = 1$, $(l + r)_p = A \cos \omega_m t + A \cos (\omega_m t + \theta)$, and $(l - r)_p = A \cos \omega_m t - A \cos (\omega_m t + \theta)$. Determine $a/_A$, the ratio of the amplitude of the pilot carrier to the amplitude of the preemphasized left or right signal, if the percentage of pilot carrier power in the composite baseband signal $s(t)$ is to be (i) 10% and (ii) 1%.

Pulse modulation and transmission

Thy friend has a friend, and thy friend's friend has a friend; be discreet.

THE TALMUD

HISTORICAL PROFILES

Alexander Graham Bell (1847–1922), inventor of the telephone, was a Scottish American scientist. The bel, the logarithmic unit introduced in Chapter 1, is named in his honor.

Bell was born in Edinburgh, Scotland, a son of Alexander Melville Bell, a well-known speech teacher. Alexander the younger also became a speech teacher after graduating from the University of Edinburgh and the University of London. After his older brother died of tuberculosis, his father decided to move to Ontario, Canada. Alexander was asked to come to Boston to work at the School of the Deaf. There he met Thomas A. Watson, who became his assistant in his electromagnetic transmitter experiment.

Throughout his life, Bell had been interested in the education of deaf people. This interest led him to invent the microphone. In 1866, he became interested in transmitting speech electrically. On March 10, 1876, Alexander sent the famous first telephone message: "Watson, come here I want you." Many inventors had been working on the idea of sending human speech by wire, but Bell was the first to succeed. The invention of the telephone grew out of improvements Bell had made to the telegraph. After inventing the telephone, Bell continued his experiments in communication. He invented the photophone – transmission of sound on a beam of light, the precursor of fiber-optics. He also invented techniques for teaching speech to the deaf. Alexander Graham Bell died in Baddek, Nova Scotia, on August 2, 1922. At the time of his burial, all telephone service stopped for one minute throughout the USA, in simple respect for his life. Since his death, the telecommunication industry has undergone an amazing revolution. Today, non-hearing people are able to use a special display telephone to communicate.

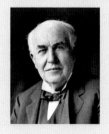

Thomas Alva Edison (1847–1931) was perhaps the greatest American inventor. He patented 1093 inventions, earning him the nickname "The Wizard of Menlo Park". These include history-making inventions such as the incandescent electric bulb, the phonograph, and the first commercial motion pictures.

Born in Milan, Ohio, the youngest of seven children, Edison received only three months of formal education because he hated school. He was home-schooled by his mother and quickly began to read on his own. In 1868, Edison read one of Faraday's books and found his calling. He moved to Menlo Park, New Jersey, in 1876, where he managed a well-staffed research laboratory. Edison worked on many projects, including sound-recording devices, the commercial phonograph, the kinetoscope, the storage battery, the electric pen, the mimeograph, and the microtasimeter. Most of his inventions came out of this laboratory. His laboratory served as a model for modern research organizations. Because of his diverse interests and the overwhelming number of his inventions and patents, Edison began to establish manufacturing companies for making the devices he invented. He designed the first electric power station to supply electric light. Formal electrical engineering education began in the mid-1880s with Edison as a role model and leader. In tribute to this important American, electric lights in the United States were dimmed for one minute on October 21, 1931, a few days after his death.

5.1 INTRODUCTION

We have considered analog communication in the last two chapters. Digital communication has several advantages over analog communication. First, digital communication is more robust than analog communication because the former can better withstand noise and distortion. Second, digital hardware implementation is flexible due to the flexibility inherent in processing digital signals. Third, it is easier and more efficient to multiplex several digital signals. Fourth, the cost of digital components continues to drop, to the point where some circuit functions cost virtually nothing in comparison to an analog alternative. Finally, digital data transmission can be used to exploit the cost effectiveness of digital integrated circuits.

In this chapter, we consider analog and digital pulse modulation systems, conversion of analog to digital signals, time-division multiplexing, and a few applications. In a pulse modulation system, the amplitude, width, or position of a pulse can vary over a continuous range in accordance with the message amplitude at the sampling instants. Analog-to-digital signal conversion consists of a three-step process – sampling, quantizing, and encoding. We begin by considering the sampling process. Later, we consider two common types of digital pulse modulation techniques: pulse code modulation (PCM) and delta modulation (DM). We then examine the concept of multiplexing. Finally, we deal with two application areas: analog-to-digital conversion and speech coding.

5.2 SAMPLING

Sampling is the process of obtaining a set of samples from a continuous signal. It is an important operation in digital communication and digital signal processing. In analog systems, signals are processed in their entirety. However, in digital systems, only samples of the signals are required for processing. The sampling can be done using a train of pulses or impulses. Here we will use an impulse sample.

Consider the continuous signal $g(t)$ shown in Figure 5.1(a). This can be multiplied by a train of impulses $\delta(t - nT_s)$, shown in Figure 5.1(b), where T_s is the *sampling interval* and $f_s = 1/T_s$ is the *sampling rate*. The samples can be considered to be obtained by passing $g(t)$ through a sampler, which is a switch that closes and opens instantaneously at the sampling instants T_s. When the switch is closed, we obtain a sample $g_s(t)$. Otherwise the output of the sampler is zero. Since $\delta(t)$ is zero everywhere except at $t = 0$, the sampled signal $g_s(t)$ can be written as

(a)

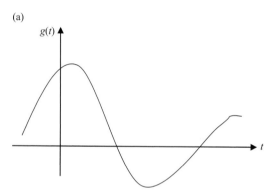

Figure 5.1. (a) Continuous (analog) signal to be sampled; (b) train of impulses; (c) sampled (digital) signal.

(b)

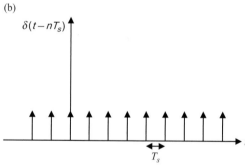

(c)

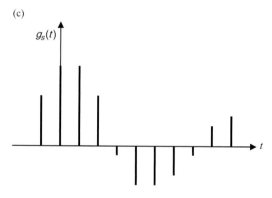

$$g_s(t) = g(t) \sum_{n=-\infty}^{\infty} \delta(t - nT_s) \tag{5.1}$$

The Fourier transform of this is

$$G_s(\omega) = \frac{1}{2\pi} \left[G(\omega) * \frac{2\pi}{T_s} \sum_{n=-\infty}^{\infty} \delta(t - nT_s) \right] = \frac{1}{T_s} \sum_{n=-\infty}^{\infty} G(\omega - n\omega_s) \tag{5.2}$$

where $\omega_s = 2\pi/T_s$. Thus, Eq. (5.2) becomes

$$G_s(\omega) = \frac{1}{T_s} \sum_{n=-\infty}^{\infty} G(\omega - n\omega_s) \tag{5.3}$$

This shows that the Fourier transform $G_s(\omega)$ of the sampled signal is a sum of translates of the Fourier transform of the original signal sampled at rate of $1/T_s$. Hence, under the following conditions,

$$1.\ G(\omega) = 0 \text{ for } |\omega| > 2\pi W, \text{ i.e. } g(t) \text{ is band-limited} \tag{5.4}$$

$$2.\ f_s = \frac{1}{T_s} = 2W \tag{5.5}$$

We notice from Eq. (5.3) that

$$G(\omega) = \frac{1}{2W} G_s(\omega) \tag{5.6}$$

Substituting Eq. (5.6) into Eq. (5.2) yields

$$G(\omega) = \frac{1}{2W} \sum_{n=-\infty}^{\infty} G(\omega - n\omega_s) \tag{5.7}$$

This shows that the Fourier transform $G(\omega)$ of a band-limited signal $g(t)$ is uniquely determined by its sample values $g(nT_s)$.

In order to ensure optimum recovery of the original signal, what must the sampling interval be? This fundamental question in sampling is answered by the *sampling theorem:*

The sampling theorem states that a band-limited signal, with no frequency component higher than W hertz, may be completely recovered from its samples taken at a frequency at least twice as high as $2W$ samples per second.

In other words, for a signal with bandwidth W hertz (with no frequency components higher than W hertz), there is no loss of information or overlapping if the sampling frequency is at least twice the highest frequency in the modulating signal. Thus,

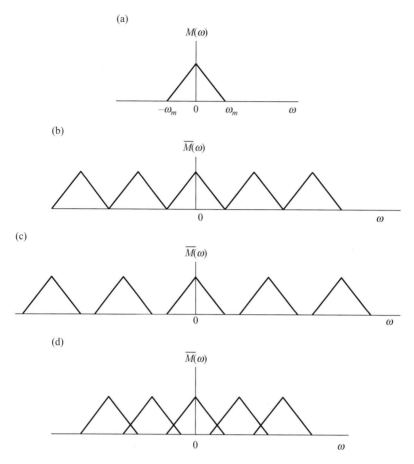

Figure 5.2. (a) Assumed spectrum; (b) sampling at the Nyquist rate; (c) sampling above the Nyquist rate; (d) sampling below the Nyquist rate.

$$\frac{1}{T_s} = f_s \geq 2W \tag{5.8}$$

The minimum sampling rate f_s is known as the *Nyquist rate* or *frequency* and the corresponding sampling interval $T_s = 1/f_s$ is the *Nyquist interval*. Thus, a band-limited analog signal $g(t)$ can be completely reconstructed from its samples $g(nT)$, if the sampling frequency is greater than $2W$ (the Nyquist rate). For example, consider a 1 kHz sinusoidal signal. Since it is band-limited to 1000 Hz, the minimum sampling rate is 2000 Hz, meaning that we take 2000 samples per second.

Consider the spectrum of $m(t)$ whose Fourier transform is in Figure 5.2(a). If the signal is sampled at the Nyquist rate $f_s = 2W$, the spectrum consists of repetitions of $M(\omega)$, as shown in Figure 5.2(b). If we sample at a rate higher than the Nyquist rate ($f_s > 2W$), we have the spectrum $\overline{M}(\omega)$ shown in Figure 5.2(c). Finally, if we sample at a rate lower than the Nyquist rate ($f_s < 2W$), there is overlapping of the spectra as shown in Figure 5.2(d). This results in aliasing and the original signal cannot be recovered.

Let us now consider the problem of reconstructing the signal $g(t)$ from the sampled signal $g_s(t)$.

It is sufficient to filter the sampled signal by an ideal low-pass filter (LPF) of gain K, bandwidth W, where $W < B < f_s - W$ and transfer function

$$H(\omega) = K\Pi\left(\frac{\omega}{4\pi B}\right)$$

With this choice, the output spectrum is

$$Y(\omega) = H(\omega)G_s(\omega) = \frac{K}{T_s}G(\omega) \tag{5.9}$$

Taking inverse Fourier transform yields

$$y(t) = \frac{K}{T_s}g(t) \tag{5.10}$$

which is the desired original signal amplified by $\frac{K}{T_s}$.

There are two major practical difficulties in reconstructing a band-limited signal from its sampled values. First, if the sampling is done at the Nyquist rate, $f_s = 2W$, to recover $g(t)$ from $g(nT_s)$ would be impossible because the ideal low-pass filter required is not realizable. However, if the sampling rate is high enough ($f_s > 2W$), the recovered signal approaches $g(t)$ more closely. In practical systems, sampling is done at a rate higher than the Nyquist. This enables reconstruction filters that are realizable and easy to build. The distance between two adjacent replicated spectra in the frequency domain is called the *guard band*. Error caused by sampling too slowly is called *aliasing* in view of the fact that higher frequencies disguise themselves as lower frequencies. Second, a signal is band-limited if its Fourier transform satisfies Eq. (5.4). A signal is time-limited if

$$g(t) = 0 \quad \text{for } |t| > T \tag{5.11}$$

i.e. $g(t)$ is of finite duration. If a signal is band-limited, it cannot be time-limited and vice versa. It turns out that all practical signals are time-limited and therefore not band-limited.

EXAMPLE 5.1
Determine the Nyquist rate and Nyquist interval for each of the following signals:

(a) $g(t) = 10 \sin 300\pi t \cos 200\pi t$
(b) $g(t) = \frac{\sin 100\pi t}{\pi t}$
(c) $g(t) = \left(\frac{\sin 100\pi t}{\pi t}\right)^2$

Solution
(a) We use the trigonometric identity

$$2 \sin x \cos y = \sin (x + y) + \sin (x - y)$$

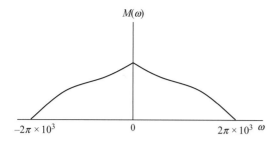

Figure 5.3. For Practice problem 5.1.

$$g(t) = 10 \sin 300\pi t \cos 200\pi t$$
$$= 5 \sin (500\pi t) + 5 \sin (100\pi t)$$

$g(t)$ is band-limited. The bandwidth is determined by the highest-frequency component. Hence, the maximum frequency is $2\pi W = 500\pi$ or $W = 250$. Hence the Nyquist rate is $2W = 500$ Hz and the Nyquist interval is $1/500 = 2$ ms.

(b) From Chapter 2, the Fourier transform of the sinc function is the rectangular pulse, i.e.

$$\frac{\sin \tau t}{\pi t} \quad \rightarrow \quad \Pi(\omega/\tau) = \begin{cases} 1, & |\omega| < \tau \\ 0, & |\omega| > \tau \end{cases}$$

We note from this that $g(t)$ is band-limited with $W = 50$ Hz. Thus, the Nyquist rate is 100 Hz and the Nyquist interval is $1/100 = 10$ ms.

(c) This signal may be regarded as the product of the signal in part (b) with itself. Since multiplication in the time domain corresponds to convolution in the frequency domain, the bandwidth of the signal is twice that of the signal in part (b), i.e. $W = 100$ Hz. Hence, the Nyquist rate is 200 Hz, while the Nyquist interval is $1/200$ s $= 5$ ms.

PRACTICE PROBLEM 5.1

Find the Nyquist rate and Nyquist interval for each of the following signals.

(a) $g(t) = 5 \cos 600\pi t + \cos 1000\pi t$.
(b) The signal $m(t)$ whose Fourier transform $M(\omega)$ is shown in Figure 5.3.

Answer: (a) 1 kHz, 1 ms. (b) 2 kHz, 0.5 ms.

5.3 ANALOG PULSE-AMPLITUDE MODULATION

Pulse modulation can be either analog or digital. Analog pulse modulation results when some attribute of a pulse varies in a one-to-one correspondence with a sample value. In analog pulse modulation, the basic idea is to use a pulse train as the carrier signal. We may choose this pulse train as square pulses, raised cosine pulses, sinc pulses or any other pulse. For simplicity, we will select square pulse trains. The parameters of the pulse train that can be varied are the amplitude,

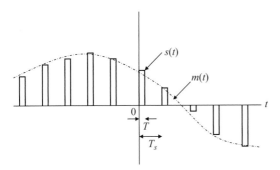

Figure 5.4. Flat-top sampling of an analog signal.

width, and position of each pulse. Varying one of these three leads to pulse-amplitude modulation (PAM), pulse-width modulation (PWM), or pulse-position modulation (PPM).

Pulse-amplitude modulation (PAM) is one in which the amplitude of individual pulses in a pulse train is varied in accordance with some characteristic of the modulating signal. In other words, the pulse train is modulated in amplitude according to the signal itself. The amplitude of the amplitude-modulated pulses conveys the information.

In PAM, the carrier signal consists of a periodic train of rectangular pulses, and the amplitudes of rectangular pulses vary according to the sample values of an analog message signal. As shown in Figure 5.4, PAM requires that we sample the message every T_s second and hold the sample for T seconds ($T < T_s$). This is known as *flat-top sampling*. The transmitted signal can be expressed as

$$s(t) = \sum_{n=-\infty}^{\infty} \underbrace{m(nT_s)}_{\text{sample}} \underbrace{h(t - nT_s)}_{\text{hold}} \tag{5.12}$$

where T_s = sampling period and $m(nT_s)$ is the sampled value of $m(t)$ obtained at $t = nT_s$; $h(t)$ is the rectangular pulse of unit amplitude and duration T, i.e.

$$h(t) = \begin{cases} 1, & 0 < t < T \\ 1/2, & t = 0, T \\ 0, & \text{otherwise} \end{cases} \tag{5.13}$$

Note that as $T \to 0$

$$\frac{1}{T} h(t) \quad \to \quad \delta(t) \tag{5.14}$$

Using the relation $h(t - nT_s) = h(t) * \delta(t - nT_s)$, the transmitted signal in Eq. (5.12) can be written as

$$s(t) = \sum_{n=-\infty}^{\infty} m(nT_s)[\delta(t - nT_s) * h(t)]$$

$$= \underbrace{\left\langle \sum_{n=-\infty}^{\infty} m(nT_s)\delta(t - nT_s) \right\rangle}_{m_{\text{sampled}}(t)} * h(t) \tag{5.15}$$

Taking the Fourier transform of this yields

$$S(\omega) = M_{\text{sampled}}(\omega)H(\omega)$$

$$= \frac{1}{T_s} \sum_{k=-\infty}^{\infty} M(\omega - \omega_s k)H(\omega) \qquad (5.16)$$

where $\omega_s = 2\pi f_s = 2\pi/T_s$ and

$$H(\omega) = T \sin c(\omega T/2)e^{-j\omega T/2} \qquad (5.17)$$

By using flat-top sampling to generate a PAM signal, we have introduced amplitude distortion and a delay of $T/2$. This distortion is called the *aperture effect*. The larger the pulse duration or aperture T, the larger the effect is.

In pulse-width modulation (PWM), sample values of the message signal are used to determine the width of the pulse signal. Thus, a PWM waveform consists of a series of pulses each having a width proportional to the sample values of the message signal. It is also known as pulse-duration modulation (PDM). With pulse-width modulation, the value of a sample of data is represented by the length of a pulse. PWM is employed in a wide variety of applications, ranging from measurement and communications to power conversion.

Pulse-position modulation (PPM) is a pulse modulation technique that uses pulses that are of uniform height and width but displaced in time from some base position according to the amplitude of the signal at the instant of sampling. PPM is also sometimes known as pulse-phase modulation. Pulse-position modulation has the advantage over pulse-amplitude modulation (PAM) and pulse-duration modulation (PDM) in that it has a lower noise immunity since all the receiver needs to do is detect the presence of a pulse at the correct time; the duration and amplitude of the pulse are irrelevant. It finds applications in wireless and optical communications.

5.4 QUANTIZATION AND ENCODING

After sampling, quantization is the next step in analog-to-digital conversion. By sampling, we transform a continuous-time signal into a discrete-time signal or sequence. The samples of the sequence can assume arbitrary values. However, in a digital implementation, the discrete-time sequence has to be represented as a digital sequence. This can be achieved through quantization.

> **Quantization** is the process of changing a continuous-amplitude signal into one with discrete amplitudes.

Given a signal $m(t)$ to be quantized, we create a new signal $m_q(t)$ (see Figure 5.5), which is an approximation to $m(t)$. Quantizers may be of the *uniform* or *nonuniform* type. In a uniform quantization, the quantization levels are uniformly spaced; otherwise the quantization is

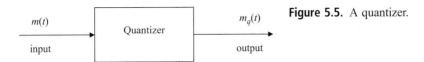

Figure 5.5. A quantizer.

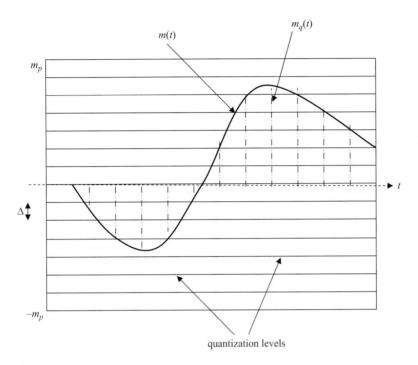

Figure 5.6. Quantization of an analog signal.

nonuniform. For uniform quantization, we assume that the amplitude of the signal $m(t)$ is confined to the range $\left(-m_p, m_p\right)$ as shown in Figure 5.6 and that the range is divided into L steps, each of step size Δ, i.e.

$$\Delta = \frac{2m_p}{L} \tag{5.18}$$

We let the amplitude of $m_q(t)$ be approximated by the midpoint value of the interval in which the sample of $m(t)$ lies. Since the quantized signal is an approximation to the original signal, the quality of the approximation can be improved by reducing the step size Δ. Thus, an analog signal can be converted to a digital one by sampling and quantizing. While sampling is chopping the signal along the time axis at the increment of T_s, quantization is slicing the signal along the amplitude at the step size of Δ.

The difference $m(t) - m_q(t)$ may be regarded as undesired signal or noise and is called the *quantization noise* $q(t)$ (or quantization error), i.e.

$$q(t) = m(t) - m_q(t) \tag{5.19}$$

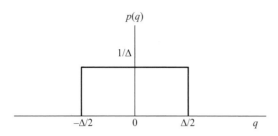

Figure 5.7. Probability density function of q.

It should be kept in mind that it is $m_q(t)$ that is rounded off, not the analog message $m(t)$. The performance of a quantizer can be characterized by the signal-to-quantization-noise ratio (SQNR) defined by

$$\text{SQNR} = 10\log_{10}\frac{\sigma^2}{D} \qquad (5.20)$$

where σ^2 is the variance of the input message $m(t)$ and D is the mean-squared quantization error. For a wide range deterministic message signal, the quantization error $q(t)$ can be assumed to be uniformly distributed over $[-\Delta/2, \Delta/2]$ as shown in Figure 5.7. Hence, the mean square error is

$$D = E[q^2] = \int_{-\infty}^{\infty} q^2 p(q)dq$$
$$= \int_{-\Delta/2}^{\Delta/2} q^2 \frac{1}{\Delta} dq = \frac{\Delta^2}{12} \qquad (5.21)$$

Substituting Eq. (5.18) into Eq. (5.21) gives

$$D = \frac{\left(\frac{2m_p}{L}\right)^2}{12} = \frac{m_p^2}{3L^2} \qquad (5.22)$$

To proceed further, we can assume that the message signal is sinusoidal and confined to the range $(-m_p, m_p)$. Then,

$$\sigma^2 = \left[\frac{m_p}{\sqrt{2}}\right]^2 = \frac{m_p^2}{2} \qquad (5.23)$$

Substituting Eqs. (5.22) and (5.23) into Eq. (5.20) gives

$$\text{SQNR} = 10\log_{10}\frac{3L^2}{2} \qquad (5.24)$$

As mentioned earlier, quantization may be nonuniform. As illustrated in Figure 5.8, the spacing between quantization levels is not uniform. The quantization levels are more closely spaced near $m(t) = 0$, but more widely spaced for the large values of $|m(t)|$ which occur infrequently. This is similar to the way a logarithm is used to view very large and very small values on the

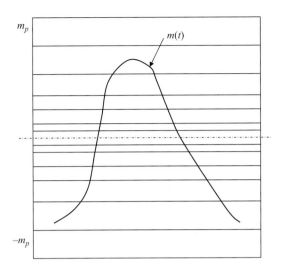

Figure 5.8. Nonuniform quantization.

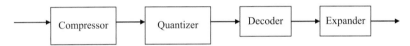

Figure 5.9. Companding process.

same set of axes. Thus, under certain conditions, nonuniform quantization may prove advantageous. For example, if the signal spends a far greater time at low levels than at higher levels, it would be expedient to have higher resolution at lower levels; the average quantization error may decrease with this approach.

The most common type of nonuniform quantization is called *companding*, which is a contraction of "compressing-expanding". The process is illustrated in Figure 5.9. The most common application of companding is in voice. All digital telephone channels employ one form of companding. Two standard companding or compression laws have been accepted by the International Telecommunication Union (ITU): the μ-law companding for North America and Japan, and the A-law companding for Europe and the rest of the world. The μ-law compander for positive amplitudes is given by

$$y(s) = \frac{\ln(1 + \mu s)}{\ln(1 + \mu)} \qquad 0 \leq s \leq 1 \tag{5.25}$$

where $s = m/m_p$ is the normalized speech signal and μ is a compression parameter often chosen as 100 or 255. The A-law compander for positive amplitudes is given by

$$y(s) = \begin{cases} \dfrac{As}{1 + \ln A}, & 0 \leq s \leq \dfrac{1}{A} \\ \dfrac{1 + \ln As}{1 + \ln A}, & \dfrac{1}{A} \leq s \leq 1 \end{cases} \tag{5.26}$$

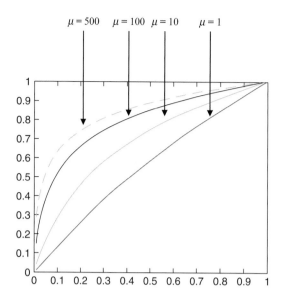

Figure 5.10. μ-law companding.

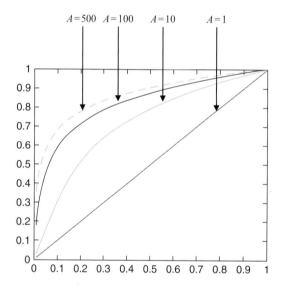

Figure 5.11. A-law companding.

where the value of $A = 87.6$ has been standardized. The functions $y(s)$ for various values of the compression parameter μ (or A) (which determines the degree of compression) are shown in Figures 5.10 and 5.11. Notice in Figure 5.10 that $y(s)$ tends to amplify small amplitudes more than larger amplitudes when $\mu > 0$. The same could be said of Figure 5.11.

Having sampled the signal and quantized it in amplitude, we arrive at a sequence of numbers which must now be encoded in binary form. Any plan for representing the discrete values as a particular arrangement of discrete events is a code. Each discrete event in the code is called a

Table 5.1. **Four-bit binary numbers**

Ordinal no.	Binary no.
0	0000
1	0001
2	0010
3	0011
4	0100
5	0101
6	0110
7	0111
8	1000
9	1001
10	1010
11	1011
12	1100
13	1101
14	1110
15	1111

codeword. The case of four bits per sample is shown in Table 5.1. We assign a distinct n binary digits (or bits) to each of the L quantization levels. Since a sequence of n bits can be arranged in 2^n distinct ways,

$$L = 2^n \quad \text{or} \quad n = \log_2 L \tag{5.27}$$

Thus, each quantized sample is encoded into n bits. For example, if there are eight quantization levels, the values can be coded into three-bit binary numbers.

Since a message signal $m(t)$ that is band-limited to W Hz requires a minimum of $2W$ samples/ s, we require a minimum channel bandwidth B Hz given by

$$B = nW \tag{5.28}$$

EXAMPLE 5.2

It is desired to design a 16-level uniform quantizer for a message signal that varies as ± 5 V. (a) Find the quantizer output value and the quantization error for the message signal amplitude of 2.2 V. (b) Determine the signal-to-quantization noise ratio.

Solution

(a) We need a uniform step size Δ. Since $L = 16$, select $\Delta = (5 + 5)/16 = 0.625$.

The amplitude 2.2 V lies between 3Δ and 4Δ so that the quantizer output is $3.5\Delta = 2.1875$.

The quantization error is $2.1875 - 2.2 = -0.0125$

(b)
$$\text{SQNR} = 10 \log_{10} \frac{3L^2}{2} = 10 \log_{10} \frac{3 \times 16^2}{2} = 25.84 \text{ dB}.$$

PRACTICE PROBLEM 5.2

Suppose we have a uniform 15-level quantizer designed for a message signal with a dynamic range of ± 10. (a) Determine the quantizer output value and the quantization error for a message signal amplitude 2.3. (b) Determine the signal-to-quantization noise ratio.

Answer: (a) 1.875, -0.425. (b) 25.84 dB.

5.5   PULSE CODE MODULATION

Pulse code modulation (PCM) is a popular digital scheme for transmitting analog data. Virtually all digital audio systems and digital telephone systems use PCM.

> **Pulse code modulation** (PCM) refers to a system in which the values of a quantized message are indicated by a series of coded pulses.

When these pulses are decoded, they indicate the standard values of the original quantized signal.

The basic operations performed in the transmitter of a PCM system are sampling, quantizing, and encoding, as shown in Figure 5.12. Among other things, the sampling enables us to use time-division multiplexing if necessary. The quantization introduces quantization noise. The encoding allows us to transmit code number rather than the sample values.

First, the message signal is sampled with a train of narrow rectangular pulses. The rate must be greater than twice the highest-frequency component of the message.

Next, the sampled version of the message is quantized. A nonuniform quantizer is often used.

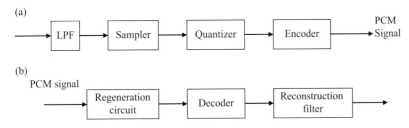

Figure 5.12. The PCM system: (a) transmitter; (b) receiver.

Last, encoding takes place. This is the process that translates the discrete set of sample values to codewords.

At the receiver end of the communications circuit, a pulse code demodulator converts the codewords (binary numbers) back into pulses having the same quantum levels as those in the modulator. These pulses are further processed to reconstruct the original analog waveform.

There are several variations of PCM. One popular variation is known as differential pulse code modulation (DPCM). DPCM is a technique for sending information about changes in the samples rather than about the sample values themselves.

Differential pulse code modulation (DPCM) is a procedure in which an analog signal is sampled and the difference between the actual value of each sample and its predicted value, derived from the previous sample or samples, is quantized, encoded, and converted to a digital signal.

The DPCM system is shown in Figure 5.13. The basic concept of DPCM is coding a difference. DPCM codewords represent differences between samples, unlike PCM, where codewords represent a sample value. If $m(k)$ is the kth sample, rather than transmitting $m(k)$, we transmit the change or difference $d(k) = m(k) - m(k-1)$. The difference between successive samples is usually smaller than the sample values themselves. This leads to fewer quantization levels and encoding with a lesser number of bits. However, DPCM employs more hardware than PCM. A typical example of a signal good for DPCM is a line in a continuous-tone (photographic) image which mostly contains smooth tone transitions. Another example would be an audio signal with a low-biased frequency spectrum.

Another variation of pulse code modulation (PCM) is the adaptive differential PCM (ADPCM), a common audio compression method. In ADPCM the quantization step size adapts to the current rate of change in the waveform which is being compressed. ADPCM requires that we only send the difference between two adjacent samples. This produces a lower bit rate and is sometimes used to effectively compress a voice signal, allowing both voice and digital data to be sent where only one would normally be sent.

(a)

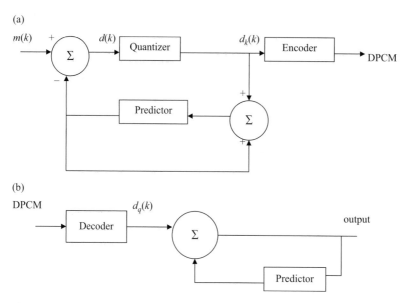

(b)

Figure 5.13. The differential PCM system: (a) transmitter; (b) receiver.

5.6   DELTA MODULATION

Delta modulation (DM) is a special case of DPCM, in which a one-bit (two-level) quantizer with magnitude $\pm\Delta$ is used. It was developed for voice telephony applications. DM output is 0 if the waveform falls in value, 1 represents a rise in value, and each bit indicates the direction in which the signal is changing (not how much), i.e. DM codes the direction of differences in signal amplitude instead of the value of the difference (DPCM).

> **Delta modulation** (DM) is a special DPCM scheme in which the difference signal $d(k)$ is encoded into just one bit.

The basic concept of delta modulation can be explained with the DM block diagram shown in Figure 5.14. Observe that the DM transmitter does not require analog-to-digital conversion. If a signal is sampled at a rate much higher than the Nyquist rate (oversampling) the correlation of adjacent samples increases. This simplifies quantization. In DM, the difference between the input signal and the latest staircase approximation is determined. Then, either a positive or a negative pulse is generated depending on the polarity of this difference.

Demodulation of DM is accomplished by integrating the output of DM to form the staircase approximation $m_q(t)$ and then passing the result through an LPF to eliminate the discrete jumps in $m_q(t)$.

The value of the step size Δ is very important in designing a DM system. Too small a step size causes sample-overload distortion, while a large value of Δ causes the modulator to follow rapid

(a)

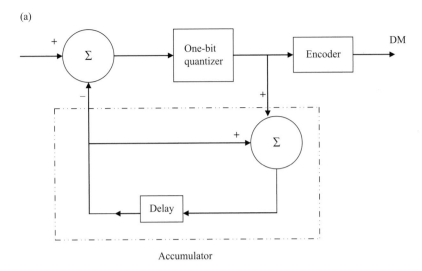

Accumulator

(b)

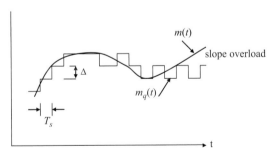

Figure 5.14. The DM system: (a) transmitter; (b) receiver.

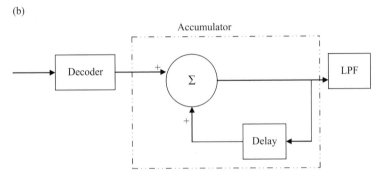

Figure 5.15. Incorrect choice of Δ causes slope overload.

change in the input signal. Figure 5.15 illustrates the consequence of incorrect step size. If Δ is too small, we get the *slope overload* condition, where the staircase approximation $m_q(t)$ cannot keep track of the rapid change in the analog signal $m(t)$. To avoid slope overload,

$$\frac{\Delta}{T_s} \geq \left|\frac{dm(t)}{dt}\right|_{max}$$

(5.29)

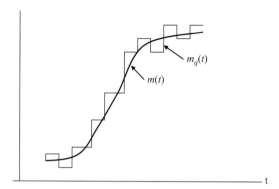

Figure 5.16. Performance of adaptive DM.

Granular noise results when the step size is too large. Granular noise is analogous to quantization noise in a PCM scheme.

There are variations of DM. One is delta-sigma modulation (DSM). This is a DM scheme which incorporates integration at its input. This is an improvement over regular DM. The integration at the input of DSM will counter the effect of the DM scheme which takes the derivative of the input sigal.

Another variation is adaptive delta modulation (ADM), in which the step size Δ is not fixed but varied according to the level of the input signal, as shown in Figure 5.16. This involves additional hardware to produce variable step sizes. By progressively increasing the step size when signal $m(t)$ is falling rapidly and causing slope overload, we allow $m_q(t)$ to catch up with $m(t)$, thereby eliminating slope overload. On the other hand, when the slope of $m(t)$ is small, reducing the step size will reduce the threshold level without increasing the granular noise (quantization noise). This leads to a greater dynamic range than for DM.

EXAMPLE 5.3

The signal $m(t) = At$ is applied to a delta modulator with step size Δ. Determine the value of A for which slope overload is avoided.

Solution

$$\frac{d}{dt}m(t) = A$$

$$\frac{\Delta}{T_s} > \left|\frac{dm(t)}{dt}\right|_{max} = A$$

In order to avoid slope overload

$$A = \frac{\Delta}{T_s}$$

PRACTICE PROBLEM 5.3

Consider $m(t) = A \cos \beta t$, where A and β are constants. Calculate the minimum step size Δ for delta modulation of this signal that is required to avoid slope overload.

Answer: $A\beta T_s$

5.7   LINE CODING

Digital data (a sequence of 1s and 0s) may be represented by several baseband data formats known as *line codes*.

> A **line code** is an assignment of a symbol or pulse to each 0 or 1 to be transmitted.

There are two basic types of line code: nonreturn-to-zero (NRZ) and return-to-zero (RZ) formats. In NRZ formats, the transmitted data bit occupies a full bit period. A long string of NRZ 1s or 0s contains no timing information since there are no level transitions. For RZ formats, the pulse width is half the full bit period. Figure 5.17 shows the waveforms of five important line codes for the binary data 1010110.

- *Unipolar nonreturn-to-zero (NRZ) signaling:* this is the simplest line code. It is also known as *on-off signaling*, where a 1 is transmitted by a pulse, while a 0 is transmitted by the absence of pulse. A major disadvantage of this line code is that there is a waste of power due to the transmitted dc level.
- *Bipolar NRZ signaling:* this is the case where a 1 is transmitted by a pulse $p(t)$, while a 0 is transmitted by pulse $-p(t)$.
- *Unipolar return-to-zero (RZ) signaling:* in this line code, a 1 is transmitted by a positive pulse that returns to zero before the end of the bit interval, while a 0 is represented by no pulse. A disadvantage of the unipolar RZ format is that long strings of 0 bits can cause loss of timing synchronization. A line code not having this problem is the Manchester line code.
- *Bipolar return-to-zero (RZ) signaling:* for this line code, a 1 is transmitted by a positive and a negative pulse alternately, with each pulse having a half-bit duration. A 0 is transmitted by no pulse.
- *Manchester (split-phase) signaling:* in this case, a 1 is transmitted by a positive pulse followed by a negative pulse so that both pulses are of equal amplitude and half-bit duration. For a 0, the polarities of the pulses are reversed. The Manchester code suppresses the dc component and this characteristic is important in some applications.

For a given application, certain factors must be considered when choosing an appropriate line code or data format. Such factors include transmission bandwidth, transparency, error detection capability, and good bit-error probability performance.

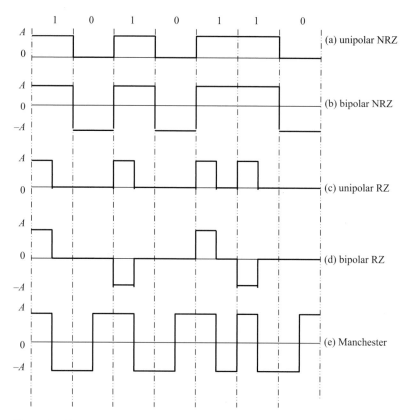

Figure 5.17. Line codes for representing binary data.

5.8   MULTIPLEXING

When the transmission channel (whether it is twisted pair, coaxial cable, optical fiber, or free space) will support a much larger bandwidth than needed for a single baseband message, it is desirable to transmit several message signals over the single channel. This is accomplished by multiplexing.

> **Multiplexing** is a technique that allows two or more messages to be transmitted over a single communication channel.

Signal multiplexing allows the transmission of numerous unrelated signals on the same channel.

The two basic multiplexing techniques are frequency-division multiplexing (FDM) and time-division multiplexing (TDM). In a sense, FDM and TDM are duals of each other. In FDM the channel frequency band is divided into smaller channels, whereas in TDM the users take turns using the entire channel for a short period. In other words, all signals operate at the same time with different frequencies in FDM, while all signals use the same frequencies but operate at different times in TDM. FDM may be used with either analog or digital signal transmission, while TDM is usually employed in the transmission of digital information because TDM can be used with analog pulse modulation systems.

5.8.1 Frequency-division multiplexing

Frequency division multiplexing (FDM) is a scheme that transmits multiple signals simultaneously over a single transmission path, such as a cable or wireless system, because subchannels can be AM, DSB, SSB, FM, ..., even a digital signal. Each signal is translated to a subchannel frequency band within the main channel. Although single sideband (SSB) modulation is widely used, any type of modulation can be used in FDM provided that the carrier spacing is sufficient to avoid spectra overlap. Numerous signals are combined to form a composite signal which is transmitted on a single communication line or channel, as typically illustrated in Figure 5.18. The figure shows the FDM of N message signals at the transmitter and their demodulation at the receiver. Each signal modulates a separate carrier so that N modulators are needed. The carriers used to frequency translate the individual signals are often referred to as *subcarriers*. The signals from the N modulators are summed and transmitted over the channel. The subchannels in the composite signal are extracted using bandpass filters. The multiplexing operation has assigned a slot in the frequency domain for each individual message. The composite signal must consist of modulated signals without overlapping spectra. In other words, the modulated message spectra are spaced out in frequency by *guard bands* (vacant bands of frequencies used to prevent channel interference), as shown in Figure 5.19. Otherwise crosstalk (a major problem of FDM) will occur.

At the receiver, the signals are separated (demultiplexed) by bandpass filters. The output of each bandpass filter is demodulated. Each demodulated signal is fed to a low-pass filter that passes the baseband message signal.

If W_k is the bandwidth of the kth frequency channel (including the guard bands), the bandwidth of the composite FDM is

$$B = \sum_{k=1}^{N} W_k \tag{5.30}$$

where N is the number of multiplexed signals.

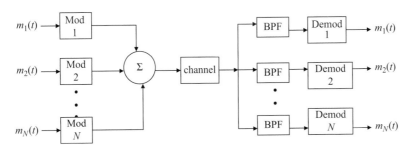

Figure 5.18. Frequency-division multiplexing.

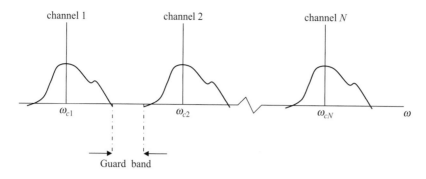

Figure 5.19. Spectrum of composite signal of the output of an FDM transmitter.

FDM is widely used in radio, TV, telephone systems, telemetry, and communication net-works. In the long-distance telephone, up to 600 voice signals are transmitted over a coaxial cable with carrier frequencies spaced 4 kHz apart. Radio and TV broadcasts employ free space (atmosphere) as the transmission channel. For example, each radio station within a certain broadcast area is assigned a frequency so that many independent channels can be sent.

Where frequency-division multiplexing is used as to allow multiple users to share a physical communication channel, it is called *frequency-division multiple access* (FDMA). FDMA is used in wireless communications. Each FDMA subscriber is assigned a specific frequency channel. No one else in the same cell or a neighboring cell can use the frequency channel while it is assigned to a user. This reduces interference, but severely limits the number of users.

5.8.2 Time-division multiplexing

One important application of sampling in communication systems is time-division multiplexing (TDM). Since the sampled waveform is "off" most of the time, the time between samples can be used for other purposes. In TDM, sample values from different signals are interlaced into a single waveform. In other words, several signals time-share the same channel. TDM is widely used in telephony, telemetry, radio, and data processing.

The simplified concept of TDM is shown in Figure 5.20. The sources are assumed to be sampled at the Nyquist rate or higher. The commutator (a time-sequential sampler) interlaces the samples to form a baseband signal. At the other end of the channel, the baseband signal is demultiplexed by the decommutator (a time-sequential distributor or sampler), which should be well synchronized with the communicator for proper operation.

We can use the Nyquist sampling theorem to determine the minimum bandwidth of a TDM system. Assume that there are N signals to be multiplexed and that each has a bandwidth of W. By the Nyquist theorem, each signal must be sampled no less than $2W$ times per second. In time interval of T, the total number of baseband samples is

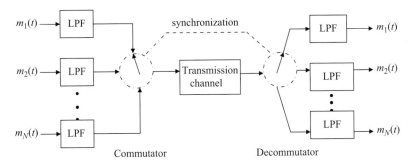

Figure 5.20. Time-division multiplexing.

$$n_s = \sum_{k=1}^{N} 2WT \qquad (5.31)$$

If the composite signal is a baseband signal of bandwidth B, the required sample rate is $2B$. In the time interval of T, we have $2BT$ samples so that

$$n_s = 2BT = \sum_{k=1}^{N} 2WT \qquad (5.32)$$

Hence,

$$B = \sum_{k=1}^{N} W = NW \qquad (5.33)$$

which is the same minimum required bandwidth obtained for FDM.

A standard TDM hierarchy has been established for digital speech transmission over telephone lines. Unfortunately, the standards adopted by North America and Japan are different from those adopted by the rest of the world. The North American hierarchy is shown in Table 5.2 and illustrated in Figure 5.21. The hierarchy starts at 64 kbps, which corresponds to PCM representation of a voice signal. This rate is known as the *digital signal zero* (DS-0). It is the fundamental building block.

- In the first level of the TDM hierarchy, 24 DS-0 bit streams are combined to form a *digital signal one* (DS-1) at 1.544 Mbps, which is 24 × 64 kbps plus a few extra bits for control purposes. The line over which the service is provided is called a T-1 line.
- In the second level of TDM, four DS-1 bit streams are combined to obtain *digital signal two* (DS-2) at 6.312 Mbps. The line handling this service is known as a T-2 line.
- In the third level of hierarchy, seven DS-2s are combined to form *digital signal three* (DS-3) at 44.736 Mbps. The line for which this service is provided is called a T-3 line.
- In the fourth level of hierarchy, six DS-3 bit streams are combined to form *digital signal four* (DS-4) at 274.176 Mbps. The corresponding line is called a T-4 line.
- In the fifth-level multiplexer, two DS-4s are combined to obtain *digital signal five* (DS-5) at 560.16 Mbps.

Table 5.2. **TDM standards for North America**

Digital signal number	Bit rate (Mbps)	No. of 64 kbps channel	Transmission media
DS-0	0.064	1	Twisted pairs
DS-1	1.544	24	Twisted pairs
DS-2	6.312	96	Twisted pairs, fiber
DS-3	44.736	672	Coax, air, fiber
DS-4	274.176	4032	Coax, fiber
DS-5	560.16	8064	Coax, fiber

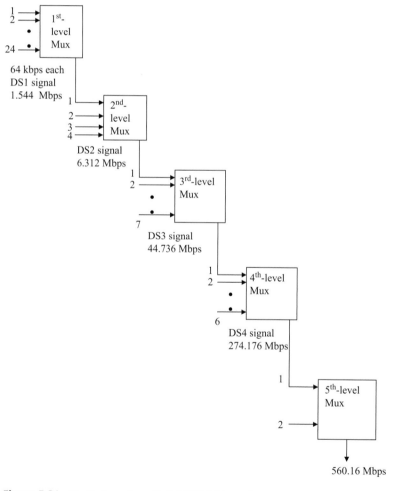

Figure 5.21. North American digital TDM hierarchy.

When time-division multiplexing is used as to allow multiple users to share a physical communications channel, it is called *time-division multiple access* (TDMA). TDMA is a digital transmission technology that allows a number of users to access a single channel without interference by allocating unique time slots to each user within each channel. The TDMA digital transmission scheme multiplexes three signals over a single channel. TDMA is employed in digital cellular telephones. The current TDMA standard for cellular divides each cellular channel into three time slots in order to increase the amount of data that can be carried. Each caller is assigned a specific time slot for transmission.

EXAMPLE 5.4

Three signals $m_1(t)$, $m_2(t)$, and $m_3(t)$ are respectively band-limited to 4.8 kHz, 1.6 kHz, and 1.6 kHz. It is desired that these signals be transmitted by time-division multiplexing. (a) Assuming that each signal is sampled at the Nyquist rate, set up a scheme for multiplexing them. (b) What must the speed of the communicator be in samples per second? (c) What is the minimum bandwidth of the channel? (d) If the commutator output is quantized and encoded with $L = 512$, what is the output bit rate?

Solution

(a) The Nyquist rates for the three signals are shown in Table 5.3. In one rotation of the commutator, we obtain one sample from each of $m_2(t)$ and $m_3(t)$ and three samples from $m_1(t)$. This implies that the commutator must have at least five poles connected to the signals as shown in Figure 5.22.

(b) From the Nyquist rates, $m_1(t)$ has 9600 samples/s, while each of $m_2(t)$ and $m_3(t)$ has 3200 samples. Thus, there is a total of 16,000 samples/s. The commutator's speed must be at least 16,000 samples/s.

(c) The mimimum channel bandwidth W is given by

$$2W = 9.6 + 3.2 + 3.2 = 16 \text{ kHz}$$

$$\text{or } W = 8 \text{ kHz}$$

(d) Since $L = 512 = 2^9 = 2^n$, the output bit rate is

$$n(16,000) = 144,000 \text{ bps} = 144 \text{ kbps}$$

Table 5.3. **For Example 5.4**

Message signal	Bandwidth	Nyquist rate
$m_1(t)$	4.8 kHz	9.6 kHz
$m_2(t)$	1.6 kHz	3.2 kHz
$m_3(t)$	1.6 kHz	3.2 kHz

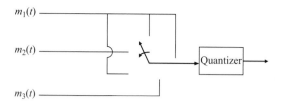

Figure 5.22. TDM scheme; for Example 5.4.

PRACTICE PROBLEM 5.4

Repeat Example 5.4 if a fourth signal $m_4(t)$ that is band-limited to 1.6 kHz is added.

Answer: (a) Commutator has six poles. (b) 19,200 samples/s. (c) 9.6 kHz. (d) 172.8 kps.

5.9 APPLICATIONS

The concepts covered in this chapter have a lot of practical applications. Such applications include analog-to-digital converters, digital audio (CD) recording, digital telephony, digital loop carriers, analog time-division switching, and channel vocoders. We will consider two of them.

5.9.1 Analog-to-digital conversion

It is often advantageous to represent analog signals in digital form. The transformation from an analog to a digital form is performed by an analog-to-digital converter (A/D or ADC).

> An **analog-to digital converter (ADC)** is a circuit that samples a continuous analog signal at specific points in time and converts the sampled values into a digital representation.

At the other end, we may need to transform the digital signal to analog form by the reverse process known as *digital-to-analog conversion* (DAC). An overwhelming variety of ADCs exist on the market today, with differing resolutions, bandwidths, accuracies, architectures, packaging, power requirements, and temperature ranges due to a variety of applications in data-acquisition, communications, instrumentation, and interfacing for signal processing, all having a host of differing requirements. There are different ways of realizing analog-to-digital conversion. These include integration, successive approximation, parallel (flash) conversion, delta modulation, pulse code modulation (PCM), and sigma-delta modulation. The successive approximation (SAR) architectures are mainly used in applications requiring high resolution and low conversion speed. The flash ADCs are very fast with typical sampling speeds ranging from 20 to 800 Megasamples/s; however, the device power consumption is high compared with

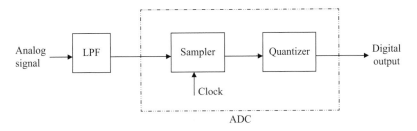

Figure 5.23. Analog-to-digital converter of a speech signal.

a SAR. The sigma-delta ADC architecture is capable of much higher resolution than the SAR. We will consider the PCM method here.

The general representation of an ADC of a speech signal is shown in Figure 5.23. As shown in Figure 5.23, the quantizing and encoding operations are often performed in the same circuit, known as the analog-to-digital converter. Since the frequency content of speech signals is limited to below 3400 Hz, the signal is first passed through a low-pass filter (LPF) before sampling. The sampling rate (8000 Hz) must be substantially higher than the highest signal frequency to avoid spurious patterns (aliasing). The analog samples are then quantized and encoded for transmission over telephone channels. The objective is to change a continuous signal into one that has discrete values. The resolution of the ADC indicates the number of discrete values it can produce. It is usually expressed in bits. For example, an ADC that encodes an analog input to one of 256 discrete values has a resolution of 8 bits since $2^8 = 256$.

5.9.2 Speech coding

In our daily communications, we encounter a lot of audio signals. The transmission of these audio signals on the telephone lines requires digitization. Speech digitization is usually accomplished in the telephone network by devices known as codecs (coder/decoders). The codecs transform the analog speech into 8-bit PCM form. Using time-division multiplexing, the channel banks combine many PCM channels into a single data stream. Standard digital sound is stored as PCM data. The audio signal is measured at a fixed sampling rate with a given precision (or resolution). On a compact disc (CD), for example, each stereo channel is measured 44,100 times a second with a precision of 16 bits, resulting in 705,600 bps. Decreasing this rate can be achieved by lowering the sampling rate or the resolution. But the best idea is to lower both values. Of course, there is a best compromise between the two.

The speech coder (which is part of a codec) can be classified into waveform coders and vocoders. The waveform coder employs algorithms to encode and decode speech such that the system output is the approximation to the input waveform. The vocoder encodes speech by extracting a set of parameters that are digitized and transmitted to the receiver. In this encoding, the spectrum of the input speech is divided up into 15 frequency bands each of bandwidth 200 Hz. The outputs of the 16 20-Hz low-pass filters are sampled, multiplexed, and A/D converted. If the sampling is at the Nyquist rate of 40 samples/s (signals of 20 Hz bandwidth) and if we employ 3 bits/sample to represent each sample, the bit rate is

$$R = 40 \frac{\text{samples/s}}{\text{filter}} \times 16 \text{ filters} \times 3 \text{ bits/sample}$$
$$= 1.9 \times 10^3 \text{ bits/s}$$

(5.34)

Many Motion Picture Experts Group (MPEG) standards have been developed for audio coding and decoding. MPEG-1 is the audio coding standard; MPEG-2 is for video and audio compression; MPEG-4 targets interactive multimedia applications and it is the multimedia compression standard for interactive applications and services. The different parts of the MPEG-7 standard collectively and individually offer a comprehensive set of multimedia description tools. MPEG-7 brings new challenges to the system expert, such as language for description representation, binary representation of description, and delivery of description either separately or jointly with the audio-visual content.

Summary

1. Sampling is the process by which a continuous-time signal is processed by measuring its magnitude at discrete time instants. If the signal is band-limited to W, it can be perfectly completely recovered from the samples if it is sampled at no less than the Nyquist rate of $f_s = 2W$.

2. Pulse-amplitude modulation (PAM) is a method in which the amplitude of the individual pulse train is varied in accordance with the signal itself. It is basically a sample-and-hold operation.

3. Quantization is a rounding-off process and it introduces an error known as quantization noise.

4. Encoding is the process of translating a discrete set of sample values into codewords.

5. Pulse code modulation (PCM) occurs when the message signal is sampled, quantized, and encoded as a sequence of binary symbols.

6. Delta modulation (DM) is usually regarded as "one-bit PCM". It is an analog-to-digital signal conversion operation in which the analog signal is approximated with a series of segments, each segment being compared to the original analog wave to indicate an increase or decrease in amplitude.

7. Multiplexing is the scheme whereby a number of separate message signals are combined together for transmission over a common channel. There two basic types: FDM and TDM. Frequency-division multiplexing (FDM) is a form of signal multiplexing where multiple baseband signals are translated to non-overlapping frequency bands and added together to create a composite signal. Time-division multiplexing (TDM) is the time-interleaving of signals from different sources so that the information from these sources can be transmitted over a single channel.

8. Two application areas of analog-to-digital converters (ADCs) and speech coding are considered. The ADC is a device that converts continuous signals to discrete values. Speech coding employs algorithms to encode and decode speech such that the system output is the approximation to the input waveform.

Review questions

5.1 A signal can be band-limited and time-limited simultaneously.
(a) True.   (b) False.

5.2 A signal is band-limited to 20 kHz. To guarantee recovery, the minimum sampling rate is:
(a) 10 kHz.   (b) 20 kHz.   (c) 40 kHz.   (d) Cannot be determined.

5.3 In telephone companies in the United States, Canada, and Japan, $\mu = 255$ is used.
(a) True.   (b) False.

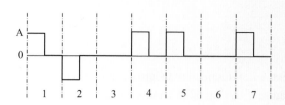

Figure 5.24. For Review question 5.9.

5.4 What causes slope overload in delta modulation?
(a) Large step size Δ.   (b) Small step size Δ.   (c) Large sampling period T_s.
(d) Small sampling period T_s.

5.5 In a delta signal modulation (DSM), the integration is performed before delta modulation.
(a) True.   (b) False.

5.6 The DS-2 consists of how many 64 kbps channels?
(a) 6.   (b) 24.   (c) 48.   (d) 96.

5.7 How many DS-3 data streams make DS-4?
(a) 2.   (b) 3.   (c) 4.   (d) 6.   (e) 10.

5.8 An ADC has 6-bit resolution. How many discrete values can it encode?
(a) 6.   (b) 12.   (c) 64.   (d) 128.

5.9 A bipolar RZ representation of a binary sequence is shown in Figure 5.24. The waveform has an error. The position of the error is:
(a) 1.   (b) 2.   (c) 4.   (d) 5.   (e) 7.

5.10 Which of these is not an analog modulation scheme?
(a) PAM.   (b) PPM.   (c) PCM.   (d) PWM.

Answers: 5.1 b, 5.2 c, 5.3 a, 5.4 b, 5.5 a, 5.6 d, 5.7 d, 5.8 c, 5.9 d, 5.10 c.

Problems

Section 5.2 Sampling

5.1 A signal is band-limited with a bandwidth of $W = 2400$ Hz. To guarantee a guard band of 800 Hz, what should the sampling rate be?

5.2 Given that $x(t)$ is band-limited, i.e. $X(\omega) = 0$ for $|\omega| > \omega_m$, show that

$$\int_{-\infty}^{\infty} [x(t)]^2 dt = T_s \sum_{n=-\infty}^{\infty} [x(nT_s)]^2$$

where $T_s = \pi/\omega_m$.

5.3 A signal $m(t)$ with bandwidth of 40 Hz is sampled at 25% above the Nyquist rate. If the resulting sampled values are

$$m(nT_s) = \begin{cases} -1, & -3 \leq n \leq 0 \\ 1, & 0 < n \leq 4 \\ 0, & \text{otherwise} \end{cases}$$

evaluate $m(t)$ at $t = 0.0125$ s.

Section 5.4 Quantization and encoding

5.4 A signal has a bandwidth of 2.4 MHz. If it is transmitted using a PCM with $L = 128$, determine the transmission bandwidth.

5.5 (a) Show that each additional bit of quantization increases the SQNR by roughly 6 dB.
(b) Obtain the SQNR for an 8-bit quantizer.

Section 5.6 Delta modulation

5.6 Given that a delta modulation has

$$m(t) = 3 \cos 40\pi t + 4 \cos 60\pi t$$

calculate the minimum sampling frequency required to prevent slope overload. Assume that $\Delta = 0.02\pi$.

5.7 A 2-kHz sinusoidal input for a DM system is sampled at five times the Nyquist rate with $\Delta = 0.15$. What is the maximum amplitude of the sinusoidal input to prevent slope overload?

Section 5.7 Line coding

5.8 (a) Using bipolar RZ, draw the pulse train for the data stream 011011101.
(b) Repeat (a) using Manchester coding.

5.9 Draw the pulse train for the data stream 1101000110 using the following line codes:
(a) unipolar NRZ   (b) bipolar NRZ   (c) unipolar RZ
(d) bipolar RZ   (e) Manchester

5.10 The 3B4B encoding scheme converts blocks of three binary digits into blocks of four binary digits according to the rule in Table 5.4. For consecutive blocks of three 1s, coded blocks 1011 and 0100 are used alternately. Similarly, for consecutive blocks of three 0s, the coded blocks 0010 and 1101 are used alternately. Find the coded stream for data 100000111111000000011010.

Section 5.8 Multiplexing

5.11 Four data channels each with an 8 kHz baseband bandwidth are to be frequency-division multiplexed. Channel 1 will be at baseband. A guard band of 25% of the bandwidth of channel 1 will be maintained between the upper edge of channel 1 and the lower edge of channel 2. In the same way, a guard band equal to 25% of the bandwidth of channel 2 will be maintained between the upper edge of channel 1 and the lower edge of channel 3, and so forth. Draw a diagram of the composite spectrum.

Table 5.4. **For Problem 5.10**

Original code	3B4B code	
	Mode 1	Mode 2
000	0010	1101
001	0011	
010	0101	
011	0110	
100	1001	
101	1010	
110	1100	
111	1011	0100

5.12 Repeat the previous problem if the four channels have the following baseband bandwidths:

Channel 1: 8 kHz
Channel 2: 10 kHz
Channel 3: 12 kHz
Channel 4: 16 kHz

5.13 (a) How much percent overhead is added to DS-1?
(b) How much percent overhead is added to DS-2?
(c) How much percent overhead is added to DS-3?

5.14 A technique known as bit stuffing is commonly used in data communications. Whenever the transmitter notices five consecutive 1s in the data, it inserts a 0 bit. The receiver does the reverse operation. Determine the data stream appearing on the following data sequences after bit stuffing.
(a) 0101111111111101   (b) 101111110111111100

5.15 The T-1 eight-bit PCM system multiplexes 24 voice channels. One framing bit is added at the beginning of each T-1 frame for synchronization purposes. The voice is sampled at 8 kHz. Determine: (a) the duration of each bit; (b) the resultant transmission rate; (c) the minimum required bandwidth.

5.16 Twenty baseband channels, each band-limited to 2.4 kHz, are sampled and multiplexed at a rate of 6 kHz. What is the required bandwidth for transmission if the multiplexed samples use a PCM system?

5.17 Time-division multiplexing is used to transmit two signals $m_1(t)$ and $m_2(t)$. The highest frequency of $m_1(t)$ is 4 kHz, while that of $m_2(t)$ is 3.2 kHz. Determine the minimum allowable sampling rate.

5.18 The outputs of 30 modems (modulator-demodulators) operating at 33.6 kbps each are time-division multiplexed onto a T-1 line. Assuming that 45 kbps is used for overhead, what percentage of the line is unused?

5.19 Eight digital lines, with each having a maximum output of 56 kbps, are combined onto a single 360-kbps line using statistical TDM. (a) How many lines can transmit simultaneously with full capacity? (b) If all the eight lines are transmitting, by how much is the data rate of each line reduced?

Section 5.9 Applications

5.20 On a compact disc (CD), each stereo channel is sampled 44,100 times a second. If the samples are quantized with a 16 bits/sample quantizer, find the number of bits in a piece of music 10 minutes long.

6 Probability and random processes

Nothing moves without a mover.

ISAAC NEWTON

TECHNICAL NOTES - Ethics

The electrical engineering curriculum in most accredited engineering schools in this country is so crowded that there is no room for courses such as engineering ethics, economics, or law. Part of the problem is that most engineering professors are not prepared to introduce engineering ethics into their classrooms.

The days when an engineer's only ethical commitment was loyalty to his or her employer have long passed. Society in general tends to hold the engineering profession to an elevated standard and expects practicing engineers to perform on a higher ethical plane. Society holds engineers accountable for their actions. And for engineers, the implications are inescapable.

As members of the engineering profession, engineers are expected to exhibit the highest standards of honesty and integrity. Engineering has a direct and vital impact on the quality of life for all people. Engineers must perform under a standard of professional behavior that requires adherence to the highest principles of ethical conduct. The ability to discern right from wrong in cases of apparent ethical dilemma is important. With ethics, there is frequently no absolute right answer, just a personal best answer. Ethics poses dilemmas, which force hard moral choices and cause us to deal with values.

Each professional engineering society has some form of ethical code. Here is a combination of some.

- Accept personal responsibility consistent with the safety, health, and welfare of the public.
- Conduct themselves responsibly, ethically, and lawfully so as to enhance the honor, reputation, and usefulness of the profession.
- Be guided in all relations by the highest standards of honesty and integrity.
- Reject bribery in all its forms.
- Perform professional services only in areas of your competence.
- Conform with state registration laws in the practice of engineering.
- Treat fairly all colleagues and co-workers regardless of their race, religion, gender, age, or national origin.

6.1   INTRODUCTION

So far in this book, we have been dealing with deterministic signals, i.e. signals which can be expressed explicitly so that they can be determined at all times. However, most of the signals we deal with in practice are random (unpredictable or erratic) and not deterministic. Random signals are encountered in one form or another in every practical communication system. For example, speech signals and television signals are random and cannot be described by equations. In fact, randomness is crucial to communication for at least two reasons. First, the very signal that is transmitted is random. Otherwise, the receiver at the end of the channel would know in advance the transmitted message and there would be no need for communication. In other words, signals must be random to convey information. Second, transmitted signals are always corrupted by noise, which is also a random signal. Thus, random signals occur in communication both as information-conveying signal and as unwanted noise signal.

A **random quantity** is one having values which are regulated in some probabilistic way.

Thus, our work with random signals and random noise must begin with the theory of probability, which is the mathematical discipline that deals with the statistical characterization of random signals and random processes.

Although the reader is expected to have had at least one course on probability theory and random variables, this chapter provides a cursory review of the basic concepts needed throughout this book. The concepts include probabilities, random variables, statistical averages or mean values, and probability models. In this chapter, we will also discuss random processes, their properties, and the basic tools used for their mathematical analysis. We also consider how the concepts developed can be applied to linear systems and demonstrated using MATLAB. A reader already versed in these concepts may skip this chapter.

6.2   PROBABILITY FUNDAMENTALS

A fundamental concept in the probability theory is the idea of an *experiment*. An experiment (or trial) is the performance of an operation that leads to results called *outcomes*. In other words, an outcome is a result of performing the experiment once. An *event* is one or more outcomes of an experiment. The relationship between outcomes and events is shown in the Venn diagram of Figure 6.1. Thus:

An **experiment** consists of making a measurement or observation.

An **outcome** is a possible result of the experiment.

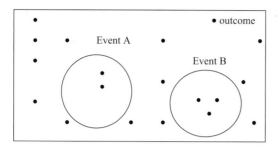

Figure 6.1. Sample space illustrating the relationship between outcomes (points) and events (circles).

An **event** is a collection of outcomes.

An experiment is said to be *random* if its outcome cannot be predicted with certainty. Thus a random experiment is one that can be repeated a number of times but yields unpredictable outcomes at each trial. Examples of random experiments are tossing a coin, rolling a die, observing the number of cars arriving at a toll booth, and keeping track of the number of telephone calls at your home. If we consider the experiment of rolling a die and regard event A as the appearance of the number 4: that event may or may not occur for every experiment.

6.2.1 Simple probability

We now define the probability of an event. The probability of event A is the number of ways event A can occur divided by the total number of possible outcomes. Suppose we perform n trials of an experiment and we observe that outcomes satisfying event A occur n_A times. We define the probability $P(A)$ of event A occurring as

$$P(A) = \lim_{n \to \infty} \frac{n_A}{n} \tag{6.1}$$

This is known as the *relative frequency* of event A. Two key points should be noted from Eq. (6.1). First, we note that the probability P of an event is always a positive number and that

$$0 \leq P \leq 1 \tag{6.2}$$

where $P = 0$ when an event is not possible (never occurs) and $P = 1$ when the event is certain (always occurs). Second, observe that for the above definition of probability to be meaningful, the number of trials n must be large.

If events A and B are disjoint or mutually exclusive, it follows that the two events cannot occur simultaneously or that the two events have no outcomes in common, as shown in Figure 6.2. In this case, the probability that either event A or B occurs is equal to the sum of their probabilities, i.e.

$$P(A \text{ or } B) = P(A) + P(B) \tag{6.3}$$

Event A   Event B   **Figure 6.2.** Mutually exclusive or disjoint events.

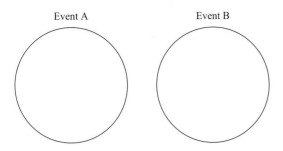

To prove this, suppose that, in an experiment with n trials, where $n \to \infty$, event A occurs n_A times, while event B occurs n_B times. Then event A or event B occurs $n_A + n_B$ times and

$$P(A \text{ or } B) = \frac{n_A + n_B}{n} = \frac{n_A}{n} + \frac{n_B}{n} = P(A) + P(B) \tag{6.4}$$

This result can be extended to the case where all possible events in an experiment are A, B, C, ..., Z, which are mutually exclusive. If the experiment is performed n times and event A occurs n_A times, event B occurs n_B times, etc. Since some event must occur at each trial,

$$n_A + n_B + n_C + \cdots + n_Z = n$$

Dividing by n and assuming n is very large, we obtain

$$P(A) + P(B) + P(C) + \cdots + P(Z) = 1 \tag{6.5}$$

which indicates that the probabilities of mutually exclusive events must add up to unity. A special case of this is when two events are complementary, i.e. if event A occurs, B must not occur and vice versa. In this case,

$$P(A) + P(B) = 1 \tag{6.6}$$

or

$$P(A) = 1 - P(B) \tag{6.7}$$

For example, in tossing a coin, the event of a head occurring is complementary to that of tail occurring. Since the probability of either event is $\frac{1}{2}$, their probabilities add up to 1.

Next, we consider when events A and B are not mutually exclusive. Two events are non-mutually exclusive if they have one or more outcomes in common, as illustrated in Figure 6.3. The probability of the union event A or B (or A + B) is

$$P(A + B) = P(A) + P(B) - P(AB) \tag{6.8}$$

where P(AB) is called the *joint probability* of events A and B, i.e. the probability of the intersection or joint event AB.

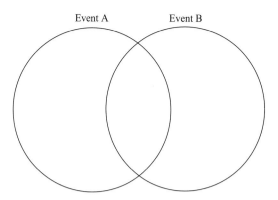

Event A   Event B

Figure 6.3. Non-mutually exclusive events.

6.2.2 Conditional probability

Sometimes we are confronted with a situation in which the outcome of one event depends on another event. The dependence of event B on event A is measured by the *conditional probability P(B/A)* given by

$$P(B/A) = \frac{P(AB)}{P(A)} \tag{6.9}$$

where $P(AB)$ is the joint probability of events A and B. The notation B/A stands for "*B given A*". If events A and B are mutually exclusive, the joint probability $P(AB) = 0$ so that the conditional probability $P(B/A) = 0$. Similarly, the conditional probability of A given B is

$$P(A/B) = \frac{P(AB)}{P(B)} \tag{6.10}$$

From Eqs. (6.9) and (6.10), we obtain

$$P(AB) = P(B/A)P(A) = P(A/B)P(B) \tag{6.11}$$

Eliminating $P(AB)$ gives

$$P(B/A) = \frac{P(B)P(A/B)}{P(A)} \tag{6.12}$$

which is a form of *Bayes' theorem.*

EXAMPLE 6.1

Three coins are tossed simultaneously. Find: (a) the probability of getting exactly two heads; (b) the probability of getting at least one tail.

Solution

If we denote HTH as a head on the first coin, a tail on the second coin, and a head on the third coin, the $2^3 = 8$ possible outcomes of tossing three coins simultaneously are the following:

HHH, HTH, HHT, HTT, THH, TTH, THT, TTT

The problem can be solved in several ways

Method 1 (intuitive approach)

(a) Let event A correspond to having exactly two heads, then

$$\text{Event } A = \{HHT, HTH, THH\}$$

Since we have eight outcomes in total and three of them are in event A, then

$$P(A) = 3/8 = 0.375$$

(b) Let B denote having at least one tail,

$$\text{Event } B = (HTH, HHT, HTT, THH, TTH, THT, TTT)$$

Hence,

$$P(B) = 7/8 = 0.875$$

Method 2 (analytic approach). Since the outcome of each separate coin is statistically independent, with head and tail equally likely,

$$P(H) = P(T) = \tfrac{1}{2}$$

(a) Event consists of mutually exclusive outcomes. Hence,

$$P(A) = P(HHT, HTH, THH)$$

$$= \left(\frac{1}{2}\right)\left(\frac{1}{2}\right)\left(\frac{1}{2}\right) + \left(\frac{1}{2}\right)\left(\frac{1}{2}\right)\left(\frac{1}{2}\right) + \left(\frac{1}{2}\right)\left(\frac{1}{2}\right)\left(\frac{1}{2}\right) = \frac{3}{8} = 0.375$$

(b) Similarly,

$$P(B) = (HTH, HHT, HTT, THH, TTH, THT, TTT)$$

$$= \left(\frac{1}{2}\right)\left(\frac{1}{2}\right)\left(\frac{1}{2}\right) + \text{in seven places} = \frac{7}{8} = 0.875$$

PRACTICE PROBLEM 6.1

Suppose two coins are tossed simultaneously. (a) List the possible outcomes. (b) What is the probability of obtaining exactly two tails? (c) Find the probability of getting at least one head.

Answer: (a) HH, HT, TH, TT. (b) 0.25. (c) 0.75.

EXAMPLE 6.2

In the switch shown in Figure 6.4, let A, B, C, and D correspond to events that switches a, b, c, and d are closed respectively. Assume that each switch acts independently but closes with probability $p = 0.5$. (a) Find the probability that the closed path exists between the end terminals. (b) Determine the probability that a closed path does not exist between the end terminals.

Table 6.1. **For Example 6.2**

Switches	Switches	
abcd	abcd	
OOOO	CCCC	←
OOOC	CCCO	←
OOOC	CCCO	←
OOCO	CCOC	←
OOCC	CCOO	←
OCOO	COCC	←
OCOC	COCO	←
OCCO	COOC	←
OCCC	COOO	

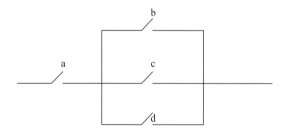

Figure 6.4. For Example 6.2; a switching circuit.

Solution

Again, this problem can be done in two ways.

Method 1 (intuitive approach). Since a switch can close (C) or open (O) and we have four switches, we have $2^4 = 16$ possible outcomes. These are shown in Table 6.1, where OCOO, for example, corresponds to switch a being open, switch b closed, and switches c and d open.

(a) In order to have a closed path, switch a must be closed and one other switch (either a, b, or c) must be closed. The outcomes that satisfy this requirement are indicated by arrows in Table 6.1, i.e. 7 out 16 equally likely outcomes. Hence.

$$P(\text{closed path}) = 7/16 = 0.4375$$

(b) Other 9 outcomes in Table 6.1 will not form a closed path. Hence,

$$P(\text{no closed path}) = 9/16 = 0.5625$$

Method 2 (analytic approach)

(a) Let $P(A)$, $P(B)$, $P(C)$, and $P(D)$ be the probabilities that switches a, b, c, and d are closed respectively. They are each $p = 0.5$. It is evident from Figure 6.4 that for a closed path to exist, switch a must be closed and either switch b, c, or d must be closed. We apply the ideas of joint probability and statistical independence.

$$P(\text{closed path}) = P[(B + C + D)A] = P(B + C + D)P(A)$$

It can be shown that

$$P(B + C + D) = P(B) + P(C) + P(D) - P(BC) - P(BD) - P(CD) + P(BCD)$$
$$= p + p + p - p^2 - p^2 - p^2 + p^3 = p(3 - 3p + p^2)$$

Hence,

$$P(\text{closed path}) = P(B + C + D)P(A)$$
$$= p^2(3 - 3p + p^2) = \frac{1}{4}\left(3 - \frac{3}{2} + \frac{1}{4}\right) = \frac{7}{16} = 0.4375$$

(b) Since the event of having a closed path and that of having no closed path are disjoint or mutually exclusive, Eq. (6.7) applies, i.e.
$P(\text{no closed path}) = 1 - P(\text{closed path}) = 1 - 0.4375 = 0.5625$,
which agrees with what we got earlier.

PRACTICE PROBLEM 6.2

In an experiment, two dice are tossed and the sum is noted. Find: (a) the probability of getting 9; (b) the probability of the sum being greater than 9; (c) the probability of the sum being either 6, 7, or 8.

Answer: (a) 1/9. (b) 1/6. (c) 4/9.

6.2.3 Random variables

Random variables are used in probability theory for at least two reasons. First, the way we have defined probabilities earlier in terms of events is awkward. We cannot use that approach for phenomena that are not susceptible to discrete outcomes and duplication. It is preferable to have numerical values for all outcomes. Second, mathematicians and communication engineers in particular deal with random processes that generate numerical outcomes. Such processes are handled using random variables.

The term "random variable" is a misnomer; a random variable is neither random nor a variable. Rather, it is a function or rule that produces numbers from the outcome of a random experiment. In other words, for every possible outcome of an experiment, a real number is assigned to the outcome. This outcome becomes the value of the random variable. We usually

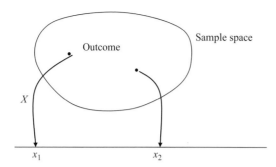

Figure 6.5. Random variable X maps elements of the sample space to the real line.

represent a random variable by an uppercase letter such as X, Y, and Z, while the value of a random variable (which is fixed) is represented by a lowercase letters such as x, y, and z. Thus, X is a function that maps elements of the sample space S to the real line $-\infty \le x \le \infty$, as illustrated in Figure 6.5.

> A **random variable** X is a single-valued real function that assigns a real value $X(x)$ to every point x in the sample space.

A random variable X may be either discrete or continuous. X is said to be a discrete random variable if it can take only discrete values. It is said to be continuous if it takes continuous values. An example of a discrete random variable is the outcome of rolling a die. An example of a continuous random variable is that of Gaussian distributed noise, to be discussed later.

Whether X is discrete or continuous, we need a probabilistic description of it in order to work with it. All random variables (discrete and continuous) have a cumulative distribution function (CDF).

> The **cumulative distribution function** (CDF) is a function giving the probability that the random variable X is less than or equal to x, for every value x.

Let us denote the probability of the event $X \le x$, where x is given, as $P(X \le x)$. The *cumulative distribution function* (CDF) of X is given by

$$F_X(x) = P(X \le x), \quad -\infty \le x \le \infty \tag{6.13}$$

for a continuous random variable X. Note that $F_X(x)$ does not depend on the random variable X, but on the assignment of X. $F_X(x)$ has the following five properties:

1.
$$F_X(-\infty) = 0 \tag{6.14a}$$

2.
$$F_X(\infty) = 1 \tag{6.14b}$$

3.
$$0 \le F_X(x) \le 1 \tag{6.14c}$$

4.
$$F_X(x_1) \le F_X(x_2), \text{if} \quad x_1 < x_2 \tag{6.14d}$$

5.
$$P(x_1 < X \le x_2) = F_X(x_2) - F_X(x_1) \tag{6.14e}$$

The first and second properties show that the $F_X(-\infty)$ includes no possible events and $F_X(\infty)$ includes all possible events. The third property follows from the fact that $F_X(x)$ is a probability. The fourth property indicates that $F_X(x)$ is a nondecreasing function. The last property is easy to prove since

$$P(X \le x_2) = P(X \le x_1) + P(x_1 < X \le x_2)$$

or

$$P(x_1 < X \le x_2) = P(X \le x_2) - P(X \le x_1) = F_X(x_2) - F_X(x_1) \tag{6.15}$$

If X is discrete, then

$$F_X(x) = \sum_{i=1}^{M} P(x_i) \tag{6.16}$$

where $P(x_i) = P(X = x_i)$ is the probability of obtaining event x_i, and N is the largest integer such that $x_N \le x$ and $N \le M$, and M is the total number of points in the discrete distribution. It is assumed that $x_1 < x_2 < x_3 < \ldots < x_M$.

It is sometimes convenient to use the derivative of $F_X(x)$, which is given by

$$f_X(x) = \frac{dF_x(x)}{dx} \tag{6.17}$$

where $f_X(x)$ is known as the *probability density function* (PDF). Note that $f_X(x)$ has the following four properties:

1.
$$f_X(x) \ge 0 \tag{6.18a}$$

2.
$$\int_{-\infty}^{\infty} f_X(x)dx = 1 \tag{6.18b}$$

3.
$$F_X(x) = \int_{-\infty}^{x} f_X(x)dx \tag{6.18c}$$

4.
$$P(x_1 \le x \le x_2) = \int_{x_1}^{x_2} f_X(x)dx \tag{6.18d}$$

Properties 1 and 2 follow from the fact that $F_X(-\infty) = 0$ and $F_X(\infty) = 1$ respectively. As mentioned earlier, $F_X(x)$ must be nondecreasing. Thus, its derivative $f_X(x)$ must always be non-negative, as stated by Property 1. Property 3 follows from Eq. (6.17), i.e. the relationship between derivative and integration. Property 4 is easy to prove. From Eq. (6.15),

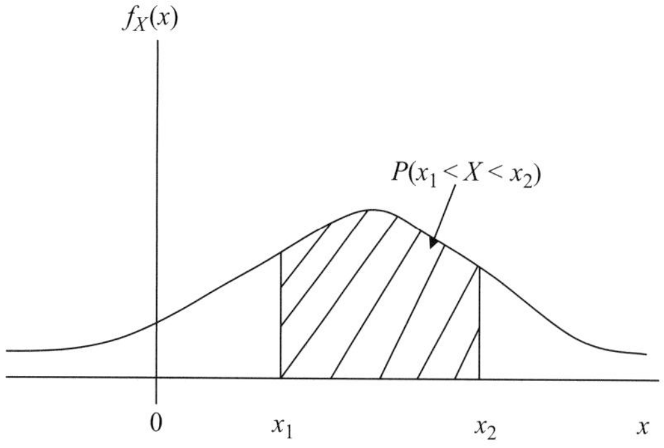

$f_X(x)$

$P(x_1 < X < x_2)$

0 x_1 x_2 x

Figure 6.6. A typical PDF.

$$P(x_1 < X \leq x_2) = F_X(x_2) - F_X(x_1)$$

$$= \int_{-\infty}^{x_2} f_X(x)dx - \int_{-\infty}^{x_1} f_X(x)dx = \int_{x_1}^{x_2} f_X(x)dx \qquad (6.19)$$

which is illustrated in Figure 6.6 for a continuous random variable. For discrete X,

$$f_X(x) = \sum_{i=1}^{M} P(x_i)\delta(x - x_i) \qquad (6.20)$$

where M is the total number of discrete events, $P(x_i) = P(x = x_i)$, and $\delta(x)$ is the impulse function. Thus,

---

The **probability density function** (PDF) of a continuous (or discrete) random variable is a function which can be integrated (or summed) to obtain the probability that the random variable takes a value in a given interval.

EXAMPLE 6.3

The CDF of a random variable is given by

$$F_X(x) = \begin{cases} 0, & x < 1 \\ \dfrac{x-1}{8}, & 1 \leq x < 9 \\ 1, & x \geq 9 \end{cases}$$

(a) Sketch $F_X(x)$ and $f_X(x)$. (b) Find $P(X \leq 4)$ and $P(2 < X \leq 7)$.

Solution

(a) In this case, X is a continuous random variable. $F_X(x)$ is sketched in Figure 6.7(a). We obtain the PDF of X by taking the derivative of $F_X(x)$, i.e.

(a)

$F_X(x)$

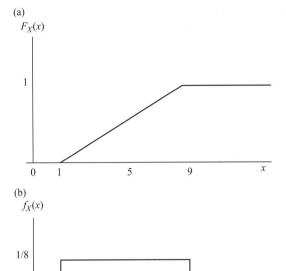

Figure 6.7. For Example 6.3: (a) CDF; (b) PDF.

(b)

$f_X(x)$

$$f_X(x) = \begin{cases} 0, & x < 1 \\ \dfrac{1}{8}, & 1 \le x < 9 \\ 0, & x \ge 9 \end{cases}$$

which is sketched in Figure 6.7(b). Notice that $f_X(x)$ satisfies the requirement of a probability because the area under the curve in Figure 6.7(b) is unity. A random number having a PDF such as shown in Figure 6.7(b) is said to be *uniformly distributed* because $f_X(x)$ is constant within 1 and 9.

(b)
$$P(X \le 4) = F_X(4) = 3/8$$

$$P(2 < x \le 7) = F_X(7) - F_X(2) = 6/8 - 1/8 = 5/8$$

PRACTICE PROBLEM 6.3

A pointer-spinning experiment leads to a uniform PFD given by

$$f_X(x) = \begin{cases} \dfrac{1}{2\pi}, & 0 < x < 2\pi \\ 0, & \text{otherwise} \end{cases}$$

(a) Sketch the corresponding $F_X(x)$. (b) Calculate $P(X > \pi/2)$ and $P(\pi/2 < X \le 3\pi/2)$.

Answer: (a) See Figure 6.8. (b) 0.75, 0.5.

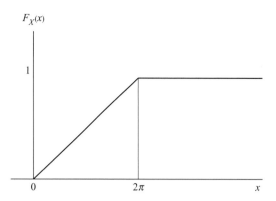

Figure 6.8. For Practice problem 6.3.

6.2.4 Operations on random variables

There are several operations that can be performed on random variables. These include the expected value, moments, variance, covariance, correlation, and transformation of the random variables. The operations are important in our study of communication systems. We begin with the mean or average values of a random variable.

Let X be a discrete random variable which takes on M values $x_1, x_2, x_3, \ldots, x_M$ that respectively occur as $n_1, n_2, n_3, \ldots, n_M$ in n trials, where n is very large. The statistical average (mean or expectation) of X is given by

$$\overline{X} = \frac{n_1 x_1 + n_2 x_2 + n_3 x_3 + \cdots + n_M x_M}{n} = \sum_{i=1}^{M} x_i \frac{n_i}{n} \qquad (6.21)$$

But by the relative-frequency definition of probability in Eq. (6.1), $n_i/n = P(x_i)$. Hence, the mean or expected value of the discrete random variable X is

$$\overline{X} = E[X] = \sum_{i=1}^{M} x_i P(x_i) \qquad (6.22)$$

where E stands for the expectation operator.

If X is a continuous random variable, we apply a similar argument. Rather than doing a summation, we replace the summation in Eq. (6.22) with integration and obtain

$$\overline{X} = E[X] = \int_{-\infty}^{\infty} x f_X(x) dx \qquad (6.23)$$

where $f_X(x)$ is the PDF of X.

In addition to the expected value of X, we are also interested in the expected value of functions of X. In general, the expected value of a function $g(X)$ of the random variable X is given by

$$\overline{g(X)} = E[g(X)] = \int_{-\infty}^{\infty} g(x) f_X(x) dx \qquad (6.24)$$

for continuous random variable X. If X is discrete, we replace the integration with summation and obtain

$$\overline{g(X)} = E[g(X)] = \sum_{i=1}^{M} g(x_i)P(x_i) \tag{6.25}$$

Consider the special case when $g(x) = X^n$. Equation (6.24) becomes

$$\overline{X^n} = E[X^n] = \int_{-\infty}^{\infty} x^n f_X(x)dx \tag{6.26}$$

$E(X^n)$ is known as the *nth moment* of the random variable X. When $n = 1$, we have the first moment $\overline{X}$ as in Eq. (6.24). When $n = 2$, we have the second moment $\overline{X^2}$ and so on.

The moments defined in Eq. (6.26) may be regarded as moments about the origin. We may also define central moments, which are moments about the mean value $m_X = E(X)$ of X, i.e.

$$E[(X - m_X)^n] = \int_{-\infty}^{\infty} (x - m_X)^n f_X(x)dx \tag{6.27}$$

It is evident that the central moment is zero when $n = 1$. When $n = 2$, the second central moment is known as the *variance* σ_X^2 of X, i.e.

$$\text{Var}(X) = \sigma_X^2 = E\left[(X - m_X)^2\right] = \int_{-\infty}^{\infty} (x - m_X)^2 f_X(x)dx \tag{6.28}$$

The square root of the variance (i.e. σ_X) is called the *standard deviation* of X. By expansion,

$$\sigma_X^2 = E\left[(X - m_X)^2\right] = E\left[X^2 - 2m_X X + m_X^2\right] = E\left[X^2\right] - 2m_X E[X] + m_X^2$$
$$= E\left[X^2\right] - m_X^2 \tag{6.29}$$

or

$$\sigma_X^2 = E\left[X^2\right] - m_X^2 \tag{6.30}$$

Note from Eq. (6.30) that if the mean $m_X = 0$, the variance is equal to the second moment $E[X^2]$.

EXAMPLE 6.4

A complex communication system is checked on a regular basis. The number of failures of the system in a month of operation has the probability distribution given in Table 6.2. (a) Find the average number and variance of failures in a month. (b) If X denotes the number of failures, determine mean and variance of $Y = X + 1$.

Table 6.2. **For Example 6.4**

No. of failures	0	1	2	3	4	5
Probability	0.2	0.33	0.25	0.15	0.05	0.02

Solution

(a) Using Eq. (6.22)

$$\overline{X} = m_X = \sum_{i=1}^{M} x_i P(x_i)$$

$$= 0(0.2) + 1(0.33) + 2(0.25) + 3(0.15) + 4(0.05) + 5(0.02)$$

$$= 1.58$$

To get the variance, we need the second moment.

$$\overline{X^2} = E(X^2) = \sum_{i=1}^{M} x_i^2 P(x_i)$$

$$= 0^2(0.2) + 1^2(0.33) + 2^2(0.25) + 3^2(0.15) + 4^2(0.05) + 5^2(0.02)$$

$$= 3.98$$

$$\text{Var}(X) = \sigma_X^2 = E[X^2] - m_X^2 = 3.98 - 1.58^2 = 1.4836$$

(b) If $Y = X + 1$, then

$$\overline{Y} = m_Y = \sum_{i=1}^{M} (x_i + 1) P(x_i)$$

$$= 1(0.2) + 2(0.33) + 3(0.25) + 4(0.15) + 5(0.05) + 6(0.02)$$

$$= 2.58$$

$$\overline{Y^2} = E(Y^2) = \sum_{i=1}^{M} (x_i + 1)^2 P(x_i)$$

$$= 1^2(0.2) + 2^2(0.33) + 3^2(0.25) + 4^2(0.15) + 5^2(0.05) + 6^2(0.02)$$

$$= 8.14$$

$$\text{Var}(Y) = \sigma_y^2 = E[Y^2] - m_Y^2 = 8.14 - 2.58^2 = 1.4836$$

which is the same as Var(X). This should be expected because adding a constant value of 1 to X does not change its randomness.

Table 6.3. **For Practice problem 6.4**

x_i	2	3	4	5	6
$P(x_i)$	0.45	0.36	0.12	0.06	0.01

PRACTICE PROBLEM 6.4

A telephone company studied the calls between two cities in the USA. Let X be the number of telephone exchanges a call passes through before reaching its destination and its probability distribution be as shown in Table 6.3. (a) Find $E(X)$ and σ_X. (b) If $Y = 2X - 1$, find $E(Y)$ and $Var(Y)$.

Answer: (a) 2.82, 0.9315. (b) 4.64, 3.4704.

6.3 SPECIAL DISTRIBUTIONS

Based on experience and usage, several probability distributions have been developed by engineers and scientists as models of physical phenomena. These distributions often arise in communication problems and deserve special attention. It is needless to say that each of these distributions satisfies the axioms of probability covered in section 6.2. In this section, we discuss four probability distributions: uniform, exponential, Gaussian, and Rayleigh distributions.

6.3.1 Uniform distribution

This distribution, also known as *rectangular distribution*, is very important for performing pseudorandom number generation used in simulations. It is also useful for describing quantizing noise that is generated in pulse code modulation. It is a distribution in which the density is constant. It models random events in which every value between a minimum and a maximum value is equally likely. A random variable X has a uniform distribution if its PDF is given by

$$f_X(x) = \begin{cases} \dfrac{1}{b-a}, & 0 \leq x \leq b \\ 0, & \text{otherwise} \end{cases} \tag{6.31}$$

which is shown in Figure 6.9. The mean and variance are given by

$$E(X) = \frac{b+a}{2} \tag{6.32a}$$

$$Var(X) = \frac{(b-a)^2}{12} \tag{6.32b}$$

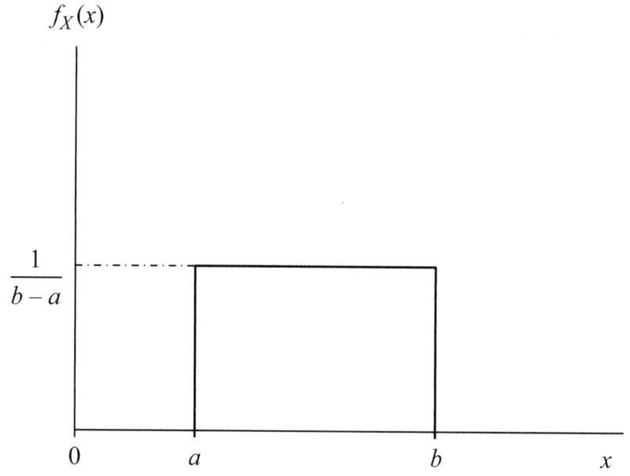

$f_X(x)$

Figure 6.9. PDF for a uniform random variable.

A special uniform distribution for which $a = 0$, $b = 1$, called the *standard uniform distribution*, is very useful in generating random samples from any probability distribution function. Also, if $Y = A \sin x$, where X is a uniformly distributed random variable, the distribution of Y is said to be *sinusoidal distribution*.

6.3.2 Exponential distribution

This distribution, also known as *negative exponential distribution*, is frequently used in simulations of queueing systems to describe the inter-arrival or inter-departure times of customers at a server. Its frequent use is due to the lack of conditioning of the remaining time on past time expended. This peculiar characteristic is known variably as Markov, *forgetfulness* or *lack of memory* property. For a given Poisson process, the time interval X between occurrences of events has an exponential distribution with the following PDF

$$f_X(x) = \lambda e^{-\lambda x} u(x) \tag{6.33}$$

which is portrayed in Figure 6.10. The mean and the variance of X are

$$E(X) = \frac{1}{\lambda} \tag{6.34a}$$

$$\mathrm{Var}(X) = \frac{1}{\lambda^2} \tag{6.34b}$$

6.3.3 Gaussian distribution

This distribution, also known as *normal* distribution, is the most important probability distribution in engineering. It is used to describe phenomena with symmetric variations above and below the mean μ. A random variable X with Gaussian distribution has its PDF of the form

$$f_X(x) = \frac{1}{\sigma\sqrt{2\pi}} \exp\left[-\frac{1}{2}\left(\frac{x-\mu}{\sigma}\right)^2\right], \quad -\infty < x < \infty \tag{6.35}$$

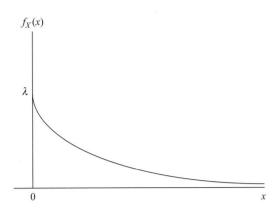

Figure 6.10. PDF for an exponential random variable.

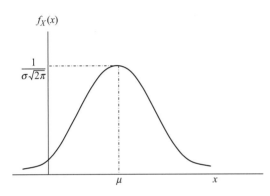

Figure 6.11. PDF for a Gaussian random variable.

where the mean

$$E(X) = \mu \tag{6.36a}$$

and the variance

$$\text{Var}(X) = \sigma^2 \tag{6.36b}$$

are themselves incorporated in the PDF. Figure 6.11 shows the Gaussian PDF. It is a common practice to use the notation $X \approx N(\mu, \sigma^2)$ to denote a normal random variable X with mean μ and variance σ^2. When $\mu = 0$ and $\sigma = 1$, we have $X = N(0,1)$, and the *normalized* or *standard normal* distribution function with

$$f_X(x) = \frac{1}{\sqrt{2\pi}} e^{-x^2/2} \tag{6.37}$$

which is widely tabulated.

It is important that we note the following points above the normal distribution which make the distribution the most prominent in probability and statistics and also in communication.

1. The binomial probability function with parameters n and p is approximated by a Gaussian PDF with $\mu = np$ and $\sigma^2 = np(1 - p)$ for large n and finite p.

2. The Poisson probability function with parameter λ can be approximated by a normal distribution with $\mu = \sigma^2 = \lambda$ for large λ.
3. The normal distribution is useful in characterizing the uncertainty associated with the estimated values. In other words, it is used in performing statistical analysis on simulation output.
4. The justification for the use of normal distribution comes from the *central limit theorem*.

The **central limit theorem** states that the distribution of the sum of n independent random variables from any distribution approaches a normal distribution as n becomes large.

Thus the normal distribution is used to model the cumulative effect of many small disturbances each of which contributes to the stochastic variable X. It has the advantage of being mathematically tractable. Consequently, many statistical analyses such as those of regression and variance have been derived assuming a normal density function. In several communication applications, we assume that noise is Gaussian distributed in view of the central limit theorem because noise is due to the sum of several random parameters.

6.3.4 Rayleigh distribution

The Rayleigh distribution is closely related to the Gaussian distribution. If Y and Z are two independent Gaussian random variables each with the same zero mean ($\mu = 0$) and variance σ^2, we can show that the random variable $X = \sqrt{Y^2 + Z^2}$ has a Rayleigh distribution with PDF given by

$$f_X(x) = \frac{x}{\sigma^2} e^{-x^2/2\sigma^2} u(x) \tag{6.38}$$

which is portrayed in Figure 6.12. Thus, Rayleigh distribution describes a random variable obtained from two Gaussian random variables. The mean and the variables of X are

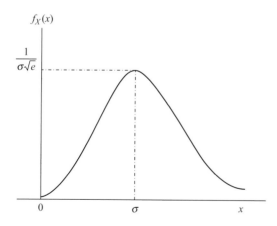

Figure 6.12. PDF of a Rayleigh random variable.

Table 6.4. **Properties of continuous probability distributions**

Name	PDF	CDF	Mean	Variance
Uniform	$f_X(x) = \dfrac{1}{b-a}$	$F_X(x) = \dfrac{x-a}{b-a}$	$\dfrac{b+a}{2}$	$\dfrac{(b-a)^2}{12}$
Exponential	$f_X(x) = \lambda e^{-\lambda x} u(x)$	$F_X(x) = 1 - e^{-\lambda x}$	$\dfrac{1}{\lambda}$	$\dfrac{1}{\lambda^2}$
Gaussian	$f_X(x) = \dfrac{1}{\sigma\sqrt{2\pi}} \exp\left[-\dfrac{1}{2}\left(\dfrac{x-\mu}{\sigma}\right)^2 \right]$	$F_X(x) = \dfrac{1}{2}\left[1 + \mathrm{erf}\left(\dfrac{x-\mu}{\sigma\sqrt{2}}\right)\right]$	μ	σ^2
Rayleigh	$f_X(x) = \dfrac{x}{\sigma^2} e^{-x^2/2\sigma^2} u(x)$	$F_X(x) = \left[1 - e^{-x^2/2\sigma^2}\right]u(x)$	$\sigma\sqrt{\dfrac{\pi}{2}}$	$\left(2 - \dfrac{\pi}{2}\right)\sigma^2$

erf(.) is the error function to be discussed in Example 6.5.

$$E(X) = \sigma\sqrt{\frac{\pi}{2}} \tag{6.39a}$$

$$\mathrm{Var}(X) = \left(2 - \frac{\pi}{2}\right)\sigma^2 \tag{6.39b}$$

Rayleigh distribution appears frequently in communications and other disciplines. For example, it is useful in describing filtered noise. A summary of the properties of the four continuous probability distributions is provided in Table 6.4.

EXAMPLE 6.5

Let X be a Gaussian random variable. (a) Find $E[X]$, $E[X^2]$, and $\mathrm{Var}(X)$. (b) Calculate $P(a < X < b)$.

Solution

(a) By definition,

$$E[X] = \int_{-\infty}^{\infty} x f_X(x)\,dx = \int_{-\infty}^{\infty} x \frac{1}{\sigma\sqrt{2\pi}} e^{-(x-\mu)^2/2\sigma^2}\,dx \tag{6.5.1}$$

Let $y = (x - \mu)/\sigma$ so that

$$E[X] = \frac{1}{\sqrt{2\pi}} \int_{-\infty}^{\infty} (\sigma y + \mu) e^{-y^2/2}\,dy = \frac{\sigma}{\sqrt{2\pi}} \int_{-\infty}^{\infty} y e^{-y^2/2}\,dy + \frac{\mu}{\sqrt{2\pi}} \int_{-\infty}^{\infty} e^{-y^2/2}\,dy \tag{6.5.2}$$

$$= 0 + \mu$$

Notice the first integral on the right-hand side is zero since the integrand is an odd function and the second integral gives μ since it represents the PDF of a Gaussian random variable $N(0,1)$. Hence,

$$E[X] = \mu \tag{6.5.3}$$

Similarly,

$$E[X^2] = \int_{-\infty}^{\infty} x^2 \frac{1}{\sigma\sqrt{2\pi}} e^{-(x-\mu)^2/2\sigma^2} dx$$

Again, we let $y = (x - \mu)/\sigma$ so that

$$E[X^2] = \frac{1}{\sqrt{2\pi}} \int_{-\infty}^{\infty} (\sigma y + \mu)^2 e^{-y^2/2} dy = \frac{1}{\sqrt{2\pi}} \int_{-\infty}^{\infty} \sigma^2 y^2 e^{-y^2/2} dy + \frac{1}{\sqrt{2\pi}} \int_{-\infty}^{\infty} 2\sigma\mu y e^{-y^2/2} dy$$

$$+ \frac{1}{\sqrt{2\pi}} \int_{-\infty}^{\infty} \mu^2 e^{-y^2/2} dy \qquad (6.5.4)$$

We can evaluate the first integral on the right-hand side by parts. The second integral is zero because the integrand is an odd function of y. The third integral yields μ^2 since it represents the PDF of a Gaussian random variable $N(0,1)$. Thus,

$$E[X^2] = \frac{\sigma^2}{\sqrt{2\pi}} \left[y e^{-y^2/2} \Big|_{-\infty}^{\infty} + \int_{-\infty}^{\infty} e^{-y^2/2} dy \right] + 2\sigma\mu(0) + \mu^2 = \sigma^2 + \mu^2 \qquad (6.5.5)$$

and

$$\text{Var}(X) = E[X^2] - E^2[X] = \sigma^2 + \mu^2 - \mu^2 = \sigma^2$$

We have established that for any real and finite number a and b, the following three integrals hold.

$$\int_{-\infty}^{\infty} \frac{1}{b\sqrt{2\pi}} \exp\left[-\frac{(x-a)^2}{2b^2} \right] dx = 1 \qquad (6.5.6a)$$

$$\int_{-\infty}^{\infty} \frac{x}{b\sqrt{2\pi}} \exp\left[-\frac{(x-a)^2}{2b^2} \right] dx = a \qquad (6.5.6b)$$

$$\int_{-\infty}^{\infty} \frac{x^2}{b\sqrt{2\pi}} \exp\left[-\frac{(x-a)^2}{2b^2} \right] dx = a^2 + b^2 \qquad (6.5.6c)$$

(b) To determine the Gaussian probability, we need the CDF of the Gaussian random variable X.

$$F_X(x) = \int_{-\infty}^{x} f_X(x) dx = \int_{-\infty}^{x} \frac{1}{\sigma\sqrt{2\pi}} e^{-(x-\mu)^2/2\sigma^2} dx = \int_{-\infty}^{\infty} \frac{1}{\sigma\sqrt{2\pi}} e^{-(x-\mu)^2/2\sigma^2} dx - \int_{x}^{\infty} \frac{1}{\sigma\sqrt{2\pi}} e^{-(x-\mu)^2/2\sigma^2} dx$$

The value of the first integral is 1 since we are integrating the Gaussian PDF over its entire domain. For the second integral, we substitute

$$z = \frac{(x - \mu)}{\sigma\sqrt{2}}, \qquad dz = \frac{dx}{\sigma\sqrt{2}}$$

and obtain

$$F_x(x) = 1 - \int_x^\infty \frac{1}{\sqrt{\pi}} e^{-z^2} dz \qquad (6.5.7)$$

We define the error function as

$$\text{erf}(x) = \frac{2}{\sqrt{\pi}} \int_0^x e^{-t^2} dt \qquad (6.5.8)$$

and the complementary error function as

$$\text{erfc}(x) = 1 - \text{erf}(x) = \frac{2}{\sqrt{\pi}} \int_x^\infty e^{-z^2} dz \qquad (6.5.9)$$

Hence, from Eqs. (6.5.7) to (6.5.9),

$$F_X(x) = \frac{1}{2}\left[1 + \text{erf}\left(\frac{x - \mu}{\sigma\sqrt{2}}\right)\right] \qquad (6.5.8)$$

and

$$P(a < x < b) = F_X(b) - F_X(a) = \frac{1}{2}\text{erf}\left(\frac{b - \mu}{\sigma\sqrt{2}}\right) - \frac{1}{2}\text{erf}\left(\frac{a - \mu}{\sigma\sqrt{2}}\right) \qquad (6.5.9)$$

Note that the definition of erf(x) varies from one book to another. Based on its definition in Eq. (6.5.8), some tabulated values are presented in Table 6.5. For example, given a Gaussian distribution with mean 0 and variance 2, we use the table to obtain

$$P(1 < x < 2) = \frac{1}{2}\text{erf}(1) - \frac{1}{2}\text{erf}(0.5) = 0.1611$$

PRACTICE PROBLEM 6.5

A random variable X has its PDF given by

$$f_X(x) = \frac{1}{\sqrt{\pi}}\exp\left(-x^2 + 2x - 4\right), \qquad -\infty < x < \infty$$

Evaluate $E[X]$, $E[X^2]$, and Var(X). Hint: Use Eq. (6.5.6).

Answer: 2, 4.5, 0.5.

Table 6.5. **Error function**

x	erf(x)	x	erf(x)
0.00	0.00000	1.10	0.88021
0.05	0.05637	1.15	0.89612
0.10	0.11246	1.20	0.91031
0.15	0.16800	1.25	0.92290
0.20	0.22270	1.30	0.93401
0.25	0.27633	1.35	0.94376
0.30	0.32863	1.40	0.95229
0.35	0.37938	1.45	0.95970
0.40	0.42839	1.50	0.96611
0.45	0.47548	1.55	0.97162
0.50	0.52050	1.60	0.97635
0.55	0.56332	1.65	0.98038
0.60	0.60386	1.70	0.98379
0.65	0.64203	1.75	0.98667
0.70	0.67780	1.80	0.98909
0.75	0.71116	1.85	0.99111
0.80	0.74210	1.90	0.99279
0.85	0.77067	1.95	0.99418
0.90	0.79691	2.00	0.99532
0.95	0.82089	2.50	0.99959
1.00	0.84270	00	0.99998
1.05	0.86244	30	1.0

EXAMPLE 6.6

This example illustrates the central limit theorem. We want to show that if $X_1, X_2, X_3, \ldots X_n$ are n dependent random variables and $c_1, c_2, c_3, \ldots, c_n$ are constants, then

$$X = c_1 X_1 + c_2 X_2 + c_3 X_3 + \cdots + c_n X_n$$

is a Gaussian random variable as n becomes large.

Solution

To make things simple, let use assume that $X_1, X_2, X_3, \ldots, X_n$ are identically distributed random variables, each with a uniform PDF as shown in Figure 6.13(a). For the sum $Y = X_1 + X_2$, the PDF of y is a convolution of the PDF in Figure 6.13(a) with itself, i.e.

$$f_Y(y) = \int_{-\infty}^{\infty} f_X(x) f_X(y - x) dx$$

(a)

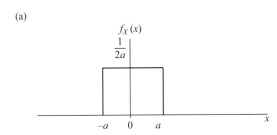

Figure 6.13. (a) PDF of uniform random variable X; (b) PDF of $Y = X_1 + X_2$; (c) PDF of $Z = X_1 + X_2 + X_3$.

(b)

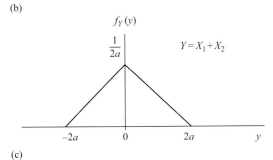

(c)

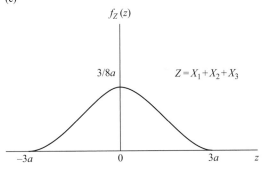

By performing the convolution, we obtain the joint PDF in Figure 6.13(b). In the same way, for the sum $Z = X_1 + X_2 + X_3$, the PDF of Z is the convolution of the PDF in Figure 6.13(a) with that in Figure 6.13(b), i.e.

$$f_Z(z) = \int_{-\infty}^{\infty} f_X(\lambda) f_Y(\lambda - z) d\lambda$$

which results in Figure 6.13(c). With only three terms, the PDF of the sum is already approaching a Gaussian PDF. According to the central limit theorem, as more terms are added, the PDF becomes Gaussian.

PRACTICE PROBLEM 6.6

If X and Y are two independent Gaussian random variables each with zero mean and the same variance σ, show that random variable $R = \sqrt{X^2 + Y^2}$ has a Rayleigh distribution. Hint: the joint PDF is $f_{XY}(x,y) = f_X(x) f_Y(y)$.

Answer: Proof

6.4 RANDOM PROCESSES

We now introduce the concept of *random* (or *stochastic*) *process* as a generalization of a random variable to include another dimension – the dimension of time. While a random variable depends on the outcome of a random experiment, a random process depends on both the outcome of a random experiment and time. In other words, if a random variable X is time-dependent, $X(t)$ is known as a *random process*. Random processes serve as models for both signals that convey information and noise. They also serve as models of systems that vary in time in a random manner.

Figure 6.14 portrays typical *realizations* or *sample functions* of a random process. From the figure, we notice that a random process is a mapping from the sample space onto an ensemble (family, set, collection) of time functions known as sample functions. Here $X(t,s_k)$ denotes the sample function or a realization of the random process for the s_k experimental outcome. It is customary to drop the s variable and use $X(t)$ to denote a random process. For a fixed time t_1, $X(t_1) = X_1$ is a random variable. Thus:

A **random process** is a family of random variables $X(t)$, indexed by the parameter t and defined on a common probability space.

It should be noted that the parameter t does not always have to represent time; it can represent any other variable such as space.

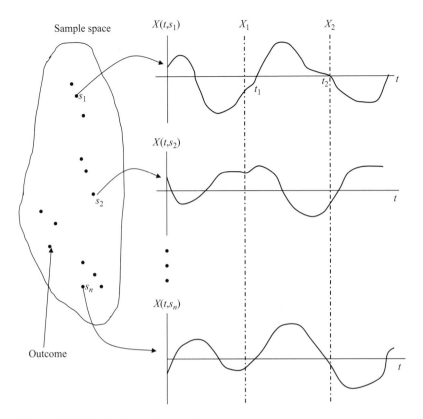

Figure 6.14. Realizations of a random process.

Random processes may be classified as:

- continuous or discrete
- deterministic or nondeterministic
- stationary or nonstationary
- ergodic or nonergodic

6.4.1 Continuous versus discrete random process

A *continuous-time random process* is one that has both a continuous random variable and continuous time. Noise in transistors and wind velocity are examples of continuous random processes. So are Wiener process and Brownian motion. A *discrete-time random process* is one in which the random variables are discrete, i.e. it is a sequence of random variables. For example, a voltage that assumes a value of either 0 or 12 V because of switching operation is a sample function from a discrete random process. The binomial counting and random walk processes are discrete processes. It is also possible to have a mixed random process which is partly continuous and partly discrete.

6.4.2 Deterministic versus nondeterministic random process

A *deterministic random process* is one for which the future value of any sample function can be predicted from a knowledge of the past values. For example, consider a random process described by

$$X(t) = A\cos{(\omega t + \Phi)} \tag{6.40}$$

where A and ω are constants and Φ is a random variable with a known probability distribution. Although $X(t)$ is a random process, one can predict its future values and hence $X(t)$ is deterministic. For a *nondeterministic random process*, each sample function is a random function of time and its future values cannot be predicted from the past values.

6.4.3 Stationary versus nonstationary random process

A *stationary random process* is one in which the probability density function of the random variable does not change with time. In other words, a random process is stationary when its statistical characteristics are time-invariant, i.e. not affected by a shift in time origin. Thus, the random process is stationary if all marginal and joint density functions of the process are not affected by the choice of time origin. A *nonstationary random process* is one in which the probability density function of the random variable is a function of time.

6.4.4 Ergodic versus nonergodic random process

An ergodic random process is one in which every member of the ensemble possesses the same statistical behavior as the entire ensemble. Thus, for ergodic processes, it is possible to determine the statistical characteristic by examining only one typical sample function, i.e. the average value and moments can be determined by time averages as well as by ensemble averages. For example, the nth moment is given by

$$\overline{X^n} = \int_{-\infty}^{\infty} x^n f_X(x)dx = \lim_{T\to\infty} \frac{1}{2T} \int_{-T}^{T} X^n(t)dt \tag{6.41}$$

This condition will only be satisfied if the process is stationary. This implies that ergodic processes are stationary as well. A nonergodic process does not satisfy the condition in Eq. (6.41). All nonstationary processes are nonergodic but a stationary process could also be nonergodic. Figure 6.15 shows the relationship between stationary and ergodic processes.

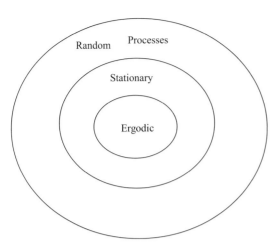

Figure 6.15. Relationship between stationary and ergodic random processes.

EXAMPLE 6.7

Classify and illustrate the random process

$$X(t) = \cos(2\pi t + \Theta)$$

where Θ is a random variable uniformly distributed on the interval $[0, 2\pi]$.

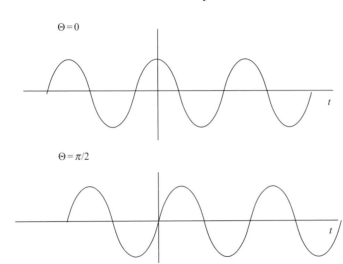

Figure 6.16. For Example 6.7; sample functions of the random process.

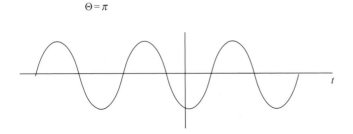

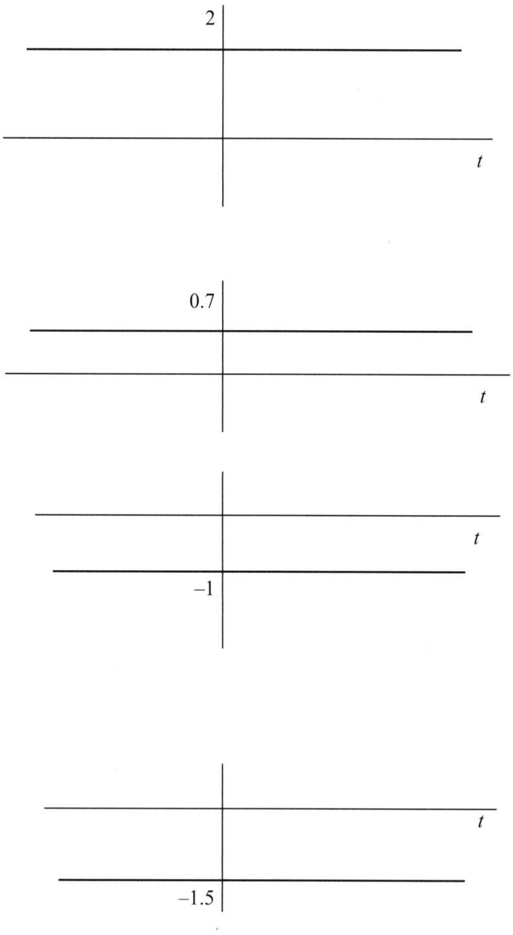

Figure 6.17. For Practice problem 6.7.

Solution

We are given an analytic expression for the random process and it is evident that it is a continuous-time and deterministic random process. Figure 6.16 displays some sample functions or realizations of the process.

PRACTICE PROBLEM 6.7

A random process is given by

$$X(t) = A,$$

where A is a random variable uniformly distributed over $[-2, 2]$. Sketch four sample functions for the random process.

Answer: See Figure 6.17.

6.5   POWER SPECTRAL DENSITIES

So far, we have characterized random processes in the time domain. We now characterize them using frequency-domain concepts. From our knowledge of Fourier transform, we know that any signal will have some form of spectral characterization. For wide-sense stationary (WSS) random processes, it is possible to define the *power spectral density* (PSD) or *spectral density function*, which provides a measure of the frequency distribution of the average power of a random process.

> The **power spectral density** (PSD) of a random process is the function that describes its power distribution with frequency.

Depending on the source, the power spectrum of a random process may be discrete, continuous, or mixed. For a random process $X(t)$, let the autocorrelation be $R_X(\tau)$. Its power spectral density $S_X(\omega)$ is the Fourier transform of $R_X(\tau)$, i.e.

$$S_X(\omega) = \mathcal{F}[R_X(\tau)] = \int_{-\infty}^{\infty} R_X(\tau)e^{-j\omega\tau}d\tau \tag{6.42}$$

or

$$R_X(\tau) = \frac{1}{2\pi}\int_{-\infty}^{\infty} S_X(\omega)e^{j\omega\tau}d\omega \tag{6.43}$$

Equations (6.42) and (6.43) constitute the *Wiener–Khinchine theorem.*

> The **Wiener–Khinchine theorem** states that the power spectral density (PSD) of a wide-sense stationary random process is the Fourier transform of the autocorrelation function.

(This strictly applies to wide-sense stationary processes.)
   Thus, $S_X(\omega)$ and $R_X(\tau)$ are Fourier transform pairs, i.e.

$$S_X(\omega) \quad \leftrightarrow \quad R_X(\tau) \tag{6.44}$$

If one is known, the other can be found exactly.
   The power density function has the following properties:

1. $S_X(-\omega) = S_X(\omega)$, i.e. it is an even function of frequency.
2. $S_X(\omega) \geq 0$, i.e. it is non-negative and real.
3. $S_X(0) = \int_{-\infty}^{\infty} R_X(\tau)d\tau$, i.e. the value of the power spectral at dc ($\omega = 0$) is the total area under the autocorrelation function.
4. $P_X = R_X(0) = E[X^2(t)] = \frac{1}{2\pi}\int_{-\infty}^{\infty} S_X(\omega)d\omega$, i.e. the average power or mean square value of a random process equals the total area under the power spectral density.

It is easy to prove these properties using Eqs. (6.42) and (6.43). Keep in mind that these properties apply only to random processes which are at least WSS.

When we are interested in WSS random processes $X(t)$ and $Y(t)$, such as the input and the output of a linear system, we can define their *cross spectral densities* (or *cross power spectral densities*) as

$$S_{XY}(\omega) = \int_{-\infty}^{\infty} R_{XY}(\tau)e^{-j\omega\tau}d\tau \tag{6.45}$$

and

$$S_{YX}(\omega) = \int_{-\infty}^{\infty} R_{YX}(\tau)e^{-j\omega\tau}d\tau \tag{6.46}$$

This implies that the cross correlation functions and the cross spectral densities form Fourier transform pairs, i.e.

$$S_{XY}(\omega) \quad \leftrightarrow \quad R_{XY}(\tau) \tag{6.47}$$

Hence,

$$R_{XY}(\tau) = \frac{1}{2\pi} \int_{-\infty}^{\infty} S_{XY}(\omega)e^{j\omega\tau}d\omega \tag{6.48}$$

$$R_{YX}(\tau) = \frac{1}{2\pi} \int_{-\infty}^{\infty} S_{YX}(\omega)e^{j\omega\tau}d\omega \tag{6.49}$$

The properties of $R_{XY}(\tau)$ and $S_{XY}(\omega)$ include the following:

1. $R_{XY}(-\tau) = R_{XY}(\tau)$
2. $|R_{XY}(\tau)| \le \sqrt{R_X(0)^2 + R_Y(0)^2} \le \frac{1}{2}[R_X(0) + R_Y(0)]$
3. $S_{XY}(\omega) = S_{XY}(-\omega) = S_{XY}^*(\omega)$
4. $S_{XY}(\omega) = 0$ if $X(t)$ and $Y(t)$ are orthogonal
5. $S_{XY}(\omega) = 2\pi m_X m_Y \delta(\omega)$ if $X(t)$ and $Y(t)$ are uncorrelated.

Finally, consider the sum of two random processes:

$$Z(t) = X(t) + Y(t) \tag{6.50}$$

where $X(t)$ and $Y(t)$ are jointly WSS. The autocorrelation of $Z(t)$ is

$$R_Z(\tau) = R_X(\tau) + R_Y(\tau) + R_{XY}(\tau) + R_{YX}(\tau) \tag{6.51}$$

Taking the Fourier transform of both sides yields

$$S_Z(\omega) = S_X(\omega) + S_Y(\omega) + S_{XY}(\omega) + S_{YX}(\omega) \tag{6.52}$$

EXAMPLE 6.8

The autocorrelation function of the random telegraph signal is given by

$$R_X(\tau) = e^{-2\alpha|\tau|}$$

where α is the mean transition rate of the signal. Determine the PSD of the signal.

Solution

From Eq. (6.42),

$$S_X(\omega) = \int_{-\infty}^{\infty} R_X(\tau) e^{-j\omega\tau} d\tau = \int_{-\infty}^{\infty} e^{-2\alpha|\tau|} e^{-j\omega\tau} d\tau$$

$$= \int_{-\infty}^{0} e^{2\alpha\tau} e^{-j\omega\tau} d\tau + \int_{0}^{\infty} e^{-2\alpha\tau} e^{-j\omega\tau} d\tau = \frac{e^{\tau(2\alpha-j\omega)}}{2\alpha - j\omega} \bigg|_{-\infty}^{0} + \frac{e^{-\tau(2\alpha+j\omega)}}{-(2\alpha + j\omega)} \bigg|_{0}^{\infty}$$

$$= \frac{1}{2\alpha - j\omega} + \frac{1}{2\alpha + j\omega} = \frac{4\alpha}{4\alpha^2 + \omega^2}$$

PRACTICE PROBLEM 6.8

A random process is described by

$$X(t) = A\cos(\omega_o t + \Theta)$$

where A is constant and Θ is uniformly distributed over the interval $(0, 2\pi)$. Find the PSD and the average power of the process.

Answer: $\frac{\pi A^2}{2}[\delta(\omega + \omega_o) + \delta(\omega - \omega_o)]$, $\quad \dfrac{A^2}{2}$

EXAMPLE 6.9

The power spectral density of a low-pass, bandlimited white noise is shown in Figure 6.18(a). Find the autocorrelation function and the average power of the noise.

Solution

This can be solved in two ways.

  Method 1: an indirect way is to express the PSD as a rectangular function

$$S_X(\omega) = N\Pi(\omega/2\omega_o)$$

From Table 2.5, we have Fourier pairs

$$\Pi(t/a) \quad \rightarrow \quad a\,\text{sinc}(\omega a/2)$$

where $a = 2\omega_o$ in our case. We now apply the duality property in Table 2.4.

(a)

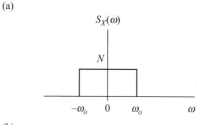

Figure 6.18. For Example 6.9.

(b)

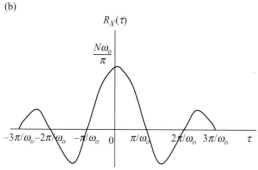

$$S_X(\omega) = N\Pi(\omega/2\omega_o) \quad \rightarrow \quad \frac{1}{2\pi} 2\omega_o N \, \mathrm{sinc}(\omega_o t)$$

Hence,

$$R_X(\tau) = \frac{N\omega_o}{\pi} \, \mathrm{sinc}(\omega_o \tau)$$

Method 2: a direct way is to use Eq. (6.43),

$$R_X(\tau) = \frac{1}{2\pi} \int_{-\infty}^{\infty} S_X(\omega) e^{j\omega\tau} d\omega = \frac{1}{2\pi} \int_{-\omega_o}^{\omega_o} N e^{j\omega\tau} d\omega = \frac{N}{2\pi} \frac{e^{j\omega\tau}}{j\tau} \Big|_{-\omega_o}^{\omega_o}$$

$$= \frac{N}{\pi\tau} \frac{e^{j\omega_o\tau} - e^{-j\omega_o\tau}}{2j} = \frac{N\omega_o}{\pi} \frac{\sin \omega_o \tau}{\omega_o \tau}$$

$$= \frac{N\omega_o}{\pi} \, \mathrm{sinc}(\omega_o \tau)$$

as obtained before. $R_X(\tau)$ is shown in Figure 6.18(b). The average power is

$$P_X = R_X(0) = \frac{N\omega_o}{\pi}$$

PRACTICE PROBLEM 6.9

The power spectral density of a random process is shown in Figure 6.19(a). Determine the autocorrelation function and sketch it. Find the average power of the process.

(a)

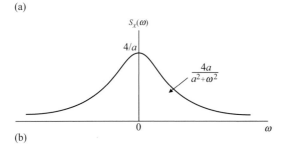

(b)

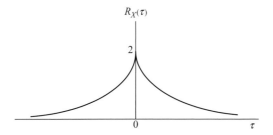

Figure 6.19. For Practice problem 6.9.

Answer: $2e^{-a|\tau|}$, see Figure 6.19(b), 2.

EXAMPLE 6.10

For random processes $X(t)$ and $Y(t)$, the cross power density is

$$S_{XY}(\omega) = \frac{4}{(a+j\omega)^3}$$

where a is a constant. Find the cross correlation function.

Solution

Let $S_{XY}(\omega) = 2F(\omega)$, where

$$F(\omega) = \frac{2}{(a+j\omega)^3}$$

From Table 2.5, the inverse Fourier transform of this is

$$f(t) = t^2 e^{-at} u(t)$$

Applying the linearity property, we obtain the inverse of $S_{XY}(\omega)$ as

$$R_{XY}(\tau) = 2\tau^2 e^{-a\tau} u(\tau)$$

PRACTICE PROBLEM 6.10

Given the cross-power spectrum density as

$$S_{XY}(\omega) = \begin{cases} a + jb\omega/\omega_o, & -\omega_o < \omega < \omega_o \\ 0, & \text{otherwise} \end{cases}$$

where ω_o, a, and b are constants. Determine the cross correlation function.

Answer: $\frac{1}{\pi\omega_o\tau^2}[(a\omega_o\tau - b)\sin(\omega_o\tau) + b\omega_o\tau\cos(\omega_o\tau)]$.

6.6   SAMPLE RANDOM PROCESSES

We have been discussing random processes in general. Specific random processes include the Poisson counting process, the Wiener process or Brownian motion, the random walking process, the Bernoulli process, and the Markov process. In this section, we consider some specific random processes that are useful in communication systems – Gaussian, white noise, and bandlimited processes.

6.6.1   Gaussian process

A random process $X(t)$ is said to be a *Gaussian process* (or *normal process*) if every linear combination of the random variable $X(t)$ is Gaussian distributed. Let $X(t_1)$, $X(t_2)$, ..., $X(t_n)$ be random variables. They have a joint Gaussian distribution if their linear combination

$$Y(t) = a_1 X(t_1) + a_2 X(t_2) + \cdots + a_n X(t_n) \tag{6.53}$$

is Gaussian distributed for every selection of the constants $a_1, a_2, \ldots, a_n$. Gaussian processes are widely used in many applications especially in communications for at least two reasons. First, using a Gaussian model to describe physical phenomena is often confirmed by experiments. Second, the unique properties of the Gaussian process make analytic results possible.

6.6.2   White noise

The term noise denotes unwanted signal. Two common types of noise in communication systems are shot noise and thermal noise. While shot noise is caused by electronic devices such as diodes and transistors, thermal noise arises from the random motion of electrons in a conductor. White noise is an idealized form of noise in which the power spectral density (PSD) does not vary with frequency. The idea of "white noise" comes from the analogous concept of "white light" which contains all the visible light frequencies. A "colored noise" is any noise that is not white. Hence, the thermal noise in a resistor and the shot noise in a transistor are colored noises.

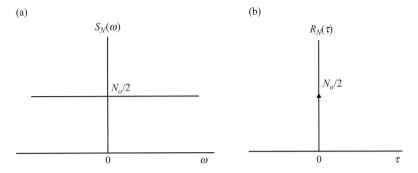

Figure 6.20. White noise: (a) its power density spectrum; (b) its autocorrelation.

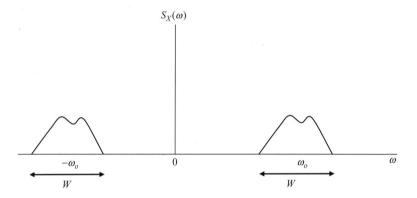

Figure 6.21. Power density spectrum of a bandlimited process.

A random process $X(t)$ is called a *white-noise process* if its PSD is constant with frequency, i.e.

$$S_X(\omega) = N_o/2 \tag{6.54}$$

where N_o is a constant. Thus a white-noise process has a flat frequency spectrum, as shown in Figure 6.20. The corresponding autocorrelation function is

$$R_X(\tau) = \frac{N_o}{2}\delta(\tau) \tag{6.55}$$

Note from Eq. (6.54) that white noise has infinite average power, which makes it physically unrealizable.

6.6.3 Bandlimited process

There are bandpass, bandlimited, and narrowband processes depending on their power spectral density (PSD). A random process is called *bandlimited* if its PSD is zero outside some frequency band, as shown in Figure 6.21. Bandlimited processes form the basis of digital systems. A bandlimited random process is said be narrowband if $W \ll \omega_o$.

6.7 APPLICATION – LINEAR SYSTEMS

There are several applications involving the foregoing theory of probability and random processes. For example, they are used in filtering, smoothing, signal processing, and modulation. A practical application will be made here on linear systems.

Consider the linear system shown in Figure 6.22, where the input $x(t)$ is an ensemble member of a random process $X(t)$. The system's response $y(t)$ is obtained by the convolution integral

$$y(t) = x(t) * h(t) = \int_{-\infty}^{\infty} x(\lambda)h(t-\lambda)d\lambda \tag{6.56}$$

where $h(t)$ is the impulse response of the system. The input–output relationship in Eq. (6.56) may be regarded as an operation of an ensemble member $x(t)$ of the random process $X(t)$ that serves as input to the linear system to produce an ensemble member $y(t)$ of a new random process $Y(t)$. Thus, $Y(t)$ may be expressed in terms of $X(t)$ as

$$Y(t) = X(t) * h(t) = \int_{-\infty}^{\infty} X(t-\lambda)h(\lambda)d\lambda \tag{6.57}$$

If $X(t)$ is wide-sense stationary, the autocorrelation function of $Y(t)$ is

$$R_Y(t, t+\tau) = E[Y(t)Y(t+\tau)]$$

$$= E\left[\int_{-\infty}^{\infty} X(t-\lambda)h(\lambda)d\lambda \int_{-\infty}^{\infty} X(t+\tau-\xi)h(\xi)d\xi\right]$$

$$= \int_{-\infty}^{\infty}\int_{-\infty}^{\infty} E[X(t-\lambda)X(t+\tau-\xi)]h(\lambda)h(\xi)d\lambda\,d\xi \tag{6.58}$$

Since $X(t)$ is WSS, Eq. (6.58) reduces to

$$R_Y(\tau) = \int_{-\infty}^{\infty}\int_{-\infty}^{\infty} R_X(t+\tau-\xi)h(\lambda)h(\xi)d\lambda\,d\xi \tag{6.59}$$

which is a double convolution of the autocorrelation function of input $X(t)$ with the system's impulse response; i.e.

$$R_Y(\tau) = h(\tau) * h(-\tau) * R_X(\tau) \tag{6.60}$$

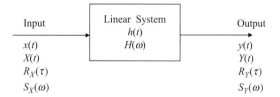

Figure 6.22. A linear system.

Taking the Fourier transform of this gives the output spectral power density as

$$S_Y(\omega) = |H(\omega)|^2 S_X(\omega) \tag{6.61}$$

where $H(\omega)$ is the Fourier transform of the impulse response $h(t)$. The average power in the output $Y(t)$ is obtained as

$$P_Y = \frac{1}{2\pi} \int_{-\infty}^{\infty} S_X(\omega)|H(\omega)|^2 d\omega \tag{6.62}$$

In a similar manner, we can show that the cross-correlation function of $X(t)$ and $Y(t)$ is

$$R_{XY}(\tau) = h(-\tau) * R_X(\tau) \tag{6.63}$$

so that

$$R_Y(\tau) = h(\tau) * R_{XY}(\tau) \tag{6.64}$$

EXAMPLE 6.11
Consider the RC low-pass filter shown in Figure 6.23. Determine the power spectral density and average power of the response $Y(t)$ when the input $X(t)$ is white noise.

Solution
In the frequency domain, the transfer function is readily obtained as

$$H(\omega) = \frac{Y(\omega)}{X(\omega)} = \frac{1/j\omega C}{R + 1/j\omega C} = \frac{1}{1 + j\omega RC}$$

For white noise, we recall that $S_X(\omega) = N_o/2$. Hence, the PSD for $Y(t)$ is

$$S_Y(\omega) = S_X(\omega)|H(\omega)|^2 = \frac{N_o/2}{1 + (\omega RC)^2}$$

The average power of $Y(t)$ can be found in two ways. One way is to use

$$P_Y = \frac{1}{2\pi} \int_{-\infty}^{\infty} S_Y(\omega)d\omega = \frac{N_o}{4\pi} \int_{-\infty}^{\infty} \frac{d\omega}{1 + (\omega RC)^2} = \frac{N_o}{4\pi} \frac{2}{RC} \tan^{-1}\omega RC \Big|_0^\omega = \frac{N_o}{4RC}$$

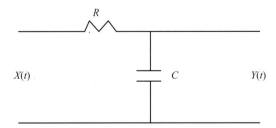

Figure 6.23. For Example 6.11; RC low-pass filter.

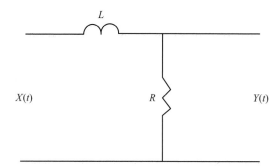

Figure 6.24. For Practice problem 6.11.

$X(t)$ R $Y(t)$

Alternatively, we can find the autocorrelation function by taking the inverse Fourier transform of $S_X(\omega)$.

$$R_Y(\tau) = \mathbb{F}^{-1}\left[\frac{N_o/2}{1 + (\omega RC)^2}\right] = \mathbb{F}^{-1}\left[\frac{N_o}{2}\left(\frac{\alpha^2}{\alpha^2 + \omega^2}\right)\right], \quad \alpha = 1/RC$$

$$= \frac{N_o}{2}\frac{\alpha}{2}e^{-\alpha|\tau|}$$

Hence,

$$P_y = R_y(0) = \frac{N_o}{4RC}$$

as obtained before.

PRACTICE PROBLEM 6.11

An RL low-pass filter is shown in Figure 6.24. If a random telegraph signal $X(t)$ passes through the filter, find the autocorrelation function of the response. Assume that

$$S_X(\omega) = \frac{4a}{4a + \omega^2}$$

where a is a constant.

Answer: $\dfrac{1}{\alpha^2 - 4a^2}\left[\alpha^2 e^{-2a|\tau|} - 2a\alpha e^{-\alpha|\tau|}\right], \quad \alpha = R/L.$

EXAMPLE 6.12

Let $X(t)$ be a differentiable WSS random process. When $X(t)$ is applied to a system, the output $Y(t)$ satisfies the stochastic differential equation

$$\dot{Y}(t) + Y(t) = X(t)$$

where $R_X = 2e^{-|\tau|}$. Find $S_Y(\omega)$.

Solution

In this case, we are not given $H(\omega)$ and we need to get it from the given differential equation. Taking the Fourier transform of both sides of the given differential equation yields

$$j\omega Y(\omega) + Y(\omega) = X(\omega) \quad \rightarrow \quad H(\omega) = \frac{Y(\omega)}{X(\omega)} = \frac{1}{1+j\omega}$$

or

$$|H(\omega)|^2 = H(\omega)H^*(\omega) = \frac{1}{1+\omega^2}$$

The autocorrelation of $X(t)$ is given by

$$S_X(\omega)^X = \mathcal{F}[R_X(\tau)] = \mathcal{F}\left[2e^{-|\tau|}\right] = \frac{2}{1+\omega^2}$$

Hence,

$$S_Y(\omega) = S_X(\omega)|H(\omega)|^2 = \frac{2}{(1+\omega^2)^2}$$

PRACTICE PROBLEM 6.12

A system is described by $Y(t) = \dot{X}^{(t)}$, where $X(t)$ and $Y(t)$ are random input and output processes. Let $R_X(\tau) = \frac{N_o}{2}\delta(\tau)$. Find $S_Y(\omega)$.

Answer: $\dfrac{N_o}{2}\delta(\omega)$.

6.8 COMPUTATION USING MATLAB

The MATLAB software can be used to reinforce the concepts learned in this chapter. It can be used to generate a random process $X(t)$ and calculate its statistics. It can also be used to plot $X(t)$ and its autocorrelation function.

6.8.1 Linear systems

MATLAB provides the command **rand** for generating uniformly distributed random numbers between 0 and 1. The uniform random-number generator can then be used to generate a random process or the PDF of an arbitrary random variable. For example, to generate a random variable X uniformly distributed over (a, b), we use

$$X = a + (a - b)U \tag{6.65}$$

where U is generated by **rand**. A similar command **randn** generates a Gaussian (or normal) distribution with mean zero and variance one.

Suppose we are interested in generating a random process

$$X(t) = 10 \cos(2\pi t + \Theta) \tag{6.66}$$

where Θ is a random variable uniformly distributed over $(0, 2\pi)$. We generate and plot $X(t)$ using the following MATLAB commands.

```
» t=0:0.01:2; % select 201 time points between 0 and 2.
» n=length(t);
» theta=2*pi*rand(1,n); % generates n=201 uniformly distributed theta
» x=10*cos(2*pi*t +theta);
» plot(t,x)
```

The plot of the random process is shown in Figure 6.25. We may find the mean and standard deviation using MATLAB commands **mean** and **std** respectively. For example the standard deviation is found using

```
» std(x)
ans =
7.1174
```

where the result is a bit off from the exact value of 6.0711. The reason for this discrepancy is that we selected only 201 points. If more points, say 10,000, are selected the two results should be very close.

If $X(t)$ generated above is an input to a Butterworth filter (a system) of order $N = 3$ in Practice problem 2.14, we want to find the output $Y(t)$. The MATLAB command **lsim** is used for finding the time response of a system to any arbitrary input signal. The format of the command is `y = lsim(num, den, x, t)`, where is $x(t)$ is the input signal and t is the time vector, $y(t)$ is the output generated, and num and den are respectively the numerator and denominator of the transfer

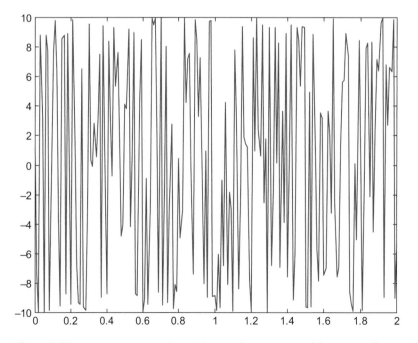

Figure 6.25. MATLAB generation of the random process $X(t) = 10\cos(2\pi t + \Theta)$.

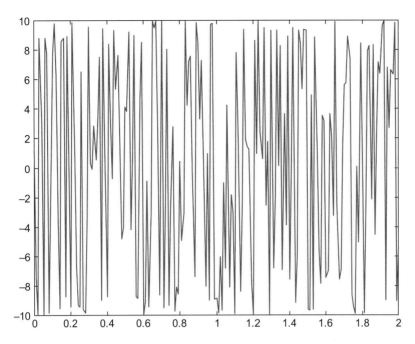

Figure 6.26. Output of the third-order Butterworth filter $Y(t)$.

function $H(s)$. For the third-order Butterworth filter, the transfer function is obtained in Practice problem 2.14 as

$$H(s) = \frac{\omega_c^3}{s^3 + 2\omega_c s^2 + 2\omega_c^2 s + \omega_c^3}, \quad \omega_c = 6.791 \times 10^7 \text{ rad/s} \quad (6.67)$$

We use the following MATLAB commands to obtain the response $y(t)$ as plotted in Figure 6.26.

```
>> t=0:0.01:2;
>> n=length(t);
>> theta=2*pi*rand(1,n);
>> x=10*cos(2*pi*t+theta);
>> wc=6.791*10^7;
>> num=wc^3; % numerator of H(s)
>> den=[ 1 2*wc 2*wc^2 wc^3] ; % denominator of H(s)
>> y=lsim(num,den,x,t);
>> plot(t,y)
```

6.8.2 Bernoulli random process

We will now use MATLAB to generate a Bernoulli random process, which is used in data communication. The process consists of random variables which assume only two states or values: $+1$ and -1 (or $+1$ and 0). In this particular case, the process may also be regarded as a *random binary process*. The probability of $X(t)$ being $+1$ is p and -1 is $q = 1-p$. Therefore, to generate a Bernoulli random variable X, we first use MATLAB **rand** to generate a random variable U that is uniformly distributed over $(0,1)$. Then, we obtain

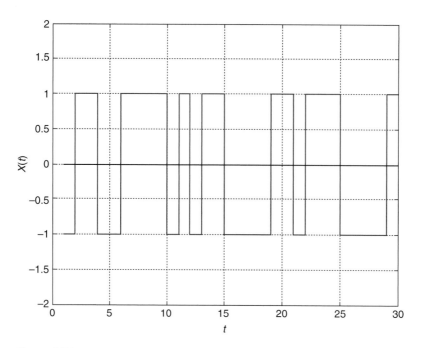

Figure 6.27. A typical sample function of a Bernoulli random process.

$$X = \begin{cases} 1, & \text{if } U \leq p \\ -1, & \text{if } U > p \end{cases} \tag{6.68}$$

i.e. we have partitioned the interval (0,1) into two segments of length p and $1 - p$. The following MATLAB program is used to generate a sample function for the random process. The sample function is shown in Figure 6.27.

```
% Generation of a Bernoulli process
% Ref: D. G. Childers, "Probability of Random Processes," Irwin, 1997, p.164
p=0.6;     % probability of having +1
q=1-p;     % probability of having -1
n=30;   % length of the discrete sequence
t=rand(1,n);     % generate random numbers uniformly distributed over (0,1)
x=zeros(length(t));   % set initial value of x equal to zero
for k=1:n
if( t(k) <= p )
x(k)=1;
else
x(k)= -1;
end
end
stairs(x);
xlabel('t')
ylabel('x(t)')
a=axis;
axis([ a(1) a(2) -2 2] );
grid on
```

Summary

1. The probability of an event is the measure of how likely it is that the event will occur as a result of a random experiment. A random experiment is one in which all the outcomes solely depend on *chance*, i.e., the outcome that will occur cannot be determined with certainty.

2. The relative-frequency definition of the probability of an event A assumes that if an experiment is repeated for a large number of times n and event A occurs n_A times,

$$P(A) = \frac{n_A}{n}$$

6. A random variable is a real-value function defined over a sample space. A discrete random variable is one which may take on only a countable number of distinct values such as 0, 1, 2, 3, ...

A continuous random variable is one which takes a continuum of possible values.

4. The cumulative distribution function (CDF) $F_X(x)$ of a random variable X is defined as the probability $P(X \leq 1)$ and $F_X(x)$ lies between 0 and 1.

5. The probability density function (PDF) $f_X(x)$ of a random variable X is the derivative of the CDF $F_X(x)$, i.e.

$$f_X(x) = \frac{dF_X(x)}{dx} \quad \leftrightarrow \quad F_X(x) = \int_{-\infty}^{\infty} x f_X(t) dt$$

Note that $f_X(x)dx$ is the probability of a random variable X lying within dx of x.

6. The mean value of a random variable X is

$$E(X) = \int_{-\infty}^{\infty} x f_X(x) dx \quad \text{if } X \text{ is continuous}$$

or

$$E(X) = \sum_{i=1}^{M} x_i P(x_i) \quad \text{if } X \text{ is discrete}$$

7. The variance of random variable X is

$$\mathrm{Var(x)} = \sigma_X^2 = E[X^2] - E^2(X)$$

where σ_X is the standard deviation of the random variable; σ_X is a measure of the width of its PDF.

8. Table 6.4 summarizes the CDF, PDF, mean, and variance of common continuous probability distributions: uniform, exponential, Gaussian, and Rayleigh.

9. The central limit theorem is the usual justification for using the Gaussian distribution for modeling. It states that the sum of independent samples from any distribution approaches the Gaussian distribution as the sample size goes to infinity.

10. A random process (also known as stochastic process) is a mapping from the sample space onto an ensemble of time functions known as sample functions. At any instant of time, the value of a random process is a random variable.

11. A continuous-time random process $X(t)$ is a family of sample functions of continuous random variables that are functions of time t, where t is a continuum of values. A random process is deterministic if future values of the sample function can be predicted from past values.

12. A random process is stationary if all its statistical properties do not change with time, i.e. $m_X(t)$ is constant and $R_X(t_1, t_2)$ depends only on $\tau = |t_2 - t_1|$. It is wide-sense stationary (WSS) if its statistical mean and variance are time-independent and the autocorrelation function depends only on τ. It is strict-sense stationary (SSS) if its statistics are invariant to the shift in the time axis.

13. The Wiener–Khinchine theorem states that the autocorrelation function $R_X(\tau)$ and the power spectral density (PSD) $S_X(\omega)$ of a stationary random process are a Fourier pair.

$$R_X(\tau) \quad \leftrightarrow \quad S_X(\omega)$$

14. Widely used random processes in communication include the Gaussian process, white noise, and bandlimited processes.

15. For a linear system with impulse response $h(t)$ and transfer function $H(\omega)$ with random input $X(t)$ and output $Y(t)$,

$$S_Y(\omega) = |H(\omega)|^2 S_X(\omega)$$

16. Some of the concepts covered in the chapter are verified using MATLAB.

Review questions

6.1 An experiment consists of rolling a die. What is the probability that the outcome is even?
(a) 0.   (b) 1/6.   (c) 1/3.   (d) 1/2.   (e) 1.

6.2 Which of the following is not a valid PDF?
(a) $f_X(x) = 3e^{-3x}u(x)$
(b) $f_Y(y) = 1, \quad -\dfrac{1}{2} < y < \dfrac{1}{2}$
(c) $f_Z(z) = u(z + 2) - u(z - 2)$
(d) $f_T(t) = \frac{1}{6}(8 - t), \quad 4 \le t \le 10$

6.3 If X is a random variable with mean 10 and variance 6 and $Y = 2X - 1$, the mean of Y is
(a) 10.   (b) 16.   (c) 19.   (d) 20.   (e) Cannot be determined.

6.4 A continuous random variable X takes equal value within its domain, the PDF of X must be:
(a) Uniform.   (b) Poisson.   (c) Gaussian.   (d) Binomial.   (e) Bernoulli.

6.5 Which of the following is not true of Gaussian distribution.
  (a) It is symmetric about the mean.
  (b) The random variable X is uniformly distributed.
  (c) The values of x near the mean are the most often encountered.
  (d) The width of the PDF curve is proportional to the standard deviation.

6.6 If the autocorrelation of a random process is unity, the power spectral density of the process is:
  (a) $\delta(\omega)$.   (b) $2\pi\delta(\omega)$.   (c) $u(\omega)$.   (d) Cannot be determined.

6.7 Let $S_X(\omega) = \frac{4}{\omega^2+4}$ for a random process $X(t)$: what is the corresponding $R_X(\tau)$?
  (a) $e^{-2t}u(t)$.   (b) $e^{-2|t|}$.   (c) $\cos 2t$.   (d) $2\pi\delta(\omega)$.

6.8 Which of the following could not be a valid power density function of a random process?
  (a) $S_X(\omega) = \frac{8}{9+\omega^2}$
  (b) $S_X(\omega) = \frac{5\omega}{1+\omega^2}$
  (c) $S_X(\omega) = 10\pi\delta(\omega)$
  (d) $S_X(\omega) = 4$

6.9 Which of the following is not true about white noise?
  (a) It has a constant spectrum.
  (b) Its spectrum contains all frequencies.
  (c) It has a finite average power.
  (d) It is physically unrealizable.

6.10 The MATLAB command **randn** is for generating a random number that is:
  (a) Gaussian distributed;
  (b) uniformly distributed over (0,1);
  (c) exponentially distributed;
  (d) Poisson distributed.

Answers: 6.1 d, 6.2 c, 6.3 c, 6.4 a, 6.5 b, 6.6 b, 6.7 b, 6.8 b, 6.9 c, 6.10 a.

Problems

Section 6.2 Probability fundamentals

6.1 Four dice are tossed simultaneously. Find the probability that at least one die shows 2.

6.2 An experiment consists of throwing two dice simultaneously. (a) Calculate the probability of having a 2 and a 5 appearing together. (b) What is the probability of the sum being 8?

6.3 A circle is split into 10 equal sectors which are numbered 1 to 10. When the circle is rotated about its center, a pointer indicates where it stops (like a wheel of fortune). Determine the probability: (a) of stopping at number 8; (b) of stopping at an odd number; (c) of stopping at numbers 1, 4, or 6; (d) of stopping at a number greater than 4.

6.4 A jar initially contains four white marbles, three green marbles, and two red marbles. Two marbles are drawn randomly one after the other without replacement. (a) Find the probability that the two marbles are red. (b) Calculate the probability that the two marbles have matching colors.

6.5 Telephone numbers are selected randomly from a telephone directory and the first digit (k) is observed. The result of the observation for 100 telephone numbers is shown below.

k	0	1	2	3	4	5	6	7	8	9
N_k	0	2	18	11	20	13	19	15	1	1

What is the probability that a phone number: (a) starts with 6; (b) begins with an odd number?

6.6 A string of 80 independent bulbs are connected in series to form the Christmas lights. The probability of a bulb being defective is 0.02. The lights are on only when all bulbs are good. Determine the probability that the lights are on.

6.7 There are 50 students in a class. Suppose 20 of them are Chinese and four of the Chinese students are female. Let event A denote "student is Chinese" and event B denote "student is female." Find: (a) $P(A)$; (b) $P(AB)$; (c) $P(B/A)$.

6.8 A bag consists of four blue marbles and six red marbles. An event consists of drawing two marbles from the bag, one at a time. Calculate:
(a) the probability of drawing a red marble given that the first draw is blue;
(b) the probability of drawing a blue marble given that the first draw is red;
(c) the probability of drawing two red marbles;
(d) the probability of drawing two blue marbles.

6.9 A box has 20 electric bulbs. If five of them are bad, what is the probability that a bulb selected randomly from the box is bad? If two bulbs are taken from the box, what is the probability that they are both bad?

6.10 A uniformly distributed random variable X has a PDF given by

$$f_X(x) = \begin{cases} k, & -2 < x < 3 \\ 0, & \text{otherwise} \end{cases}$$

(a) Determine the value of constant k.
(b) Find $F_X(x)$.
(c) Calculate $P(|X| \leq 1)$ and $P(X > 1)$.

6.11 A continuous random variable X has the following PDF

$$f_X(x) = \begin{cases} kx, & 1 < x < 4 \\ 0, & \text{otherwise} \end{cases}$$

(a) Find the value of constant k.
(b) Obtain $F_X(x)$.
(c) Evaluate $P(X \leq 2.5)$.

6.12 Given a PDF

$$f_Z(z) = \mu e^{-z/3} u(z)$$

Determine the value of μ that will make the PDF valid.

6.13 A random variable has a PDF given by

$$f_X(x) = \begin{cases} \dfrac{1}{2\sqrt{x}}, & 0 < x < 1 \\ 0, & \text{otherwise} \end{cases}$$

Find the corresponding $F_X(x)$ and $P(0.5 < X < 0.75)$.

6.14 Given the function

$$f_X(x) = \frac{x^n}{n!} e^{-x}, \quad 0 < x < \infty, \quad n > 0$$

Show that $f_X(x)$ is a PDF.

6.15 A Cauchy random variable X has PDF

$$f_X(x) = \frac{1}{\pi(1 + x^2)}, \quad -\infty < x < \infty$$

Find the corresponding CDF.

6.16 A random variable with PDF

$$f_X(x) = \begin{cases} \dfrac{1}{8}, & 4 < x < 12 \\ 0, & \text{otherwise} \end{cases}$$

Determine: (a) $E[X]$, $E[X^2]$, and Var (X); (b) $P(3 \le X \le 10)$.

6.17 The *skew* is defined as the third moment taken about the mean, i.e.

$$\text{skew}(X) = E\left[(X - m_x)^3\right] = \int_{-\infty}^{\infty} (x - m_x)^3 f_X(x) dx$$

Given that a random variable X has a PDF

$$f_X(x) = \begin{cases} \dfrac{1}{6}, & 4 < x < 10 \\ 0, & \text{otherwise} \end{cases}$$

find skew(X).

6.18 A random variable T represents the lifetime of an electronic component. Its PDF is given by

$$f_T(t) = \frac{t}{\alpha^2} \exp\left[-\frac{t^2}{\alpha^2}\right] u(t)$$

where $\alpha = 10^3$. Find $E[T]$ and Var(T).

Hint: $\displaystyle\int_0^\infty x^2 e^{-x^2} dx = \frac{\sqrt{\pi}}{4}, \quad \int_0^\infty x^3 e^{-x^2} dx = \frac{1}{2}$

6.19 A random variable X has the PDF

$$f_X(x) = \frac{e^{-(x-2)^2/32}}{\sqrt{32\pi}}, \quad -\infty < x < \infty$$

Calculate $P(4 < X < 10)$.

Section 6.3 Special distributions

6.20 A uniform random variable X has $E[X] = 1$ and Var(X) = 1/2. Find its PDF and determine $P(X > 1)$.

6.21 An independent random variable is uniformly distributed with the PDF shown in Figure 6.28. Calculate its mean and variance.

6.22 A continuous random variable X may take any value with equal probability within the interval range 0 to α. Find $E[X]$, $E[X^2]$, and Var(X).

6.23 A random variable X with mean 3 follows an exponential distribution. (a) Calculate $P(X < 1)$ and $P(X > 1.5)$. (b) Determine λ such that $P(X < \lambda) = 0.2$.

6.24 A zero-mean Gaussian random variable has a variance of 9. Find a such $P(|X| > a) < 0.01$.

6.25 A Gaussian random variable X has a mean of 1. If the probability that X lies between 2 and 4 is 0.1, find Var(X).

6.26 Let X have a Gaussian distribution with mean μ and variance σ^2. Use the table or MATLAB to find:
(a) $P(\mu - \sigma < X < \mu + \sigma)$
(b) $P(\mu - 2\sigma < X < \mu + 2\sigma)$
(c) $P(\mu - 3\sigma < X < \mu + 3\sigma)$

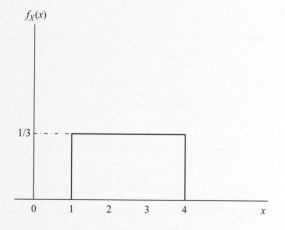

Figure 6.28 For Problem 6.21.

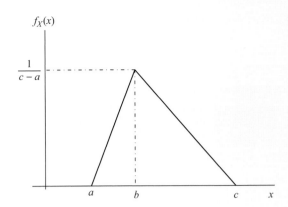

$f_X(x)$

Figure 6.29 For Problem 6.31.

6.27 A measurement of a noise voltage produces a Gaussian random signal with zero mean and variance 2×10^{-11} V^2. Find the probability that a sample measurement exceeds 4 μV.

6.28 Show that the CDF of a Rayleigh random variable is

$$F_X(x) = \left(1 - e^{-x^2/2\sigma^2}\right) u(x)$$

6.29 A random variable is Rayleigh distributed with mean 6. (a) Find Var(X). (b) Calculate $P(X \geq 10)$.

6.30 A random variable X is Rayleigh distributed with PDF

$$f_X(x) = \frac{x}{16} \exp\left(-x^2/32\right) u(x)$$

Find $P(1 < X < 2)$.

6.31 A random variable has triangular PDF as shown in Figure 6.29. Find $E[X]$ and Var(X).

Section 6.4 Random processes

6.32 Let $X(t) = A \cos 2\pi t$ be a random process with A being a random variable uniformly distributed over $(-1, 1)$. Sketch three realizations of the random process.

6.33 If $X(t) = A \sin 4t$, where A is a random variable uniformly distributed between 0 and 2, (a) sketch four sample functions of $X(t)$; (b) find $E[X(t)]$ and $E[X^2(t)]$.

Section 6.5 Power spectral densities

6.34 Determine which of the following are valid PSD functions. State at least one reason why or why not.

(a) $S_X(\omega) = 5 + \delta(\omega - 2)$

(b) $S_X(\omega) = \frac{2}{9+\omega^2}$

(c) $S_X(\omega) = 10 \, \text{sinc}^2(4\omega)$

(d) $S_X(\omega) = e^{-4\omega^2} \cos^2\omega$

(a)

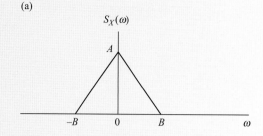

Figure 6.30. For Problem 6.37.

(b)

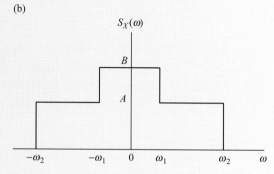

6.35 A stationary random process $X(t)$ has a PSD as

$$S_X(\omega) = 5\pi\delta(\omega) + \frac{2(\omega^2 + 10)}{\omega^4 + 4\omega^2 + 3}$$

(a) Obtain the autocorrelation function of $X(t)$.
(b) Calculate the average power of $X(t)$.

6.36 Find the correlation function for each of the following PSDs:
(a) $S_X(\omega) = \frac{\omega^2}{9+\omega^2} + 4\pi\delta(\omega)$
(b) $S_X(\omega) = \frac{30}{\omega^4+13\omega^2+36}$

6.37 Determine the autocorrelation function for each of the PSDs in Figure 6.30.

6.38 A random process $X(t)$ has

$$R_X(\tau) = 6e^{-4|\tau|}\cos 5\pi\tau$$

(a) Find $E[X(t)]$ and $E[X^2(t)]$.
(b) Determine $\text{Var}[X(t)]$.
(c) Calculate the average power of $X(t)$.
(d) Obtain the PSD.

6.39 Determine $S_X(\omega)$ if:
(a) $R_X(\tau) = 5\delta(\tau)$
(b) $R_X(\tau) = e^{-2\tau^2}\cos \omega_o\tau$
(c) $R_X(\tau) = 2e^{-\tau^2}$
(d) $R_X(\tau) = 4\frac{\sin 2\pi\tau}{2\pi\tau}$

6.40 The autocorrelation function of a random process $X(t)$ is

$$R_X(\tau) = 4 + 6e^{-2|\tau|}$$

Determine: (a) the PSD of $X(t)$; (b) the average power of $X(t)$.

6.41 Two independent stationary random processes $X(t)$ and $Y(t)$ have their PSDs as

$$S_X(\omega) = \frac{\omega^2}{\omega^2 + 4} \quad \text{and} \quad S_Y(\omega) = \frac{4}{\omega^2 + 4}$$

respectively. If $Z(t) = X(t) - Y(t)$, find:
(a) the PSD of $Z(t)$, i.e. $S_Z(\omega)$.
(b) the cross-PSD $S_{XY}(\omega)$.
(c) the cross-PSD $S_{YZ}(\omega)$.

6.42 A random process $X(t)$ has $E[X^2] = 3$ and PSD $S_X(\omega)$. Find $E[Y^2]$ if:
(a) $S_Y(\omega) = 2S_X(\omega)$.
(b) $S_Y(\omega) = S_X(3\omega)$.
(c) $S_Y(\omega) = S_X(\omega/4)$.

6.43 A random process $X(t)$ has its PSD as

$$S_X(\omega) = \frac{124 + 9\omega^2}{(4 + \omega^2)(25 + \omega^2)}$$

By using partial fractions, find the corresponding autocorrelation function.

Section 6.7 Application – linear systems

6.44 A stationary random process $X(t)$ is applied to a network having an input response

$$h(t) = 2te^{-3t}u(t)$$

The cross-correlation of the input $X(t)$ with the output $Y(t)$ is

$$R_{XY}(\tau) = 2\tau e^{-3\tau}u(\tau)$$

Determine the autocorrelation of $Y(t)$ and its average power.

6.45 A random process $X(t)$ is applied to a system having

$$h(t) = 5te^{-2t}u(t)$$

Given that $E[X(t)] = 3$, find $E[Y(t)]$.

6.46 A filter has an impulse response $h(t) = e^{-2t}u(t)$. If the input process $X(t)$ to the filter has

$$S_X(\omega) = \frac{A}{1 + (\omega B)^2}$$

where A and B are constants, find the autocorrelation function and power spectra density of the output.

6.47 Consider the circuit in Figure 6.31. Find $S_Y(\omega)$ in terms of $S_X(\omega)$.

6.48 If a stationary random process $X(t)$ with $R_X(\tau) = 5e^{-2|\tau|}$ is applied to the circuit in Figure 6.32, find $S_X(\omega)$ and $S_Y(\omega)$.

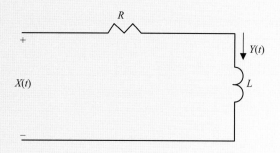

Figure 6.31. For Problem 6.47.

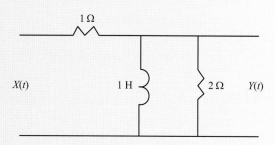

Figure 6.32. For Problem 6.48.

Section 6.8 Computation using MATLAB

6.49 Use MATLAB to generate a random process $X(t) = A \cos(2\pi t)$, where A is a Gaussian random variable with mean zero and variance one. Take $0 < t < 4$ s.

6.50 Repeat the previous problem if A is a random variable uniformly distributed over $(-2, 2)$.

6.51 Given that the autocorrelation function $R_X(\tau) = 2 + 3e^{-\tau^2}$, use MATLAB to plot the function for $-2 < \tau < 2$.

6.52 Use MATLAB to generate a random process

$$X(t) = 2\cos\left(2\pi t + B[n]\frac{\pi}{4}\right)$$

where $B[n]$ is a Bernoulli random sequence taking the values of $+1$ and -1 with $p = 0.6$. Take $0 < t < 3$ s.

6.53 Use MATLAB to generate a Bernoulli process $X(t)$ that assumes values of 1 and 0 with $p = 0.5$. This special case of the Bernoulli process is called the binary white noise process.

7 | Noise in analog communications

Learn all you can from the mistakes of others. You won't have time to make them all yourself.

HISTORICAL PROFILES

Harry Nyquist (1889–1976), American physicist, engineer, a prolific inventor who made fundamental theoretical and practical contributions to telecommunications.

Born in Nilsby, Sweden, Nyquist was the fourth child of eight. He migrated to the United States in 1907 and received a PhD in physics in 1917 at Yale University. During his 37 years of service with Bell Systems, he received 138 US patents and published 12 technical papers. His many important contributions include the invention of the vestigial sideband transmission system, the mathematical explanation of thermal noise, the Nyquist sampling theorem, the Nyquist stability theorem, and the well-known Nyquist diagram for determining the stability of feedback systems. Before his death, Nyquist received many honors for his outstanding work in communications.

Guglielmo Marconi (1874–1937), an Italian electrical engineer, invented the wireless telegraph, which paved the way for the invention of radio.

He was born in Bologna to a rather wealthy Italian father and Irish mother and educated privately in Bologna, Florence, and Leghorn. Even as a boy he took a keen interest in electrical science and studied the works of Maxwell, Hertz, and others. Marconi was determined to turn Hertz's laboratory demonstration into a practical means of communication. In 1895 he began laboratory experiments at the family home near Bologna, where he succeeded in sending wireless signals over a distance of one and a half miles, thus becoming the inventor of the first practical system of wireless telegraphy. The Italian government was not interested in Marconi's work, but the British Admiralty was, and it installed Marconi's radio equipment in some of its ships. Marconi continued to refine and expand upon his inventions in the next few years, and then turned toward the business aspects of his work. Between 1902 and 1912 he patented several new inventions.

He received many honors, including a share in the Nobel Prize for physics in 1909. Although he continued to perform experiments in the new field of radio, which evolved from wireless telegraphy, his later efforts were mainly directed to affairs of state.

7.1 INTRODUCTION

Noise can be found everywhere in nature, particularly in electrical devices. Noise is present in all analog devices, but the digital domain is relatively noise-free. In this chapter, we will investigate the performance of analog communication systems in the presence of noise, while the next chapter is devoted to the performance of digital systems in the presence of noise. The performance is measured using the signal-to-noise (power) ratio (SNR) for analog communication systems or the probability of error for digital communication systems.

High noise levels, resulting in low SNR, can render a communication system unusable. We will assume that modulators, demodulators, filters, and other subsystems are ideal and that the performance of the system is degraded by additive noise at the receiver's input.

The chapter begins by taking a closer look at the sources and types of noise. We then examine the behavior of baseband systems in the presence of noise. The results for such systems serve as a basis for comparison with other systems including linear modulation systems (AM, DSB, SSB, and VSB) and nonlinear modulation schemes (PM and FM). In all cases, our objective is to investigate the performance of the system given that the received signal is contaminated by noise. We compare these modulation systems in terms of their performance. We finally discuss how MATLAB can be used to evaluate the performance of the systems considered in this chapter.

7.2 SOURCES OF NOISE

Noise is any signal which interferes with and corrupts the desired signal. As illustrated in the typical communication system model of Figure 7.1, noise is an inevitable component of any communication system and its effects can degrade the overall performance of the communication system. For this reason, techniques for minimizing the effect of noise are often sought. A typical effect of noise $n(t)$ on received signal $r(t)$ is shown in Figure 7.2. By definition,

> **Noise** is any unwanted electrical or magnetic signal that corrupts a desired signal.

By nature, noise is a random process. Although there are different kinds of noises, they can be grouped into two broad categories: internal noise and external noise. While internal noise is generated by components associated with the signal itself, external noise results from natural or man-made electrical or magnetic phenomena and contaminate the signal as it is being transmitted. Noise produces undesirable effects and degrades the performance of communication

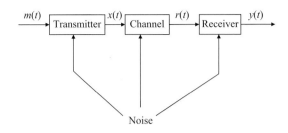

Figure 7.1. Model of a typical communication system.

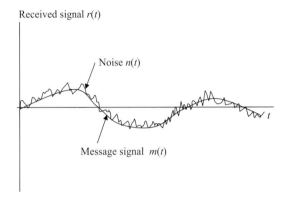

Figure 7.2. Message signal $m(t)$ and noise $n(t)$.

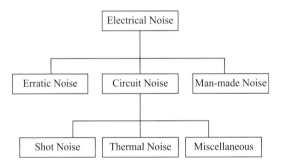

Figure 7.3. Classification of noise.

systems because it limits the ability to correctly recover or identify the sent message and therefore limits information transfer.

Noise sources can be internal or external. Internal or intrinsic noise is due to thermal agitation and other physical behavior of the system components at the molecular level. External sources include alternators, battery chargers, arcs, lightning bolts, ignition systems, power-line switching, fluorescent lights, welding machines, electrical motors, cell or cordless phones, radio/radar transmitters, and computers.

As shown in Figure 7.3, electrical noise can also be classified as erratic noise, man-made noise, or circuit noise. Erratic noise is due to atmospheric disturbances such as lightning discharges, the Sun, and other natural electrical disturbances. Man-made noise is caused by machines or ignition systems which produce an electrical spark. Circuit noise is generated by resistors and active circuit elements such as transistors and op amps. Electrical noise can be annoying when it causes audible or visual interference in radios and televisions. The most common types of electrical noise are *thermal* noise and *shot* noise. Besides these, we recall our discussion of the idealized form of noise known as *white noise* in section 6.6.

7.2.1 Thermal noise

Thermal noise is perhaps the most important type of noise in most communication systems. It is produced by the thermally excited random motion of electrons in a conductor such as a metallic resistor. Thermal noise appears in virtually all circuits that contain resistive devices and is temperature-dependent. The resulting random voltage $v_n(t)$ across the open-circuited terminals of a resistor R at temperature T is known as *thermal noise*.

Thermal noise is caused by thermal agitation of electrons within a conducting medium such as a resistor.

The mean square value of voltage $V_n(t)$ has been confirmed by theory and experiment as

$$V_n^2 = 4kTRB \tag{7.1}$$

where k = Boltzmann's constant = 1.38×10^{-23} joules per kelvin (J/K)
T = absolute temperature in kelvins (K)
R = resistance in ohms (Ω)
B = observation bandwidth in hertz (Hz)

The power density spectrum of $V_n(t)$ is

$$S_n(\omega) = 2kTR \tag{7.2}$$

Equations (7.1) and (7.2) are valid for $\omega \leq 2\pi \times 10^{13}$ rad/s. Owing to this broad range of frequency (about 10,000 GHz), thermal noise behaves as white so that $\eta/2 = 2kTR$ for a resistive thermal source. Thus, a noisy resistor may be modeled by the Thévenin equivalent consisting of a voltage generator of value V_n^2 in series with a noiseless resistor, as shown in Figure 7.4.

7.2.2 Shot noise

Shot (or quantum) noise is the time-dependent variation in electric current caused by the discreteness of the electronic charge. Although we usually think of current as being continuous, it is actually discrete. Shot noise results from the fact that the current is not a continuous flow of electronic charge but the sum of discrete pulses in time, each corresponding to the transfer of an electron through the conductor.

(a) (b)

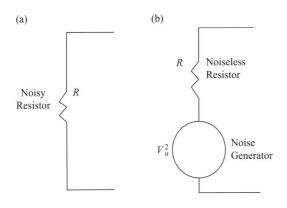

Figure 7.4. A resistor with thermal noise: (a) noisy resistor; (b) its equivalent circuit.

Shot noise is due to random fluctuations in the motion of charge carriers (or electric current) in a conductor.

The term "shot noise" was coined in vacuum-tube days when electrons would strike the metal anode and produce a sound similar to pouring a bucket of shots over a metal surface. Shot noise is well known to occur in electronic devices such as diodes and transistors. It is absent in a metallic resistor because the electron-photon scattering smoothes out current fluctuations that result from the discreteness of the electrons, leaving only thermal noise.

Shot noise depends on the direct current flow and the bandwidth of the device. The mean square value of current $I_n(t)$ is

$$I_n^2 = 2eI_{dc}B \qquad (7.3)$$

where e = electron charge = 1.6×10^{-19} coulombs (C)
B = observation bandwidth in hertz (Hz)
I_{dc} = direct current flowing through the device

Shot noise has an approximately flat PSD like thermal noise. For this reason, shot noise may be considered as "white noise". As shown in Figure 7.5, a noisy diode may be modeled by the Norton equivalent consisting of a current source of value I_n^2 in parallel with a resistor.

In the sections that follow, we will lump the two effects (thermal and shot noise) in the communication system into one source for the purpose of analysis. As illustrated in Figure 7.6, we shall assume an *additive white Gaussian noise* (AWGN) channel model to make the analysis tractable. With that assumption, we may express the received signal $R(t)$ as

$$R(t) = m(t) + n(t) \qquad (7.4)$$

where $n(t)$ is a zero-mean, white Gaussian process with PSD

$$S_n(\omega) = 2\pi\left(\frac{\eta}{2}\right) \text{ watts/rad/s}, \quad -\infty < \omega < \infty \qquad (7.5)$$

(a)   (b)

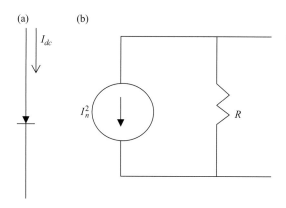

Figure 7.5. A model of shot noise: (a) a noisy diode; (b) its equivalent circuit.

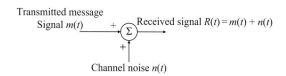

Figure 7.6. Model of additive noise channel.

It is customary to design systems in such a way that they will perform well in the presence of AWGN. For the remaining portion of this chapter, we shall discuss how various communication systems behave in the presence of an AWGN channel model.

EXAMPLE 7.1

A 2-MΩ resistor is maintained at a temperature of 23 °C. (a) Find the thermal noise voltage and the equivalent current if the bandwidth of interest is 500 kHz. (b) Calculate the mean noise power.

Solution

(a) We first convert the temperature from Celsius to Kelvin, i.e.

$$T = 23 + 273 = 296 \text{ K}$$

$$V_n^2 = 4kTRB = 4 \times 1.38 \times 10^{-23} \times 296 \times 2 \times 10^6 \times 500 \times 10^3$$
$$= 1.634 \times 10^{-8}$$

or the rms value is

$$V_n = 1.278 \times 10^{-4} = 127.8 \text{ μV}$$

which is quite small and would be insignificant for applications in which the signal level is of the order of volts or millivolts. The equivalent current is obtained as

$$I_n^2 = \frac{V_n^2}{R^2} = \frac{4kTRB}{R^2} = \frac{1.634 \times 10^{-8}}{4 \times 10^{12}} = 0.4085 \times 10^{-20}$$

or the rms value is

$$I_n = 0.6391 \times 10^{-10} = 63.91 \text{ pA}$$

which happens to be the same as V_n/R.

(b) The average power is

$$P = \frac{V_n^2}{R} = kTB = \frac{1.278 \times 10^{-4}}{2 \times 10^6} = 63.9 \text{ pW}$$

which is very small.

PRACTICE PROBLEM 7.1

A 2-A direct current flows through a 5-kΩ resistor and a temperature-limiting diode which has a limiting bandwidth of 15 kHz. Determine the mean square of the noise current and corresponding voltage.

Answer: 9.6×10^{-15} A^2, 2.4×10^{-7} V^2.

7.3   BASEBAND SYSTEMS

Baseband systems are important because they serve as a standard against which other systems are compared. A baseband communication system is one in which there is no modulation or demodulation; the signal is transmitted directly. This mode of transmission is appropriate for twisted pairs, coaxial cables, or optical fiber.

A baseband system is shown in Figure 7.7, where it is evident that the transmitter and receiver are ideal low-pass (or baseband) filters (LPF) with bandwidth B Hz. The LPF at the transmitter restricts the input signal spectrum to a specific bandwidth, while the LPF at the receiver eliminates the out-of-band noise (noise outside the band) as shown in Figure 7.8.

We assume that the message signal $m(t)$ is a zero-mean ergodic random process band-limited to $W \ (= 2\pi B)$. If the channel is distortionless over the message band, the average signal power S_o at the output of the receiver is the same as the average signal power S_i at the input of the receiver. Thus,

$$S_o = S_i \tag{7.6}$$

The average noise power at the output of the receiver is

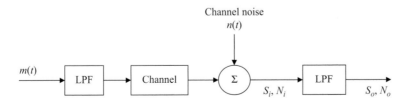

Figure 7.7. A baseband system.

(a)

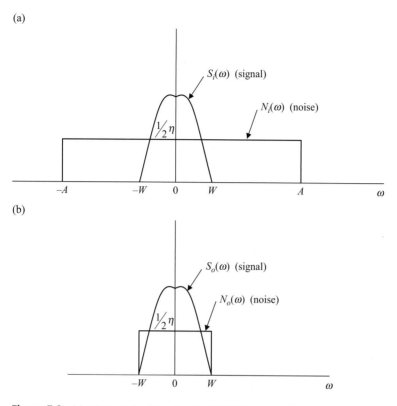

(b)

Figure 7.8. (a) PSDs at the filter input; (b) PSDs at the filter output.

$$N_o = E\left[n_o^2(t)\right] = \frac{1}{2\pi} \int_{-W}^{W} S_n(\omega)d\omega \tag{7.7}$$

where $S_n(\omega)$ is the power spectral density (PSD) of noise signal $n(t)$. To simplify the calculations, if we assume white noise, $S_n(\omega) = \eta/2$ (W/Hz). For this case,

$$N_o = \frac{1}{2\pi} \int_{-W}^{W} \frac{\eta}{2} d\omega = \frac{\eta W}{2\pi} = \eta B \tag{7.8}$$

The output SNR is

$$\left(\frac{S}{N}\right)_o = \frac{S_o}{N_o} = \frac{S_i}{\eta B} \tag{7.9}$$

If we let

$$\gamma = \frac{S_i}{\eta B} \tag{7.10}$$

then

$$\boxed{\frac{S_o}{N_o} = \gamma = \frac{S_i}{\eta B}} \tag{7.11}$$

This value of SNR serves as a basis of comparing the output SNR of other communication systems.

EXAMPLE 7.2

A received signal consists of the message signal $4 \cos 2\pi \times 10^4 t$ and white noise with power spectral density 0.002 W/Hz. The received signal is passed through a bandpass filter with a passband between 2000 and 2500 Hz. Find the SNR at the output of the filter.

Solution

In the steady state, we can assume that the entire message signal $m(t)$ appears at the output so that

$$S_o = \frac{A^2}{2} = \frac{4^2}{2} = 8 \text{ W}$$

The noise power is

$$N_o = 2 \left(\frac{1}{2\pi} \int_{\omega_1}^{\omega_2} S_n(\omega) d\omega \right) = 2 \int_{f_1}^{f_2} S_n(f) df, \quad \omega = 2\pi f$$

$$= 2 \int_{2000}^{2500} 0.002 \, df = 2 \text{ W}$$

Hence, the output SNR is

$$\frac{S_o}{N_o} = \frac{8}{2} = 4 \quad \text{or} \quad 10 \log_{10} 4 = 6.021 \text{ dB}$$

PRACTICE PROBLEM 7.2

In a given communication system, noise of spectral density $S_n(f) = e^{-2|f|}$ is added to a signal $m(t) = 0.3 \sin 4\pi t$ and the total received signal forms the input to an ideal bandpass filter with a passband from 1.5 to 3.0 Hz. Find the SNR in dB at the output of the filter.

Answer: 8.88 dB.

EXAMPLE 7.3

The PSDs of a signal and noise are shown in Figures 7.9(a) and (b) respectively.

The sum of the signal and noise is passed through a bandpass filter whose transfer function is shown in Figure 7.9(c). Find the SNR at the input and at the output of the filter.

Solution

We obtain the powers by finding areas under the corresponding PSDs, i.e.

$$S_i = 2 \int_{50}^{120} S_m(f)df = 2(120 - 50) \times 20 = 2800 \text{ W}$$

$$N_i = 2 \int_{0}^{400} S_n(f)df = 2(400 - 0) \times 1 = 800 \text{ W}$$

The input SNR is

$$\frac{S_i}{N_i} = \frac{2800}{800} = 3.5 \text{ or } 5.441 \text{ dB}$$

We recall from Chapter 6 that given a linear system with transfer function $H(f)$, the PSD of the output can be obtained from the PSD of the input by using

$$S_Y(f) = |H(f)|^2 S_x(f)$$

Hence, when the signal and noise are passed through the bandpass filter with the transfer function in Figure 7.9(c), we obtain the PSDs of the output signal and the output noise by multiplying the input PSDs by the square of the filter's $H(f)$. Doing that produces the PSDs in Figure 7.10 for the output signal and output noise. Thus,

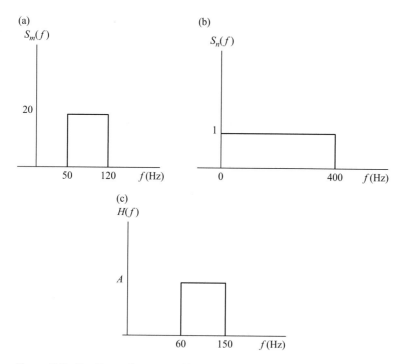

Figure 7.9. For Example 7.3: (a) PSD of signal $m(t)$; (b) PSD of noise $n(t)$; (c) transfer function of bandpass filter.

(a)

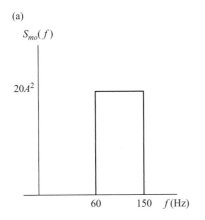

$S_{mo}(f)$

$20A^2$

60   150 f(Hz)

(b)

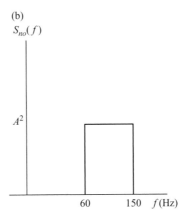

$S_{no}(f)$

A^2

60   150 f(Hz)

Figure 7.10. For Example 7.3: (a) PSD of the signal at the output; (b) PSD of the noise at the ouput.

$$S_o = 2 \int_{60}^{150} S_m(f)df = 2(150 - 60) \times 20A^2 = 3600A^2 \text{ W}$$

$$N_o = 2 \int_{60}^{150} S_n(f)df = 2(150 - 60) \times A^2 = 180A^2 \text{ W}$$

The output SNR is

$$\frac{S_o}{N_o} = \frac{3600A^2}{180A^2} = 20 \text{ or } 13.01 \text{ dB}$$

i.e. there is SNR improvement of $13.01 - 5.441 = 8.57$ dB. Note the result is not affected by A since both the signal and noise are multiplied by A.

PRACTICE PROBLEM 7.3

Find the output SNR in Example 7.3 if the bandpass filter passes from 50 to 100 Hz instead of 60 to 150 Hz.

Answer: 13.01 dB.

7.4 AMPLITUDE-MODULATION SYSTEMS

We now examine the effect of noise on various kinds of amplitude-modulated (AM) signals. We want to find the SNR of the output of the receiver that demodulates AM signals for DSB-SC, SSB-SC, and AM systems.

7.4.1 DSB system

The double-sideband suppressed carrier (DSB-SC) demodulator is illustrated in Figure 7.11. The transmitted DSB signal is assumed to be given by

$$x(t) = Am(t)\cos\omega_c t \tag{7.12}$$

where $m(t)$ is the message signal. The input signal power is

$$S_i = E[x^2(t)] = \frac{1}{2}E[A^2 m^2(t)] = \frac{1}{2}A^2 S_m \tag{7.13}$$

The noise can be expanded into its direct and quadrature components as

$$n(t) = n_c(t)\cos\omega_c t + n_s(t)\sin\omega_c t \tag{7.14}$$

Thus the received signal is given by

$$R(t) = Am(t)\cos\omega_c t + n_c(t)\cos\omega_c t + n_s(t)\sin\omega_c t \tag{7.15}$$

We demodulate the signal by first multiplying the received signal $R(t)$ by a locally generated sinusoid $2\cos\omega_c t$ (synchronous demodulation) and then passing the resulting signal through an LPF. The multiplication yields

$$\begin{aligned} 2R(t)\cos\omega_c t &= 2Am(t)\cos\omega_c t\cos\omega_c t + 2n_c(t)\cos\omega_c t\cos\omega_c t \\ &\quad + 2n_s(t)\sin\omega_c t\cos\omega_c t \\ &= Am(t) + Am(t)\cos2\omega_c t + n_c(t) + n_c(t)\cos2\omega_c t + n_s(t)\sin2\omega_c t \end{aligned} \tag{7.16}$$

The LPF passes only the low-frequency components and rejects the double-frequency components so that the output becomes

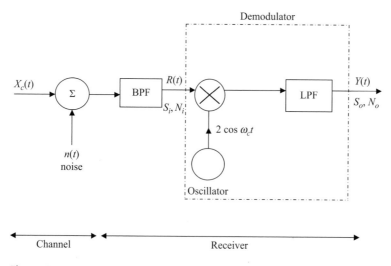

Figure 7.11. A DSB system.

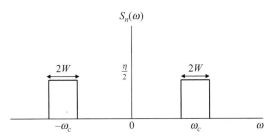

Figure 7.12. DSB demodulator input noise spectrum.

$$Y(t) = Am(t) + n_c(t) \tag{7.17}$$

indicating that the message signal and noise are additive at the receiver output. The output message signal power is

$$S_o = E\left[A^2 m^2(t)\right] = A^2 E\left[m^2(t)\right] = A^2 S_m = 2S_i \tag{7.18}$$

The noise power is given by

$$N_o = E\left[n_c^2(t)\right] = E\left[n^2(t)\right] \tag{7.19}$$

Again, if we assume white noise with double-sided power spectral density $S_n(\omega) = \eta/2$ (watts/Hz) with its spectrum centered at ω and a bandwidth of $2W$, as illustrated in Figure 7.12, then

$$N_o = \frac{1}{2\pi}(2)\int_{-W}^{W} \frac{\eta}{2} d\omega = \frac{\eta W}{\pi} = 2\eta B \tag{7.20}$$

The output SNR is

$$\left(\frac{S}{N}\right)_o = \frac{S_o}{N_o} = \frac{2S_i}{2\eta B} \tag{7.21}$$

or

$$\boxed{\frac{S_o}{N_o} = \gamma} \tag{7.22}$$

This indicates that the output SNR of the DSB is the same as that for a baseband system. Therefore, the DSB system has the same noise performance as a baseband system.

7.4.2 SSB system

The single-sideband system is portrayed in Figure 7.13. The SSB signal is given by

$$X(t) = m(t)\cos\omega_c t \pm m_h(t)\sin\omega_c t \tag{7.23}$$

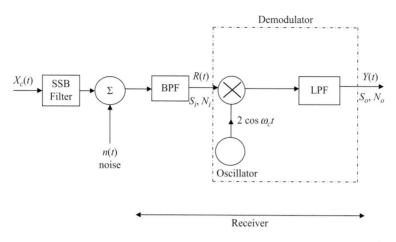

Figure 7.13. An SSB system.

where $m_h(t)$ is the Hilbert transform of $m(t)$: the plus sign represents the lower sideband, and the minus sign represents the upper sideband. The input signal power is

$$S_i = E\left[X^2(t)\right] = \frac{1}{2}E\left[m^2(t)\right] + \frac{1}{2}E\left[m_h^2(t)\right] = E\left[m^2(t)\right] = S_m \tag{7.24}$$

since

$$E[m(t)m_h(t)] = 0 \quad \text{and} \quad E[m^2(t)] = E\left[m_h^2(t)\right]$$

The received signal is

$$\begin{aligned} R(t) &= X(t) + n(t) \\ &= [m(t) + n_c(t)]\cos\omega_c t + [m_h(t) + n_s(t)]\sin\omega_c t \end{aligned} \tag{7.25}$$

Parallel to our discussion of DSB, we multiply this signal by a locally generated sinusoid $2\cos\omega_c t$ and then pass the resulting signal through an LPF. The quadrature components of both the signal and noise are rejected so that the demodulator output is

$$Y(t) = m(t) + n_c(t) \tag{7.26}$$

Hence the output signal power is

$$S_o = E\left[m^2(t)\right] = S_m = S_i \tag{7.27}$$

The output noise power is given by

$$N_o = E\left[n_c^2(t)\right] = E\left[n^2(t)\right] \tag{7.28}$$

Again, we assume white noise with double-sided PSD $S_n(\omega) = \eta/2$ (watts/Hz). The power spectral densities of both $n(t)$ and $n_c(t)$ are shown in Figure 7.14. Hence.

$$N_o = \frac{1}{2\pi}\int_{-W}^{W}\frac{\eta}{2}d\omega = \frac{\eta W}{2\pi} = \eta B \tag{7.29}$$

(a)

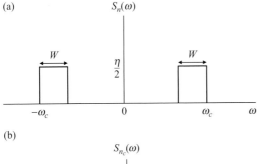

Figure 7.14. PSD of: (a) $n(t)$; (b) $n_c(t)$.

(b)

The output SNR is

$$\left(\frac{S}{N}\right)_o = \frac{S_o}{N_o} = \frac{S_i}{\eta B} \tag{7.30}$$

or

$$\frac{S_o}{N_o} = \gamma \tag{7.31}$$

Notice that the SNR in the SSB system is identical to that of a DSB system. However, the transmission bandwidth for the DSB system is twice that for the SSB system. Thus, for a given transmitted power and transmission bandwidth, the baseband, DSB-SC, and SSB-SC systems all give the same output SNR.

7.4.3 AM synchronous demodulation

We will consider both synchronous demodulation and envelope demodulation. The case of synchronous demodulation is similar to DSB-SC except that the carrier is not suppressed but added. The modulated signal is given by

$$x(t) = A[1 + \mu m(t)] \cos \omega_c t \tag{7.32}$$

where $m(t)$ is the message signal (or baseband signal that amplitude-modulates the carrier $A \cos \omega_c t$) and μ is the modulation index (of the AM signal) which determines the percentage modulation

$$0 < \mu < 1 \tag{7.33}$$

The input signal power is

$$S_i = E[x^2(t)] = \frac{1}{2}E\left[\{A[1 + \mu m(t)]\}^2\right]$$
$$= \frac{1}{2}E\left[\{A^2 + 2A^2\mu m(t) + A^2\mu^2 m^2(t)\}^2\right] = \frac{1}{2}A^2\left(1 + \mu^2 S_m\right) \tag{7.34}$$

where $E[m(t)] = 0$ since $m(t)$ is assumed to have a zero mean. The signal received by the demodulator is

$$R(t) = x(t) + n(t)$$
$$= A[1 + \mu m(t)]\cos\omega_c t + n_c(t)\cos\omega_c t + n_s(t)\sin\omega_c t \tag{7.35}$$

We follow the same argument made for the DSB system. After multiplying by $2\cos\omega_c t$, low-pass-filtering, and dc blocking, we get the output signal to be

$$Y(t) = A\mu m(t) + n_c(t) \tag{7.36}$$

The output signal power is

$$S_o = E\left[\mu^2 A^2 m^2(t)\right] = \mu^2 A^2 S_m = \frac{2\mu^2 S_m}{1 + \mu^2 S_m}S_i \tag{7.37}$$

and the output noise power is given by

$$N_o = E\left[n_c^2(t)\right] = E\left[n^2(t)\right] = 2\eta B \tag{7.38}$$

Hence, the output SNR is obtained

$$\left(\frac{S}{N}\right)_o = \frac{S_o}{N_o} = \frac{2\mu^2 S_m S_i}{1 + \mu^2 S_m}\frac{1}{2\eta B} = \frac{\mu^2 S_m}{1 + \mu^2 S_m}\left(\frac{S_i}{\eta B}\right)$$

or

$$\boxed{\frac{S_o}{N_o} = \frac{\mu^2 S_m}{1 + \mu^2 S_m}\gamma} \tag{7.39}$$

Since $\mu^2 S_m \leq 1$,

$$\frac{S_o}{N_o} \leq \frac{\gamma}{2} \tag{7.40}$$

This shows that the output SNR in AM is at least 3 dB worse than that in DSB-SC and SSB-SC depending on the modulation index μ. In practice, the degradation of an AM system is usually larger than 3 dB. This inferior performance is due to the wastage of transmitted power caused by transmitting the carrier along with the message. Of course, the motivation for employing AM is not noise performance; it is rather taking advantage of the fact that AM allows the use of simple, economical envelope detection for demodulation.

7.4.4 AM envelope detection

Envelope detection is the common way of demodulating an AM signal partly because an envelope detector is cheap. This is also the reason it is widely used in AM broadcast receivers.
    As usual, the modulated signal is

$$x(t) = A[1 + \mu m(t)] \cos \omega_c t \tag{7.41}$$

with power

$$S_i = E\left[x^2(t)\right] = \frac{1}{2}A^2\left(1 + \mu^2 S_m\right) \tag{7.42}$$

while the noise is given by

$$n(t) = n_c(t) \cos \omega_c t + n_s(t) \sin \omega_c t \tag{7.43}$$

The received signal at the detector input is

$$\begin{aligned} R(t) &= x(t) + n(t) \\ &= \{A[1 + \mu m(t)] + n_c(t)\} \cos \omega_c t + n_s(t) \sin \omega_c t \end{aligned} \tag{7.44}$$

This can be written in polar form as

$$R(t) = E_R(t) \cos \{\omega_c t - \varphi_R(t)\} \tag{7.45}$$

where the envelope $E_R(t)$ and the phase $\varphi_R(t)$ are given by

$$E_R(t) = \sqrt{\{A[1 + \mu m(t)] + n_c(t)\}^2 + n_s^2(t)} \tag{7.46}$$

and

$$\varphi_R(t) = \tan^{-1}\left[\frac{n_s(t)}{\{A[1 + \mu m(t)] + n_c(t)\}}\right] \tag{7.47}$$

$E_R(t)$ is the output of the detector. We will consider two extreme cases of small noise and large noise. We will also consider the intermediate case.

Case 1: Small noise
In this case, the signal dominates. In other words, we have large signal-to-noise ratio (SNR $\gg$ 1).
If

$$|A[1 + \mu m(t)]| \gg |n(t)| = \sqrt{n_c(t)^2 + n_s^2(t)} \text{ for all } t, \tag{7.48}$$

then

$$|A[1 + \mu m(t)]| \gg |n_c(t)| \text{ or } |n_s(t)| \text{ for all } t \tag{7.49}$$

Under this condition, we may approximate Eq. (7.46) as

$$E_R(t) = A[1 + \mu m(t)] + n_c(t) \tag{7.50}$$

The term A, the dc component, is blocked by a capacitor so that the output of the envelope detector is

$$Y(t) = A\mu m(t) + n_c(t) \tag{7.51}$$

which is the same as Eq. (7.36) obtained for the synchronous detector. Hence, we obtain an identical result in Eq. (7.39), namely

$$\boxed{\frac{S_o}{N_o} = \frac{\mu^2 S_m}{1 + \mu^2 S_m} \gamma} \tag{7.52}$$

Thus, for a large SNR (when the noise is small in comparison with the signal), the envelope detector performs the same way as the synchronous detector.

Case 2: Large noise

In this case, the noise dominates. In other words, the noise is large compared with the signal so that SNR $\ll 1$. The reverse of Eqs. (7.48) and (7.49) takes place, namely

$$|n(t)| = \sqrt{n_c(t)^2 + n_s^2(t)} \gg |A[1 + \mu m(t)]| \text{ for all } t, \tag{7.53}$$

then

$$|n_c(t)| \text{ or } |n_s(t)| \gg |A[1 + \mu m(t)]| \text{ for all } t \tag{7.54}$$

Hence, we may approximate Eq. (7.46) as

$$
\begin{aligned}
E_R(t) &= \sqrt{\{A[1 + \mu m(t)] + n_c(t)\}^2 + n_s^2(t)} \\
&= \sqrt{A^2[1 + \mu m(t)]^2 + 2A[1 + \mu m(t)]n_c(t) + n_c^2(t) + n_s^2(t)} \\
&\simeq \sqrt{[n_c^2(t) + n_s^2(t)]\left[1 + \frac{2An_c(t)}{n_c^2(t) + n_s^2(t)}\right][1 + \mu m(t)]}
\end{aligned}
\tag{7.55}
$$

where the condition in Eq. (7.53) has been applied. If we employ the approximation

$$\sqrt{1 + x} \approx 1 + \frac{x}{2} \tag{7.56}$$

to Eq. (7.55), we obtain

$$
\begin{aligned}
E_R(t) &= E_n(t)\left[1 + \frac{An_c(t)}{E_n^2(t)}\sqrt{[1 + \mu m(t)]}\right] \\
&= E_n(t) + \frac{n_c(t)}{E_n(t)}A\sqrt{[1 + \mu m(t)]}
\end{aligned}
$$

or

$$E_R(t) = E_n(t) + A\sqrt{[1 + \mu m(t)]}\cos\phi_n(t) \tag{7.57}$$

where $E_n(t)$ and $\phi_n(t)$ are the envelope and phase of the noise $n(t)$ respectively, i.e.

$$E_n(t) = \sqrt{n_c^2(t) + n_s^2(t)}$$ (7.58)

and

$$\phi_n(t) = -\tan^{-1}\frac{n_s(t)}{n_c(t)}$$ (7.59)

We observe from Eq. (7.57) that there is no term which is proportional to the message signal $m(t)$.

In fact, the signal is multiplied by a random noise term, $\cos\phi_n(t)$. Under this condition, the signal is buried in the noise and cannot be recovered from the envelope detector. Therefore, it is meaningless to talk about the output SNR for this case.

Case 3: Threshold

There is an intermediate or transition case where the message signal power is approximately equal to the input noise power. As the SNR decreases from high to low value, a *threshold* is reached.

> The **threshold** is a point above which the effect of noise is negligible and below which the performance of the system deteriorates rapidly.

Although the threshold effect does not occur in coherent demodulation, it is not unique to envelope detection; it is also found in frequency modulation. For this reason, coherent demodulation is often preferred when the SNR is low.

A detailed analysis of the threshold effect is complex. We shall merely state the formula for the SNR as

$$\frac{S_o}{N_o} \simeq 0.916A^2\mu^2 S_m\gamma^2$$ (7.60)

Threshold takes place when γ is of the order of 10 or less.

7.4.5 Square-law detector

The linear envelope detector described by Eq. (7.46) is hard to analyze over a wide range of SNR due to the square root operation involved. The square-law detector can be realized by replacing the envelope detector in Figure 7.15 with a squaring device followed by a low-pass

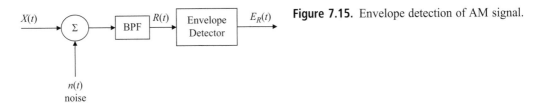

Figure 7.15. Envelope detection of AM signal.

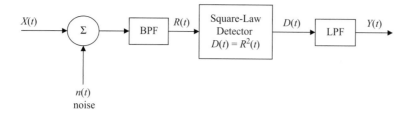

Figure 7.16. Square-law detection of AM signal.

filter, as shown in Figure 7.16. In other words, the square-law detector squares its input $R(t)$ and then passes low-frequency components. From Eq. (7.44), the output of the square-law detector is

$$D(t) = R^2(t) = \{x(t) + n(t)\}^2$$
$$= \{A[1 + \mu m(t)] \cos \omega_c t + n(t)\}^2$$
$$= A^2 \cos^2 \omega_c t + 2A^2 \mu m(t) \cos \omega_c t + A^2 \mu^2 m^2(t) \cos^2 \omega_c t$$
$$+ 2n(t)A \cos \omega_c t + 2n(t)A\mu m(t) \cos \omega_c t + n^2(t) \qquad (7.61)$$

Since $\cos^2 x = \frac{1}{2}(1 + \cos 2x)$, we can simplify Eq. (7.61) to become

$$D(t) = \frac{1}{2}A^2 + \frac{1}{2}A^2 \cos 2\omega_c t + 2A^2 \mu m(t) \cos \omega_c t + \frac{1}{2}A^2 \mu^2 m^2(t) + \frac{1}{2}A^2 \mu^2 m^2(t) \cos 2\omega_c t$$
$$+ 2n(t)A \cos \omega_c t + 2n(t)A\mu m(t) \cos \omega_c t + n^2(t) \qquad (7.62)$$

We now drop terms whose PSDs lie outside the LPF. The first term on the right-hand side is the dc component, which is blocked assuming that the detector output is ac-coupled. The spectrum of the message signal $m(t)$ extends to W ($=2\pi B$) so that the spectrum of $m^2(t)$ extends to $2W$. Hence, we drop the $m(t) \cos 2\omega_c t$ and $m^2(t) \cos 2\omega_c t$ because their PSDs extend over the range $2\omega_c \pm W$ and $2\omega_c \pm 2W$ respectively. Thus, at the output of the LPF, we obtain

$$Y(t) = \frac{1}{2}A^2 \mu^2 m^2(t) + 2A^2 \mu m(t) \cos \omega_c t$$
$$+ 2n(t)A \cos \omega_c t + 2n(t)A\mu m(t) \cos \omega_c t + n^2(t) \qquad (7.63)$$

Assuming that $|\mu m(t)| \ll 1$, which is required in order to avoid envelope distortion, we get

$$Y(t) \simeq 2A^2 \mu m(t) \cos \omega_c t + 2n(t)A \cos \omega_c t + n^2(t) \qquad (7.64)$$

The output signal power is

$$S_o = E\left[\{2A^2 \mu m(t) \cos \omega_c t\}^2\right] = A^4 \mu^2 E\left[m^2(t)\right] = A^4 \mu^2 S_m \qquad (7.65)$$

The output noise power is

$$N_o = E\left[\left\{2n(t)A\cos\omega_c t + n^2(t)\right\}^2\right]$$

$$= E\left[4n^2(t)A^2\cos^2\omega_c t + 2n^3(t)A\cos\omega_c t + n^4(t)\right]$$

$$= 2A^2 E[n^2] + 0 + E[n^4(t)]$$

$$= N_1 + N_2 \qquad (7.66)$$

where N_1 and N_2 are due to the terms involving $E[n^2(t)]$ and $E[n^4(t)]$ respectively. Again, if we assume white noise so that $S_n(\omega) = \eta/2$, then

$$N_1 = 2A^2 E[n^2] = 2A^2 \frac{1}{2\pi}\int_{-W}^{W} S_n(\omega)d\omega = \frac{A^2 W\eta}{\pi} = 2A^2\eta B \qquad (7.67)$$

To find N_2, we utilize the fact that the PSD due to $n^2(t)$ can be expressed in terms of the PSD of $n(t)$ using

$$S_{n^2}(\omega) = 2\pi E\left[n^2(t)\right]\delta(\omega) + \frac{1}{\pi}S_n(\omega) * S_n(\omega) \qquad (7.68)$$

where the last term involves the convolution of $S_n(\omega) = \eta/2$ by itself. Both $S_n(\omega)$ and $S_{n^2}(\omega)$ are shown in Figure 7.17. Assuming that the dc term is blocked, we simply find the area of the shaded portion (which is two trapezoids back to back) to get N_2. Thus,

$$N_2 = E[n^2(t)] = \frac{1}{2\pi}\int_{-W}^{W} S_{n^2}(\omega)d\omega$$

$$= \frac{1}{2\pi}(\text{area of shaded portion}) = \frac{1}{2\pi}2\frac{1}{2}W\left(2\eta^2 B + \eta^2 B\right)$$

$$= \frac{W}{2\pi}\left(3\eta^2 B\right) = 3\,\eta^2 B^2 \qquad (7.69)$$

where the formula for finding the area of a trapezoid (Area $= \frac{1}{2}[a+b]h$) has been applied. Hence,

$$N_o = N_1 + N_2 = 2A^2\eta B + 3\eta^2 B^2 \qquad (7.70)$$

From Eqs. (7.65) and (7.70), we obtain the output SNR as

$$\boxed{\frac{S_o}{N_o} = \frac{A^4\mu^2 S_m}{2A^2\eta B + 3\eta^2 B^2}} \qquad (7.71)$$

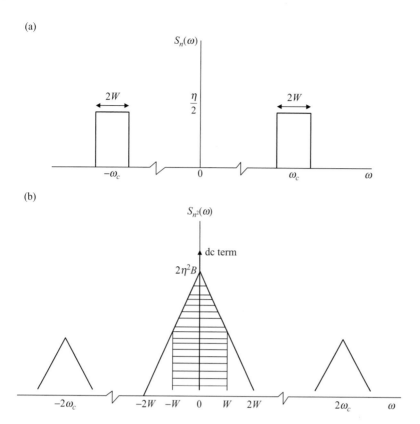

Figure 7.17. Noise PSDs: (a) $S_n(\omega)$; (b) $S_{n^2}(\omega)$.

With $|\mu m(t)| \ll 1$, the input power is obtained from Eq. (7.42) as

$$S_i = \frac{1}{2}A^2 \qquad \text{or} \qquad A^2 = 2S_i \tag{7.72}$$

Substituting for A in Eq. (7.71) yields

$$\frac{S_o}{N_o} = \frac{4\mu^2 S_i S_m}{4S_i + 3\eta B}\left(\frac{S_i}{\eta B}\right) = \frac{\mu^2 S_m}{1 + \frac{3}{4\gamma}}\gamma \tag{7.73}$$

If we consider the two special cases of high-input SNR ($\gamma \gg 1$) and low-input SNR ($\gamma \ll 1$), we obtain

$$\frac{S_o}{N_o} = \begin{cases} \mu^2 S_m \gamma, & \gamma \gg 1 \\ \frac{4}{3}\mu^2 S_m \gamma^2, & \gamma \ll 1 \end{cases} \tag{7.74}$$

This indicates that the square-law detector experiences the threshold effect just like the envelope detector. The performance of the square-law detector deteriorates rapidly below threshold.

EXAMPLE 7.4

An audio signal $m(t)$ with a bandwidth of 12 kHz is transmitted with additive white noise with PSD $\frac{\eta}{2} = 10^{-8}$ W/Hz. It is required that the output SNR be at least 50 dB. Assuming $S_m = 0.4$ and a 30 dB power loss, calculate the necessary transmission bandwidth B_T and the average required transmitted power S_t for:

(a) DSB AM,
(b) SSB AM,
(c) conventional AM (envelope detection) with $\mu = 1$.

Solution

(a) For DSB AM,

$$B_T = 2B = 24 \text{ kHz}$$

Since $50 \text{ dB} = 10\log_{10}x \quad \rightarrow \quad x = 10^{50/10} = 10^5$,

$$\frac{S_o}{N_o} = \gamma = \frac{S_i}{\eta B} = \frac{S_i}{2(10^{-8})(12 \times 10^3)} = 10^5 \quad \rightarrow \quad S_i = 24 \text{ W}$$

Since there is a power loss of 30 dB or 10^3, the required power transmitted is

$$S_t = 10^3 S_i = 24 \text{ kW}$$

(b) For SSB,

$$B_T = B = 12 \text{ kHz}$$

Since the SNR is the same,

$$S_t = 10^3 S_i = 24 \text{ kW}$$

(c) For the envelope detector,

$$B_T = 2B = 24 \text{ kHz}$$

$$\frac{S_o}{N_o} = \frac{\mu^2 S_m}{1 + \mu^2 S_m}\gamma = \frac{1(0.4)}{1 + 0.4}\frac{S_i}{2(10^{-8})(12 \times 10^3)} = 10^5$$

or

$$S_i = 84 \text{ W}$$

The required power transmitted is

$$S_t = 10^3 S_i = 84 \text{ kW}$$

PRACTICE PROBLEM 7.4

A communication system transmits a message with a bandwidth of 1 MHz. If the transmitted power is 30 kW, the channel attenuation is 60 dB, and the noise PSD is 0.5 nW/Hz, determine the output SNR in dB if the modulation scheme is: (a) DSB; (b) SSB; (c) conventional AM with $\mu = 1$ and $S_m = 0.2$

Answer: (a) 14.44 dB; (b) 14.44 dB; (c) 6.99 dB.

EXAMPLE 7.5

A message signal $m(t)$ with an amplitude distribution of $f_m(t) = \frac{1}{2}e^{-|t|}$ is demodulated by an envelope detector. Assuming that the modulation index $\mu = 1$, find the value of γ at the threshold if the threshold occurs when $E_n < A$ with a probability of 0.99.

Solution

$$S_m = E\left[m^2(t)\right] = \int_{-\infty}^{\infty} t^2 f_m(t)dt = \frac{1}{2}\int_{-\infty}^{\infty} t^2 e^{-|t|}dt = \int_{0}^{\infty} t^2 e^{-t}dt = 2$$

$$P(E_n < A) = 0.99 \quad \rightarrow \quad P(E_n \geq A) = 0.01$$

We recall that the noise envelope $E_n(t) = \sqrt{n_c^2(t) + n_s^2(t)}$, where n_c and n_s are Gaussian processes with variance σ_n^2, it follows that E_n is a Rayleigh process. Hence,

$$0.01 = P(E_n \geq A) = \int_{A}^{\infty} \frac{E_n}{\sigma_n^2} e^{-E_n^2/2\sigma_n^2} dE_n = e^{-A^2/2\sigma_n^2}$$

or $\frac{A^2}{2\sigma_n^2} = -\ln(0.01) = 4.6052$.

Since we assume a zero-mean Gaussian noise and a bandwidth of $2B$,

$$\sigma_n^2 = E[n(t)] = 2\left(\frac{\eta}{2}\right)2B = 2\eta B$$

so that

$$A^2 = 4.6052(4\eta B)$$

For envelope detection with $\mu = 1$, Eq. (7.42) gives

$$S_i = E[x_c(t)] = \frac{1}{2}A^2\left(1 + \mu^2 S_m\right) = \frac{3}{2}A^2$$

Thus, at threshold,

$$\gamma_{Th} = \frac{S_i}{\eta B} = \frac{\frac{3}{2}(4.6052)(4\eta B)}{\eta B} = 27.63 \text{ or } 14.41 \text{ dB}$$

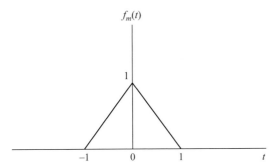

Figure 7.18. For Practice problem 7.5; PDF of $m(t)$.

PRACTICE PROBLEM 7.5

Determine γ_{Th} in an AM envelope detector with $\mu = 0.5$ if the threshold occurs when $E_n > A$ with probability 0.005. Assume that the message signal $m(t)$ has an amplitude distribution as shown in Figure 7.18.

Answer: 10.43 dB.

7.5 ANGLE-MODULATION SYSTEMS

Having considered the presence of noise in linear modulation systems, we are prepared to examine the effect of noise in nonlinear or angle-modulation (FM and PM) systems. Consider the angle-modulation system shown in Figure 7.19. As usual, we assume that the input to the predetection filter is a modulated carrier

$$x(t) = A \cos \left[\omega_c t + \varphi(t) \right] \tag{7.75}$$

where

$$\varphi(t) = \begin{cases} k_p m(t) & \text{for PM} \\ k_f \displaystyle\int_{-\infty}^{t} m(\lambda)d\lambda & \text{for FM} \end{cases} \tag{7.76}$$

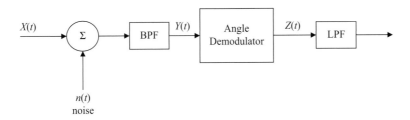

Figure 7.19. Angle-modulation system.

and $m(t)$ is the normalized message signal, k_p is the phase-deviation constant and k_f is the frequency deviation constant. The power in the input signal is

$$S_i = E[x^2(t)] = \frac{A^2}{2} \tag{7.77}$$

We will apply a common approach for analyzing frequency modulation (FM) and phase modulation (PM) cases and later separate the results by using the appropriate $\varphi(t)$. The noise $n(t)$ is given in quadrature form as

$$n(t) = n_c(t) \cos \omega_c t + n_s(t) \sin \omega_c t = E_n(t) \cos [\omega_c t + \phi_n(t)] \tag{7.78}$$

At the input of the demodulator,

$$
\begin{aligned}
Y(t) &= X(t) + n(t) \\
&= A \cos [\omega_c t + \varphi(t)] + E_n(t) \cos [\omega_c t + \phi_n(t)]
\end{aligned} \tag{7.79}
$$

where $n(t)$ is zero-mean white noise with

$$S_i(f) = \begin{cases} \eta/2, & \text{for } |f - f_c| < B_T/2 \\ 0, & \text{otherwise} \end{cases} \tag{7.80}$$

and $E_n(t)$ and ϕ_n are the envelope and phase of $n(t)$. The transmission bandwidth B_T of the angle modulated signal is

$$B_T = \begin{cases} 2(k_p + 1)B, & \text{for PM} \\ 2(D + 1)B, & \text{for FM} \end{cases} \tag{7.81}$$

where $D = \Delta f / B = k_f / (2\pi B)$ is the deviation ratio of the FM system. (For tone-modulated FM, the deviation ratio is also called the modulation index.) We can write $Y(t)$ in Eq. (7.79) in polar form as

$$Y(t) = R(t) \cos [\omega_c t + \psi(t)] \tag{7.82}$$

where

$$\psi(t) = \varphi(t) + \tan^{-1} \left(\frac{E_n(t) \sin (\phi_n - \varphi)}{A + E_n(t) \cos (\phi_n - \varphi)} \right) \tag{7.83}$$

Assuming that the signal power is much larger than the noise power, i.e. $E_n(t) \ll A$ and using the approximation $\tan^{-1} x \simeq x$, the second term (the noise term) on the right hand side of Eq. (7.83) can be approximated so that

$$\psi(t) \simeq \underbrace{\varphi(t)}_{\substack{\text{signal} \\ \text{term}}} + \underbrace{\frac{E_n(t)}{A} \sin [\phi_n(t) - \varphi(t)]}_{\substack{\text{noise} \\ \text{term}}} \tag{7.84}$$

Notice that the noise term depends on the signal term $\varphi(t)$ due to the nonlinear nature of angle modulation. The demodulator processes the signal $Y(t)$ and its output depends on whether the system is PM or FM.

7.5.1 PM system

For the phase-modulation system, the demodulator detects the phase of input signal $Y(t)$ so that the demodulator output is

$$Z(t) = \psi(t) = \varphi(t) + \frac{E_n(t)}{A} \sin\left[\phi_n(t) - \varphi(t)\right] \tag{7.85}$$

As far as computing the SNR is concerned, we can replace $\phi_n(t) - \varphi(t)$ with $\phi_n(t)$ without affecting the result. Thus

$$Z(t) = \varphi(t) + \frac{E_n(t)}{A} \sin\phi_n(t)$$
$$= k_p m(t) + \frac{n_s}{A} \tag{7.86}$$

where Eqs. (7.76) and (7.78) have been applied. The power of the signal term is

$$S_o = E\left[k_p^2 m^2(t)\right] = k_p^2 E\left[m^2(t)\right] = k_p^2 S_m \tag{7.87}$$

and the power of the noise term is

$$N_o = E\left[\frac{n_s^2}{A^2}\right] = \frac{1}{A^2} E\left[n_s^2\right] = \frac{2\eta B}{A^2} = \frac{\eta B}{S_i} \tag{7.88}$$

where A^2 has been replaced by $2S_i$ according to Eq. (7.77). Thus,

$$\frac{S_o}{N_o} = \frac{k_p^2 S_i S_m}{\eta B} = k_p^2 \left(\frac{S_i}{\eta B}\right) S_m$$

or

$$\boxed{\frac{S_o}{N_o} = k_p^2 S_m \gamma} \tag{7.89}$$

7.5.2 FM system

For frequency modulation, the output of the demodulator is obtained from Eqs. (7.76) and (7.84) as

$$Z(t) = \frac{d\psi}{dt} = k_f m(t) + \frac{n_s'}{A} \tag{7.90}$$

Y(t) → [$H(\omega) = j\omega$] → Z(t) = Y'(t)   **Figure 7.20.** Transfer function of a differentiator.

The power in the signal term is

$$S_o = E\left[k_f^2 m^2(t)\right] = k_f^2 E\left[m^2(t)\right] = k_f^2 S_m \qquad (7.91)$$

To find the power due to the noise term, we notice that the noise is passed through an ideal differentiator whose transfer function is $j\omega$ as shown in Figure 7.20 because

$$Z(t) = \frac{dY(t)}{dt} \quad \overset{\text{Fourier Transform}}{\longrightarrow} \quad Z(\omega) = j\omega Y(\omega) \qquad (7.92a)$$

or

$$H(\omega) = \frac{Z(\omega)}{Y(\omega)} = j\omega \qquad (7.92b)$$

Hence the PSD of n_s' is

$$S_{n_s'} = |H(\omega)|^2 S_{n_s} = \begin{cases} \omega^2 \eta, & |\omega| \le 2\pi B = W \\ 0, & \text{otherwise} \end{cases} \qquad (7.93)$$

and

$$
\begin{aligned}
N_o &= E\left[\frac{n_s'(t)^2}{A^2}\right] = \frac{1}{A^2} E\left[n_s'(t)^2\right] \\
&= \frac{1}{A^2}\frac{1}{2\pi}\int_{-W}^{W} S_{n_s'}(\omega)d\omega = \frac{1}{2\pi A^2}\int_{-W}^{W} \omega^2 \eta \, d\omega \\
&= \frac{\eta W^3}{3\pi A^2} = \frac{\eta W^3}{6\pi S_i}
\end{aligned} \qquad (7.94)
$$

where A^2 has been replaced by $2S_i$ according to Eq. (7.77). Thus,

$$\frac{S_o}{N_o} = \frac{6\pi k_f^2 S_m S_i}{\eta W^2(2\pi B)} = \frac{3k_f^2 S_m}{W^2}\left(\frac{S_i}{\eta B}\right) = \frac{3k_f^2 S_m}{W^2}\gamma \qquad (7.95)$$

But $\Delta\omega = |k_f m(t)| = k_f$ since $|m(t)| \le 1$.

$$\frac{S_o}{N_o} = 3\left(\frac{\Delta\omega}{W}\right)^2 S_m \gamma$$

or

$$\boxed{\frac{S_o}{N_o} = 3D^2 S_m \gamma} \qquad (7.96)$$

where D is the frequency deviation ratio of the FM system.

We should note the following points:

(1) In general, FM is superior to PM when $S_m(\omega)$ is concentrated at high frequencies, whereas PM is superior to FM when $S_m(\omega)$ is concentrated at low frequencies. For most practical signals $m(t)$, $S_m(\omega)$ is concentrated at low frequencies, making PM superior to FM.

(2) In FM, the effect of noise is pronounced at higher frequencies. In other words, signal components at higher frequencies will suffer more from noise than lower-frequency components.

(3) There is no restriction on the magnitude of the output SNR for FM, as there was for PM. However, the output SNR cannot be increased indefinitely by simply increasing the frequency deviation ratio D and thus increasing the transmission bandwidth. A trade-off must be made between output SNR and bandwidth.

(4) Comparing PM and FM with AM is important at this point. In AM, increasing the transmitted power directly increases the power of the output signal at the receiver because the message is in the amplitude of the signal. In angle modulation, increasing the transmitted power does not increase the demodulated message power because the message is in the phase. To increase the SNR, we increase the modulator sensitivity (k_p for PM and k_f for FM) without increasing the transmitted power. For this reason, FM systems are used in most low-power applications such as in satellite communications.

7.5.3 Threshold effects

We recall that the expressions for the output SNR for PM and FM were based on the assumption that the signal power was much greater than the noise power (i.e. $E_n(t) \ll A$). When the reverse is the case, the receiver cannot distinguish between the signal and noise. The message cannot be recovered. The threshold effect occurs when the noise and signal levels are comparable. This is an effect we observed in an AM system with envelope detection. It is pronounced in FM.

The SNR at the output of the FM detector can be obtained by investigating the statistical characteristics of the noise spikes. Since noise is random, we can only determine average values of the quantities of interest. It can be shown that at threshold, the value of γ is

$$\gamma_{\text{Th}} = 20(D + 1) \tag{7.97}$$

where D is the frequency deviation ratio of the FM signal. Using Eq. (7.97), we can determine the minimum allowed D to ensure that the system works above threshold. Substituting Eq. (7.96) into Eq. (7.97) yields

$$\boxed{\frac{S_o}{N_o} = 60D^2(D + 1)S_m} \tag{7.98}$$

which is the relationship between output SNR and the lowest possible frequency deviation ratio.

EXAMPLE 7.6

A message signal $m(t)$ with $S_m = 1/2$ is transmitted over a channel with $B = 20$ kHz and additive white noise with $\eta/2 = 1$ nW/Hz. If the power loss in channel is 40 dB and it is required that the output SNR be more than 35 dB, calculate the transmission bandwidth and the necessary average transmitted power for: (a) PM with $k_p = 4$; (b) FM with $D = 4$.

Solution

(a) From Eq. (7.81),

$$B_T = 2(k_p + 1)B = 2(4 + 1) \times 20 \text{ kHz} = 200 \text{ kHz}$$

Since,

$$35 \text{ dB} \quad \rightarrow \quad 10^{35/10} = 3162.3$$

$$40 \text{ dB} \quad \rightarrow \quad 10^{40/10} = 10^4$$

From Eq. (7.89),

$$\frac{S_o}{N_o} = k_p^2 S_m \gamma = k_p^2 S_m \frac{S_i}{\eta B} = (4)^2 \left(\frac{1}{2}\right) \frac{S_i}{2 \times 10^{-9} \times 20 \times 10^3} \geq 3162.3$$

or

$$S_i \geq 0.0158$$

Thus,

$$S_t = S_i(10^4) = 15.8 \text{ W}.$$

(b) For this case,

$$B_T = 2(D + 1)B = 2(4 + 1) \times 20 \text{ kHz} = 200 \text{ kHz}$$

From Eq. (7.96),

$$\frac{S_o}{N_o} = 3D^2 S_m \gamma = 3D^2 S_m \frac{S_i}{\eta B} = 3(4)^2 \left(\frac{1}{2}\right) \frac{S_i}{2 \times 10^{-9} \times 20 \times 10^3} \geq 3162.3$$

or

$$S_i \geq 0.0053$$

Thus,

$$S_t = S_i(10^4) = 53 \text{ W}$$

PRACTICE PROBLEM 7.6

An FM system with $\Delta f = 80$ kHz and $B = 20$ kHz uses sinusoidal modulation, i.e. $m(t) = \cos \omega_c t$. (a) Calculate the output SNR. (b) What is the improvement (in dB) of the FM system over the baseband system?

Answer: (a) 24γ; (b) 13.8 dB.

7.6 PREEMPHASIS AND DEEMPHASIS FILTERS

In commercial FM broadcasting and reception, it is noticed that the noise PSD is largest in the frequency range where the signal PSD is smallest. In other words, the higher frequencies contribute more to the noise than the lower frequencies. To improve the noise performance, we employ preemphasis/deemphasis filtering. Preemphasis allows the higher frequencies to be increased in amplitude before being used to modulate the carrier. As shown in Figure 7.21, the preemphasis filtering is done before modulation (before noise is introduced). Since the filtering operation distorts the signal, an inverse operation must take place at the receiver using deemphasis filtering. This way, the useful signal is restored to its original form but the noise, which was added after the preemphasis, is now reduced. Thus,

> **Preemphasis** is a filtering process used in FM broadcasting to boost the high-frequency components of the message signal $m(t)$ (before modulation) in order to improve the overall SNR.

> **Deemphasis** is the low-pass filtering process used in FM reception to restore (after detection) amplitude-frequency characteristics of the preemphasized signal.

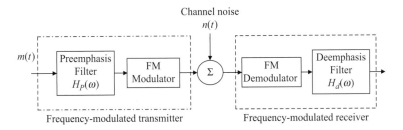

Figure 7.21. Preemphasis/deemphasis filtering in FM communication systems.

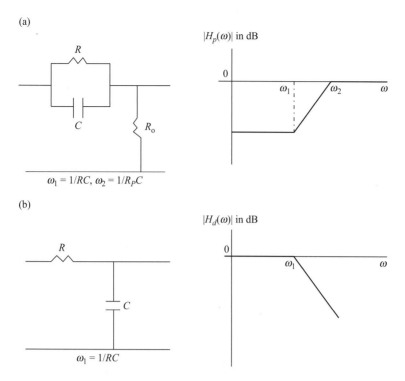

(a)

$\omega_1 = 1/RC, \ \omega_2 = 1/R_pC$

(b)

$\omega_1 = 1/RC$

Figure 7.22. Typical realization of: (a) a preemphasis filter; (b) a deemphasis filter.

Typical *RC* circuit realizations of preemphasis and deemphasis filters and their Bode magnitude plots are shown in Figure 7.22, where R_p is the parallel combination of R and R_o. The transfer function of the preemphasis is of the form

$$H_p(\omega) = K \frac{(1 + j\omega/\omega_1)}{(1 + j\omega/\omega_2)} \tag{7.99}$$

where K is a constant, while that of the deemphasis is

$$H_d(\omega) = \frac{1}{1 + j\omega/\omega_1} \tag{7.100}$$

The values of the components are based on the proper choice of ω_1 and ω_2. The choice of

$$f_1 = \omega_1/2\pi = 1/2\pi RC \tag{7.101}$$

is standardized in the USA as the frequency where the signal PSD is down by 3 dB; this is 2.1 kHz for broadcasting. This is known as 75-μs preemphasis because the RC time constant of the filter is 75 μs. The choice of

$$f_2 = \omega_2/2\pi = 1/2\pi R_pC, \quad R_p = \frac{RR_o}{R + R_o} \tag{7.102}$$

is made to be well above the audio range, say $f_2 \geq 30$ kHz.

We can determine the improvement in the output SNR due to the preemphasis/deemphasis filtering by calculating the decrease in the noise power. Without the deemphasis filter, the output noise power is given by Eq. (7.94), namely

$$N_o = \frac{\eta W^3}{3\pi A^2} \tag{7.103}$$

In the presence of the deemphasis filter, the output noise power is

$$
\begin{aligned}
N'_o &= \frac{1}{2\pi} \int_{-W}^{W} S_{n_o}(\omega) |H_d(\omega)|^2 d\omega = \frac{1}{\pi} \int_0^{W} \frac{\omega^2 \eta}{A^2} |H_d(\omega)|^2 d\omega \\
&= \frac{\eta}{A^2 \pi} \int_0^{W} \frac{\omega^2 d\omega}{1 + (\omega/\omega_1)^2} \\
&= \frac{\eta \omega_1^3}{A^2 \pi} \left(\frac{W}{\omega_1} - \tan^{-1} \frac{W}{\omega_1} \right)
\end{aligned}
\tag{7.104}
$$

Thus, we obtain the noise improvement factor as

$$\Gamma = \frac{N_o}{N'_o} = \frac{1}{3} \frac{(W/\omega_1)^3}{W/\omega_1 - \tan^{-1}(W/\omega_1)} \tag{7.105}$$

The improvement derived from using preemphasis/deemphasis can be significant in a noisy environment. Although the scheme can be used in AM, it is not effective because the effect of noise on AM is uniform throughout the spectrum of the modulating signal.

EXAMPLE 7.7

In a commercial FM broadcasting system, $B = 15$ kHz and $f_1 = 2.1$ kHz. Find the improvement in output SNR with preemphasis/deemphasis filtering.

Solution

$$W/\omega_1 = B/f_1 = 15/2.1 = 7.14.$$

so that

$$
\begin{aligned}
\Gamma &= \frac{1}{3} \frac{(W/\omega_1)^3}{W/\omega_1 - \tan^{-1}(W/\omega_1)} = \frac{1}{3} \frac{7.14^3}{7.14 - \tan^{-1} 7.14} \\
&= 21.25 \text{ or } 13.27 \text{ dB}
\end{aligned}
$$

Since the output SNR typically varies from 40 to 50 dB, the improvement is significant.

PRACTICE PROBLEM 7.7

Design RC preemphasis and deemphasis filters such that $R = 75$ kΩ.

Answer: $C = 1$ nF, $R_o = 5.709$ kΩ.

7.7   COMPARISON OF ANALOG MODULATION SYSTEMS

At this point, we may now compare the various analog continuous-wave (CW) systems we have discussed in this chapter. The comparison is based on transmission bandwidth B_T, the output SNR normalized to γ, threshold point γ_{Th} if applicable, dc response or low-frequency response, and system complexity. Table 7.1 shows the comparison.

The following points should be noted.

1. All linear modulation schemes (baseband, AM, DSB-SC, SSB) have the same SNR performance as the baseband system, except for the wasted carrier power in conventional AM.
2. From system viewpoint, AM is the least complex modulation, while suppressed carrier VSB is the most complex.
3. Of the linear modulation schemes suppressed carrier methods are better than the conventional AM – SNRs are better and there is no threshold effect.
4. When bandwidth is a major concern, single sideband and vestigial sideband are the choice.
5. The nonlinear/angle modulation schemes (PM and FM) provide significant improvement in terms of noise performance, provided that the input signal is above the threshold.

Table 7.1. **Comparison of analog modulation schemes**

Type	B_T	$(S/N)_o/\gamma$	γ_{Th}	dc	Complexity	Comments
Baseband	B	1	–	No	Minor	No modulation
AM	$2B$	$\dfrac{\mu^2 S_m}{1 + \mu^2 S_m}$	20	No	Minor	Envelope detection
DSB	$2B$	1	–	Yes	Major	Synchronous detection
SSB	B	1	–	No	Moderate	Synchronous detection
VSB	$B+$	1	–	Yes	Major	Synchronous detection
PM	$2(k_p + 1)B$	$k_p^2 S_m$	≥ 20	Yes	Moderate	Phase detection
FM	$2(D + 1)B$	$3D^2 S_m$	≥ 20	Yes	Moderate	Frequency detection

B = message bandwidth, $S_m = <m^2(t)>$, $\gamma = \frac{S_i}{\eta B}$.

6. Whether FM or PM is superior in noise performance depends on the message signal spectrum. For most practical signals, the noise performance of PM is superior to FM.
7. When an FM system is compared to an AM system with a modulation index of 1 operating under similar noise conditions, Table 7.1 shows that the FM signal has a signal to noise ratio which is $3D^2$ better than the AM system, where D is the modulation index or deviation ratio for the FM signal.
8. When power conservation is an issue, FM is better than other schemes. For this reason, FM is used a lot in low-power applications.

The choice of a particular modulation scheme depends on several factors. Since there are no universal solutions to all communication problems, communication engineers must be open-minded and be willing to review all alternatives available to them.

7.8   COMPUTATION USING MATLAB

As usual, MATLAB is a handy tool for computation. It can be used for demonstrating the performance of the various communication systems covered in this chapter. Here we use it to

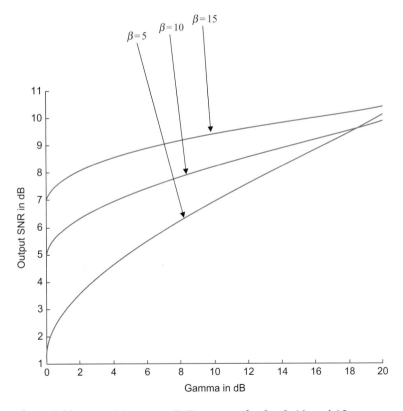

Figure 7.23. Plot of the output SNR versus γ for $\beta = 5$, 10, and 15.

evaluate the performance of an FM detector by examining the statistical characteristics of impulses called spikes or "clicks". The analysis is complicated and we simply state the result. For a sinusoidal message signal (i.e. $m(t) = \cos \omega_c t$), the output SNR is given by

$$\frac{S_o}{N_o} = \frac{\frac{3}{2} \beta^2 \gamma}{1 + \frac{12 \beta \gamma}{\pi} \exp \left[- \frac{\gamma}{2(1 + \beta)} \right]} \qquad (7.106)$$

where $\gamma = S_i / \eta B$ as usual and β is the modulation index or frequency deviation ratio. Using the following MATLAB program, the output SNR as in Eq. (7.106) is plotted against γ for $\beta = 5$, 10, and 15 in Figure 7.23. The plots illustrate the threshold phenomenon and demonstrate that the FM system degrades rapidly as γ falls below certain levels. The threshold value is dictated by the modulation index β.

```
% MATLAB code to generate FM performance curves
beta = [5 10 15];
gammadB = 0:0.1:20;
hold on    % hold the current plot for other plots
for k=1:length(beta)
  for j=1:length(gammadB)
  num=1.5*beta(k)^2*gammadB(j);
  fac =exp(-0.5*gammadB(j)/(1+beta(k)));
  den=1 + 12*beta(k)*gammadB(j)*fac/pi;
  snr = num/den;
  snrdB(j)=10*log10(snr);
  end
  plot(gammadB,snrdB)
end
hold off    % release
xlabel('gamma in dB')
ylabel('Output SNR in dB')
```

Summary

1. Noise is a random signal which interferes with the transmission and reception of information through a communication system. Noise is inevitable in communication systems. Low noise levels are usually tolerable, but high noise levels can render a system unusable.

2. Thermal noise is internally generated noise, which is linearly dependent on temperature. The mean square value of voltage $V_n(t)$ across resistor R is

$$V_n^2 = 4kTRB$$

where k is Boltzmann's constant, T is absolute temperature in kelvins (K), and B is the observation bandwidth.

3. Shot noise is caused by fluctuations in dc current flowing through an electronic device such as diodes and transistors. The mean square value of current $I_n(t)$ due to shot noise is

$$I_n^2 = 2eI_{dc}B$$

where e is the electron charge, B is the observation bandwidth, and I_{dc} is the direct current flowing through the device. Both thermal and shot noise have a white power spectrum (i.e. the noise power does not depend on frequency over a very wide frequency range).

4. The output SNR of a baseband communication system corrupted by additive white noise with the single-sided PSD of η is $\gamma = \frac{S_i}{\eta B}$, where B is the signal bandwidth and S_i is the signal power. Both DSB and SSB systems have the same output SNR as baseband.

5. Both AM synchronous demodulation and envelope demodulation (for small noise) give same output SNR, namely,

$$\frac{S_o}{N_o} = \frac{\mu^2 S_m}{1 + \mu^2 S_m}\gamma$$

where μ is the modulation index and $S_m = \langle m^2(t) \rangle$.

6. For the square-law detector,

$$\frac{S_o}{N_o} = \begin{cases} \mu^2 S_m \gamma, & \gamma \gg 1 \\ \dfrac{4}{3}\mu^2 S_m \gamma^2, & \gamma \ll 1 \end{cases}$$

As a nonlinear system, the square-law detector exhibits a threshold effect.

7. For a PM system,

$$\frac{S_o}{N_o} = k_p^2 S_m \gamma$$

where k_p is the phase-deviation constant.

8. For an FM system,

$$\frac{S_o}{N_o} = 3D^2 S_m \gamma$$

where $D = k_f / W$ is the deviation ratio and $W = 2\pi B$.

9. Commercial broadcast FM uses the preemphasis/deemphasis filtering technique to improve the output SNR. The message is preemphasized before modulation so that we can deemphasize the noise/interference relative to the message after demodulation.

10. The performance characteristics of the various analog modulation schemes discussed in this chapter are summarized in Table 7.1. The output or postdetection SNRs provided in this table constitute the most important result in this chapter.

11. MATLAB serves as a tool for evaluating the performance of the various communication systems covered in this chapter.

Review questions

7.1 Which of the following is not characteristic of noise?
(a) It is inevitable.
(b) Its effect on the message signal is measured by the signal-to-noise ratio (SNR).
(c) It is additive.
(d) It can have a severe effect on communication systems.

7.2 Which of the following is not a source of man-made noise?
(a) Fluorescent lights.　　(b) Transistors.　　(c) Ignition systems.
(d) Radio/radar transmitters.　　(e) Computers.

7.3 What kind of noise is caused by the random fluctuations of electric current?
(a) White noise.　　(b) Colored noise.　　(c) Thermal noise.　　(d) Shot noise.

7.4 Which communication system does not require modulation/demodulation?
(a) Baseband.　　(b) DSB.　　(c) VSB.　　(d) PM.　　(e) FM.

7.5 Which modulation is not linear in nature?
(a) Conventional AM.　　(b) DSB.　　(c) SSB.　　(d) FM.

7.6 An angle-modulated signal is described by

$$x(t) = 100 \cos \left(2\pi \times 10^7 t + 0.002 \cos 2\pi \times 10^4 t \right)$$

If this represents a PM signal, what is k_p?
(a) 1000.　　(b) 100.　　(c) 2π.　　(d) 0.002.

7.7 An FM system with $\Delta f = 60$ kHz and $B = 20$ kHz uses tone modulation. Its frequency deviation ratio is:
(a) 60.　　(b) 20.　　(c) 30.　　(d) 10.　　(e) 3.

7.8 Which modulation scheme does not exhibit a threshold?
(a) FM.　　(b) Square-law detector.　　(c) DSB.　　(d) AM envelope detector.

7.9 Which of these is not true of a preemphasis network?
  (a) It reduces the amount of noise.
  (b) It improves the output SNR.
  (c) It provides a gain for the components of $m(t)$ at high frequencies.
  (d) It is a filter whose transfer function is constant for low frequencies and behaves like a differentiator at the higher frequencies.
  (e) It can be used in conventional AM to provide a performance improvement.
7.10 Which modulation scheme is appropriate for low-power applications?
  (a) AM.   (b) FM.

Answers: 7.1 c, 7.2 b, 7.3 d, 7.4 a, 7.5 d, 7.6 d, 7.7 e, 7.8 c, 7.9 a, 7.10 b.

Problems

Section 7.2 Sources of noise

7.1 What are the differences between thermal and shot noise?
7.2 The transistors used in a receiver have an average input resistance of 2 kΩ. Calculate the thermal noise for the receiver with: (a) a bandwidth of 300 kHz and a temperature of 37 °C; (b) a bandwidth of 5 MHz and a temperature of 350 K.
7.3 A TV receiver operates on channel 2 ($54 < f < 60$ MHz) at an ambient temperature of 27 °C. If the input resistance is 1 kΩ, calculate the rms noise voltage at the input.
7.4 Determine the rms noise voltage arising from thermal noise in two resistors, 40 Ω and 60 Ω, at T = 250 K if the resistors are connected: (a) in series; (b) in parallel. Assume a bandwidth of 2 MHz.
7.5 Determine the rms value of the noise current flowing through a temperature-limited diode if $I_{dc} = 15$ mA and $B = 2$ MHz.
7.6 A noisy diode operates in the frequency range $f_c = 10^6 \pm 100$ kHz. If the direct current is 0.1 A, find the rms noise current flowing through the diode.

Section 7.3 Baseband systems

7.7 A baseband communication system has a distribution channel and power spectra density of white noise $\eta/2$ of 10^{-8} W/Hz. The signal transmitted has a 5 kHz bandwidth. If an RC low-pass filter with a 3 dB bandwidth of 10 kHz is used at the receiving end to limit noise power, determine the output noise power.

Section 7.4 AM systems

7.8 An AM system operates above the threshold with a modulation index of 0.4. If the message signal is $10 \cos 10\pi t$, calculate the output SNR in dB relative to the baseband performance γ.
7.9 An AM system operates with modulation index of $\mu = 0.5$. The modulating signal is a sinusoid (i.e. $m(t) = \cos\omega t$). (a) Find the output SNR in terms of γ. (b) Calculate the improvement in dB in the output SNR if μ is increased from 0.5 to 0.9.

Section 7.5 Angle-modulation systems

7.10   An angle-modulated signal is described by

$$x(t) = 20\cos\left(2\pi \times 10^6 t + 50\cos 100\pi t\right)$$

(a)   If the modulation is PM with $k_p = 5$, what is $m(t)$?

(b)   If $x(t)$ represents an FM signal, with $k_f = 5$, what is $m(t)$?

Section 7.6 Preemphasis and deemphasis

7.11   What are preemphasis and deemphasis? Why are they not used for AM?

7.12   Two FM systems are described by 75-µs and 25-µs deemphasis. Assuming that $B = 15$ kHz, compare their performance.

Section 7.8 Computation Using MATLAB

7.13   An envelope detector with a tone modulation operates above threshold. Use MATLAB to plot the output SNR in dB as a function of γ for $\mu = 0.4$, 0.6, and 0.8.

7.14   Use MATLAB to plot the noise improvement factor in Eq. (7.105) as a function of $\alpha = W/\omega_1$ for $0.1 \le \alpha \le 10$.

7.15   The transfer function of a preemphasis filter is given by

$$H(\omega) = \frac{\left(\omega^2 + 56.8 \times 10^6\right)\omega^4}{\left(\omega^2 + 6.3 \times 10^6\right)\left(\omega^2 + 0.38 \times 10^9\right)\left(\omega^6 + 9.58 \times 10^{26}\right)}$$

Use MATLAB to plot $|H(\omega)|$ on a dB scale with $1 < \omega < 1000$ rad/s.

8 Noise in digital communication systems

Let him that would move the world, first move himself.

SOCRATES

TECHNICAL NOTES – Global positioning system

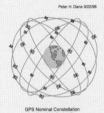

Peter H. Dana 9/22/98

GPS Nominal Constellation
24 Satellites in 6 Orbital Planes
4 Satellites in each Plane
20,200 km Altitudes, 55 Degree Inclination

From time immemorial, people have been trying to figure out a reliable way to tell where they are. The global positioning system (GPS) is a satellite-based navigation system. It is the only system today able to show you your exact position on the earth at any time, in any weather, anywhere. It is one of history's most exciting and revolutionary developments, and new applications for it are constantly being discovered.

Operationally, GPS is divided into three segments: the space segment, the control segment, and the user segment. The space-based segment consists of 24 satellites. They travel on polar orbits around the Earth once every 12 hours at an altitude of about 20,000 km. The satellites transmit signals that can be detected by anyone with a GPS receiver. The control segment continuously tracks the satellites and updates their orbital parameters and clocks. This consists of the master control station in Colorado Springs and four unmanned stations in Hawaii, Kwajalein, Diego Garcia, and Colorado. The user segment consists of GPS receivers, which can be hand-carried or installed on aircraft, ships, tanks, submarines, cars, and trucks. The receivers detect the satellite transmissions and perform calculations to determine the position, altitude, and time outputs. Using the receiver, you can determine your location with great precision.

GPS is funded by and controlled by the US Department of Defense (DOD). While there are many thousands of civil users of GPS world-wide, the system was designed for and is operated by the US military. (The Global Navigation Satellite System (GLONASS), being deployed by the Russian Federation, has much in common with the US GPS.) GPS has become important for nearly all military operations and weapons systems. For any extended-range weapon system, precise and continuous positioning information is invaluable. GPS is rapidly becoming a key element of the basic infrastructure of the world's economy and holds the promise of dramatic increases in productivity. GPS, married to modern digital communications, will bring enormous productivity improvements to the world's shipping, airline, and transportation industries, as well as dramatically improve the efficiency of police, ambulance, and fire dispatching. There will probably be a time soon when every car on the road can be equipped with a GPS receiver.

8.1   INTRODUCTION

In the previous chapter, we considered analog communication systems in the presence of additive white Gaussian noise (AWGN) channels and evaluated the signal-to-noise (power) ratio (SNR) for various types of analog system. In this chapter, we consider the behavior of digital communication systems in the presence of AWGN channels. The assumption of AWGN channels both simplifies the analysis and corresponds to a large number of communication systems. Digital signal transmission/reception is different from analog signal transmission/reception in many respects. First, instead of reproducing a waveform, we are often concerned with determining the presence or absence of a pulse. Second, we may know in advance the shape of the pulse but not its amplitude or when it will arrive. For these reasons, the concept of signal-to-noise ratio used for analog systems is somewhat irrelevant here. We use the *probability of error* (also known as *bit error rate*) as a performance measure for different kinds of modulation signal on an AWGN channel.

The principal components of a digital communication system are shown in Figure 8.1. The source of information generates a stream of information. The state of the information source is associated with symbols $c_k, k = 1, 2, ..., M$ by the coder. The modulator uniquely associates a waveform $s_k(t)$ with c_k. The channel adds an additive white noise $n(t)$ with zero mean and constant variance to the signal. The demodulator recovers $s_k(t)$ from the channel output.

We begin by looking at matched-filter receiver. We then consider baseband signals, which are digital signals intended for direct transmission. Such signals involve using modulation schemes that do not require a sinusoidal carrier to perform frequency translation. We will discuss coherent and noncoherent detections leading to a comparison of modulation schemes. Next, we discuss *M*-ary communication and spread-spectrum systems. Finally, we use MATLAB to plot and compare the performance of various modulation methods covered in the chapter.

8.2   MATCHED-FILTER RECEIVER

The matched filter is the heart of the optimum detector. A matched filter is one that maximizes the filter output SNR when the input is corrupted by white Gaussian noise.

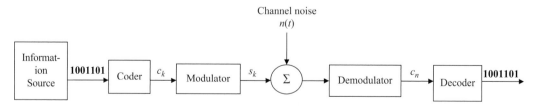

Figure 8.1. A typical digital communication system.

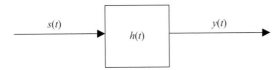

s(t)

h(t)

y(t)

Figure 8.2. Input and output of a matched filter.

A **matched filter** is a system that maximizes the output SNR.

Consider the system shown in Figure 8.2. If $s(t)$ is any physical signal, then a filter which is matched to $s(t)$ is, by definition, one with impulse response

$$h(t) = ks(t_o - t) \tag{8.1}$$

where k is a constant and t_o is the delay. Since the arbitrary constant k multiplies both the signal and noise, its value is irrelevant and we may choose $k = 1$ for convenience. This leads to

$$h(t) = s(t_o - t) \tag{8.2a}$$

This states that a matched (optimal) filter is characterized by an impulse response that is a time-reversed and delayed version of the input signal $s(t)$. Taking the Fourier transform of Eq. (8.2a) yields

$$H(\omega) = S(-\omega)e^{-j\omega t_o} = S^*(\omega)e^{-j\omega t_o} \tag{8.2b}$$

where $S^*(\omega)$ is the complex conjugate of the Fourier transform of signal $s(t)$. Matched filters find applications in digital communication systems and radar systems.

It is needless to say that the output of the filter is

$$y(t) = h(t) * s(t) \tag{8.3}$$

If $s(t)$ is confined to the interval $0 \le t \le T$, Eq. (8.2a) may be written as

$$h(t) = \begin{cases} s(T - t), & 0 \le t \le T \\ 0, & \text{otherwise} \end{cases}$$

Substituting this in Eq. (8.3) gives

$$y(t) = \int_0^t h(\tau)s[T - (t - \tau)]d\tau$$

When $t = T$,

$$y(t) = \int_0^T h(\tau)s(\tau)d\tau$$

indicating that $y(t)$ is the *correlation* of $h(t)$ and $s(t)$.

It is readily shown that for a signal $s(t)$ corrupted by white Gaussian noise $n(t)$, the maximum signal-to-noise ratio at the output of the matched filter is

$$\left(\frac{S}{N}\right)_{\text{max}} = \frac{\displaystyle\int_0^T s^2(t)\,dt}{\eta/2} = \frac{2E}{\eta} \tag{8.4}$$

where E is the signal energy and $\eta/2$ is the two-sided noise spectral density. It is assumed that $s(t)$ is confined to the interval $0 \le t \le T$.

EXAMPLE 8.1

Find the impulse response of the matched filter for the two signals shown in Figure 8.3.

Solution

(a) Using Eq. (8.2a),

$$h_1(t) = s_1(t_o - t)$$

which means that we get $h_1(t)$ by first time-reversing $s_1(t)$ and then shifting (or delaying) the time-reversed signal by t_o, as shown in Figure 8.4(a).

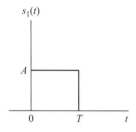

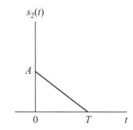

Figure 8.3. For Example 8.1.

(a)

Figure 8.4. For Example 8.1.

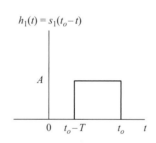

(b)

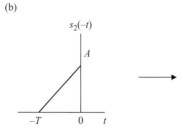

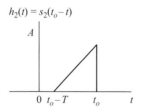

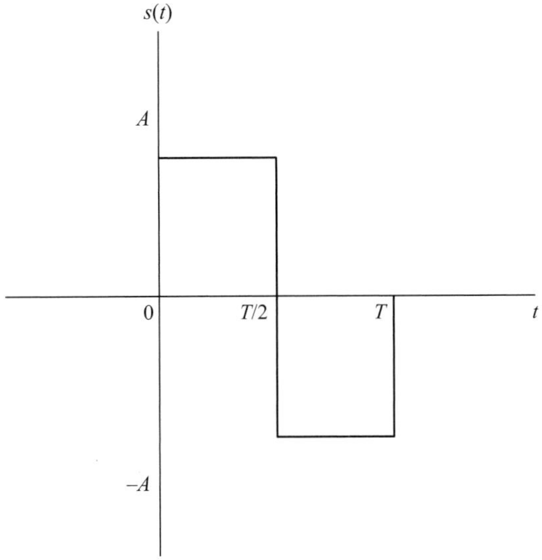

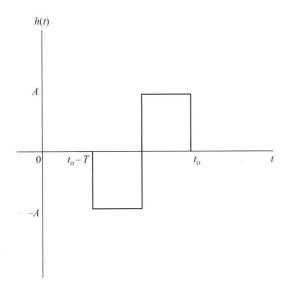

Figure 8.5. For Practice problem 8.1.

Figure 8.6. For Practice problem 8.1.

(b) Similarly,

$$h_2(t) = s_2(t_o - t)$$

which is obtained as shown in Figure 8.4(b).

PRACTICE PROBLEM 8.1

Obtain the impulse response of the matched filter for the signal shown in Figure 8.5.

Answer: See Figure 8.6.

8.3 BASEBAND BINARY SYSTEMS

These are binary systems without modulation. If T (in seconds/bit) is the time it takes to transmit one bit of data, then the transmission rate or bit rate (bits/second) is

$$R = \frac{1}{T} \tag{8.5}$$

The transmitted signal over a bit interval $(0, T)$ is

$$s(t) = \begin{cases} s_1(t), & \text{for binary 1} \\ s_2(t), & \text{for binary 0} \end{cases} \tag{8.6}$$

where $0 \leq t \leq T$, $s_1(t)$ is the waveform used if a binary 1 is transmitted and $s_2(t)$ is the waveform used if binary 0 is transmitted. The binary signal $s(t)$ plus noise $n(t)$ at the receiver input produces a baseband waveform $r(t)$. Both $s(t)$ and $r(t)$ are shown in Figure 8.7 for different binary signaling. We now consider each of these binary signalings.

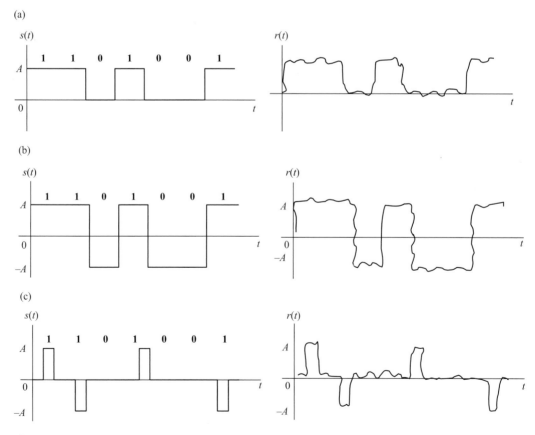

Figure 8.7. Baseband binary signalings: (a) unipolar; (b) polar; (c) bipolar.

8.3.1 Unipolar signaling

This is also known as *on-off signaling*. For this case,

$$s(t) = \begin{cases} s_1(t) = A, \text{for binary 1} \\ s_2(t) = 0, \quad \text{for binary 0} \end{cases} \tag{8.7}$$

for $0 \le t \le T$ and $A > 0$. This signal along with the noise $n(t)$ is present at the receiver so that

$$r(t) = s(t) + n(t) \tag{8.8}$$

where $n(t)$ is additive white Gaussian noise with zero mean and variance $\sigma^2 = \eta/2$.

We first evaluate the performance of the receiver when an LPF, with unity gain and bandwidth $B > 2/T$, is used. Making $B > 2/T$ allows the signal to be preserved while noise is reduced. The received signal at a given time $t = t_o$ is

$$r(t_o) = \begin{cases} A + n(t_o), \text{if signal is present} \\ n(t_o), \quad \text{if signal is absent} \end{cases} \tag{8.9}$$

Since $n(t)$ is random, we need probability to describe the process of making the correct decision. In order to calculate the probability of error in making the correct decision, we need to find the following decision rule that compares r with a threshold μ:

$$\begin{aligned} \text{Signal present} &: r(t_o) > \mu \\ \text{Signal absent} &: r(t_o) < \mu \end{aligned} \tag{8.10}$$

where the intermediate case of $r(t_o) = \mu$ is ignored because it occurs with probability zero. Intuitively, we expect the threshold value μ to be at some intermediate level, $0 < \mu < A$.

The probability density function (PDF) of r when the no signal is present is

$$p_0(r) = \frac{1}{\sqrt{2\pi}\sigma} e^{-r^2/2\sigma^2} \tag{8.11a}$$

and the PDF of r when the signal is present is

$$p_1(r) = \frac{1}{\sqrt{2\pi}\sigma} e^{-(r-A)^2/2\sigma^2} \tag{8.11b}$$

These PDFs are depicted in Figure 8.8.

There are two ways errors can occur: s_1 is received when s_2 is sent or s_2 is received when s_1 is sent. Their corresponding probabilities are equivalent to the shaded areas in Figure 8.8 and are given by:

$$P_{e0} = P(r > \mu | s_2 \text{ sent}) = \int_{\mu}^{\infty} \frac{1}{\sqrt{2\pi}\sigma} e^{-r^2/2\sigma^2} dr \tag{8.12a}$$

$$P_{e1} = P(r < \mu | s_1 \text{ sent}) = \int_{-\infty}^{\mu} \frac{1}{\sqrt{2\pi}\sigma} e^{-(r-A)^2/2\sigma^2} dr \tag{8.12b}$$

(a)

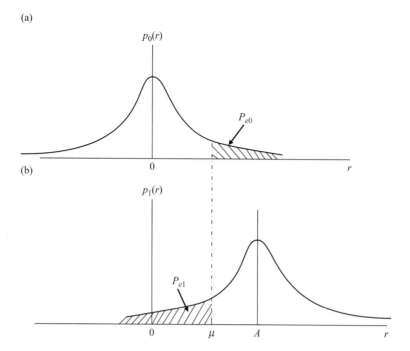

(b)

Figure 8.8. Probability density functions with decision threshold and error probabilities.

Notice from Figure 8.8 that increasing the threshold μ decreases P_{e0} but increases P_{e1} or vice versa. The bit error rate (BER) or probability of error is

$$P_e = P(r > \mu|s_2 \text{ sent})P(s_2 \text{ sent}) + P(r < \mu|s_1 \text{ sent})P(s_1 \text{ sent})$$

or

$$P_e = P_0 P_{e0} + P_1 P_{e1} \tag{8.13}$$

where P_0 and P_1 are the source digit probabilities of zeros and ones, respectively; i.e. P_1 is the probability that a binary 1 is sent. In a long string of message, 0s and 1s are equally likely to be transmitted so that $P_0 = P_1 = \frac{1}{2}$. From Figure 8.8, the sum of the two shaded areas will be minimum if

$$\mu = A/2$$

With this threshold, the two shaded areas in Figure 8.8 are equal due to symmetry. Hence,

$$P_e = \frac{1}{2}(P_{e0} + P_{e1}) = P_{e0} = P_{e1} \tag{8.14}$$

Thus,

$$P_e = \int_{\mu}^{\infty} \frac{1}{\sqrt{2\pi}\sigma} e^{-r^2/2\sigma^2} dr$$

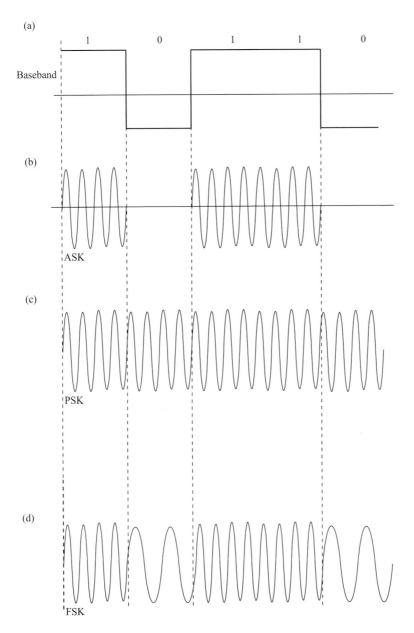

Figure 8.9. Digital modulation schemes.

If we change variable $z = r/\sigma$, we get

$$P_e = \int_{\mu/\sigma}^{\infty} \frac{1}{\sqrt{2\pi}} e^{-z^2/2} dz = Q\left(\frac{\mu}{\sigma}\right)$$

where $Q(\cdot)$ is a form of complementary error function related to erfc($\cdot$), i.e..

$$Q(x) = \frac{1}{\sqrt{2\pi}} \int_x^\infty e^{-y^2/2} dy$$

or

$$Q(x) = \frac{1}{2} \text{erfc}\left(\frac{x}{\sqrt{2}}\right) \tag{8.15}$$

For large values of x ($x \gg 1$), $Q(x)$ can be approximated as

$$Q(x) \simeq \frac{1}{x\sqrt{2\pi}} e^{-x^2/2}, \quad x \gg 1 \ (x > 4) \tag{8.16}$$

But $\mu = A/2$.

$$P_e = Q\left(\frac{A}{2\sigma}\right) \tag{8.17}$$

This is the minimum error probability for unipolar signaling in Gaussian noise for equally likely source digits. To express this result in terms of more familiar quantities, we let the average power be

$$S = \frac{1}{2}(0)^2 + \frac{1}{2}(A)^2 = \frac{A^2}{2}$$

and the average noise power be

$$\sigma^2 = 2(\eta/2) = \eta$$

Substituting these in Eq. (8.17) yields

$$P_e = Q\left(\sqrt{\frac{S}{2\eta}}\right) \text{(lowpass filter)} \tag{8.18}$$

We now consider the case of a matched filter. The matched filter is designed to give the optimum SNR for a given signal in the presence of white noise. The matched filter needs to be matched to the difference signal $s_1(t) - s_2(t)$. Thus, Eq. (8.17) becomes

$$P_e = Q\left(\frac{a_1 - a_2}{2\sigma}\right) \tag{8.19}$$

where a_1 or a_2 is the mean value of $r(t)$ depending on whether $s_1(t)$ or $s_2(t)$ was sent. Due to the nature of the $Q(x)$, to maximize P_e requires that we minimize x. Thus, the receiving filter is matched to input when the argument of Q in Eq. (8.19) is minimized.

$$\frac{(a_1 - a_2)^2}{\sigma^2} = \frac{E_d}{\eta/2} = \frac{2E_d}{\eta} \tag{8.20}$$

where E_d is the energy of the difference signal at the input of the receiving filter, i.e.

$$E_d = \int_0^T [s_1(t) - s_2(t)]^2 dt \tag{8.21}$$

Thus, from Eqs. (8.19) and (8.20),

$$P_e = Q\left(\sqrt{\frac{E_d}{2\eta}}\right) \tag{8.22}$$

For unipolar signaling,

$$E_d = \int_0^T (0^2 + A^2) dt = A^2 T$$

so that

$$P_e = Q\left(\sqrt{\frac{E_d}{2\eta}}\right) = Q\left(\sqrt{\frac{A^2 T}{2\eta}}\right) \tag{8.23}$$

if we define E_b as the *average energy per bit*. Since the energy for a binary 1 is $A^2 T$, while the energy for a binary 0 is 0, the average energy per bit is

$$E_b = \frac{1}{2}(A^2 T) + \frac{1}{2}(0) = \frac{A^2 T}{2}$$

Thus, Eq. (8.23) becomes

$$\boxed{\begin{aligned} P_e &= Q\left(\sqrt{\frac{E_b}{\eta}}\right) = Q(\sqrt{\gamma_b}) \\ &\simeq \frac{1}{\sqrt{2\pi\gamma_b}} e^{-\gamma_b/2}, \quad \gamma_b \gg 1 \quad (\gamma_b > 4) \quad \text{(matched filter)} \end{aligned}} \tag{8.24}$$

The parameter $\gamma_b = E_b/\eta$ is the normalized energy per bit or the average energy required to transmit one bit of data over a white noise channel. It is often called the *signal-to-noise ratio* per bit (or simply SNR per bit). It is customary to express BER in terms of γ_b so that we can compare the performance of different digital communication systems.

Two things should be noted from Eq. (8.24). First, the BER depends only on the SNR per bit and not on any other characteristics of the signals and noise. Second, the SNR per bit γ_b is also the output SNR of the matched filter demodulator.

8.3.2 Polar signaling

This is regarded as the most efficient scheme. For this case,

$$s(t) = \begin{cases} s_1(t) = +A, \text{ for binary 1} \\ s_2(t) = -A, \text{ for binary 0} \end{cases}$$ (8.25)

where $0 \leq t \leq T$. This is also called *antipodal signaling* because $s_2(t) = -s_1(t)$. We follow the same steps as for unipolar signaling except that $S = (A/2)^2$. For a matched filter, the difference signal energy is

$$E_d = \int_0^T [s_1(t) - s_2(t)]^2 dt = \int_0^T [A - -A]^2 dt = 4A^2 T$$

while the average energy per bit is

$$E_b = \frac{1}{2}(A^2 T) + \frac{1}{2}(A^2 T) = A^2 T$$

Substituting these in Eq. (8.22) gives

$$P_e = Q\left(\sqrt{\frac{E_d}{2\eta}}\right) = Q\left(\sqrt{\frac{4A^2 T}{2\eta}}\right)$$

or

$$\boxed{\begin{aligned} P_e &= Q\left(\sqrt{\frac{2E_b}{\eta}}\right) = Q(\sqrt{2\gamma_b}) \\ &\simeq \frac{1}{\sqrt{4\pi\gamma_b}} e^{-\gamma_b}, \quad \gamma_b \gg 1 \quad (\gamma_b > 4) \quad \text{(matched filter)} \end{aligned}}$$ (8.26)

8.3.3 Bipolar signaling

This is the case where binary 0 is represented by a zero level, while binary 1s are represented by alternating positive or negative values, i.e.

$$s(t) = \begin{cases} s_1(t) = \pm A, \text{ for binary 1} \\ s_2(t) = 0, \text{ for binary 0} \end{cases}$$ (8.27)

where $0 \leq t \leq T$. This is similar to the unipolar case except that we need to make some modifications. In this case, the decision rule now involves two thresholds, $+\mu$ and $-\mu$. For a matched filter, $E_d = A^2 T = 2E_b$ as in unipolar signaling. Thus,

$$\boxed{\begin{aligned} P_e &= 1.5Q\left(\sqrt{\frac{E_b}{\eta}}\right) = 1.5Q(\sqrt{\gamma_b}) \\ &\simeq \frac{1.5}{\sqrt{2\pi\gamma_b}} e^{-\gamma_b/2}, \quad \gamma_b \gg 1 \quad (\gamma_b > 4) \quad \text{(matched filter)} \end{aligned}}$$ (8.28)

where Eq. (8.16) has been applied. This indicates that P_e for bipolar signaling is 50% greater than that for unipolar signaling. In section 8.8, we will use MATLAB to compare the performance of the three cases.

EXAMPLE 8.2

A bit stream has a signal energy of 4×10^{-5} J and Gaussian white noise with PSD $\eta/2 = 0.5 \times 10^{-5}$ W/Hz. Assume that a matched-filter receiver is used, determine the bit error probability for the following signaling: (a) unipolar; (b) polar.

Solution

$$\text{Let} \quad x = \sqrt{\frac{E_b}{\eta}} = \sqrt{\frac{4 \times 10^{-5}}{2 \times 0.5 \times 10^{-5}}} = 2$$

Since $Q(x)$ is related to the complementary error function, one can use Eq. (8.15) along with MATLAB or the table in Appendix C to find $Q(x)$.
(a) Using unipolar signaling,

$$P_e = Q\left(\sqrt{\frac{E_b}{\eta}}\right) = Q(2) = 0.0228$$

which implies that 228 out of every 10,000 bits are errored.
(b) With polar signaling,

$$P_e = Q\left(\sqrt{\frac{2E_b}{\eta}}\right) = Q(2\sqrt{2}) = 0.0023$$

This means that 23 out every 10,000 bits are in error. This confirms the fact that polar signaling is more efficient than unipolar signaling.

PRACTICE PROBLEM 8.2

Rework the problem in Example 8.1 if bipolar signaling is used.

Answer: 0.0342.

EXAMPLE 8.3

A polar binary signal is a +5 V or −5 V pulse in the interval $(0, T)$. If a white Gaussian noise with PSD $\eta/2 = 10^{-7}$ W/Hz is added to the signal, find the maximum bit rate that can be sent with a probability of error 10^{-7}.

Solution

From the $Q(x)$ table,

$$Q(x) = 10^{-7} \quad \rightarrow \quad x = 5.2$$

For polar signaling,

$$P_e = Q\left(\sqrt{\frac{2E_b}{\eta}}\right)$$

where $E_b = A^2 T$

$$x = 5.2 = \sqrt{\frac{2E_b}{\eta}} = \sqrt{\frac{2A^2 T}{\eta}} \quad \rightarrow \quad T = 5.2^2 \frac{\eta}{2A^2}$$

or

$$T = 5.2^2 \frac{\eta}{2A^2} = 27.04 \times \frac{2 \times 10^{-7}}{2 \times (5)^2} = 1.0816 \times 10^{-7}$$

Hence, the maximum bit rate is

$$R = \frac{1}{T} = 9.2456 \times 10^6 \text{ b/s} = 9.25 \text{ Mbps}$$

PRACTICE PROBLEM 8.3

A binary system transmits 1.5 Mbps with a BER of 10^{-4}. Given that the channel noise spectrum is $\eta/2 = 10^{-8}$ W/Hz, calculate signal level A required using polar signaling.

Answer: 0.909.

8.4 COHERENT DETECTION

The error probability P_e for the optimum detector does not depend on the shape of the pulse but only on the energy of the pulse. Therefore, the error probability of a modulated scheme is identical to that of the corresponding baseband scheme of the same energy.

A digital signal can modulate the amplitude, frequency, or phase of a sinusoidal carrier wave leading respectively to amplitude-shift keying (ASK), frequency-shift keying (FSK), or phase-shift keying (PSK). With a rectangular baseband pulse used as the modulating waveform, the three schemes are shown in Figure 8.9. We now consider the cases of ASK, PSK, and FSK separately. In each case, we should keep in mind that the signal plus the white Gaussian noise is present at the matched-filter receiver.

8.4.1 Amplitude-shift keying

This is also known as *on-off keying* (OOK). For this case,

$$s(t) = \begin{cases} s_1(t) = A \cos \omega_c t, & \text{for binary 1} \\ s_2(t) = 0, & \text{for binary 0} \end{cases} \tag{8.29}$$

where $0 \leq t \leq T$ and T is an integer times $1/f_c = 2\pi/\omega_c$. Since the performance depends only on the pulse energy, the performance of a matched-filter receiver is given by Eq. (8.22). The difference signal energy is

$$E_d = \int_0^T [A \cos \omega_c t - 0]^2 dt = \frac{A^2 T}{2}$$

while the average energy per bit is

$$E_b = \frac{1}{2}\left(\frac{A^2 T}{2}\right) + \frac{1}{2}(0) = \frac{A^2 T}{4}$$

Therefore,

$$P_e = Q\left(\sqrt{\frac{E_d}{2\eta}}\right) = Q\left(\sqrt{\frac{A^2 T}{4\eta}}\right)$$

or

$$\boxed{P_e = Q\left(\sqrt{\frac{E_b}{\eta}}\right)} \qquad (8.30)$$

Comparing this with Eq. (8.24), we note that the performance of ASK is essentially the same as that for baseband unipolar signaling.

8.4.2 Phase-shift keying

For this case,

$$s(t) = \begin{cases} s_1(t) = A \cos \omega_c t, & \text{for binary 1} \\ s_2(t) = -A \cos \omega_c t, & \text{for binary 0} \end{cases} \qquad (8.31)$$

where $0 \leq t \leq T$. Due to the fact that $s_2(t) = -s_1(t)$, this is also known as *phase-reversal keying* (PRK). For matched-filter receiver, the performance is obtained from Eq. (8.22), where the difference signal energy is

$$E_d = \int_0^T [A \cos \omega_c t - A \cos \omega_c t]^2 dt = 2A^2 T$$

while the energy per bit is

$$E_b = \frac{1}{2}\left(\frac{A^2 T}{2}\right) + \frac{1}{2}\left(\frac{A^2 T}{2}\right) = \frac{A^2 T}{2}$$

Consequently, the BER is

$$P_e = Q\left(\sqrt{\frac{E_d}{2\eta}}\right) = Q\left(\sqrt{\frac{2A^2T}{2\eta}}\right)$$

or

$$\boxed{P_e = Q\left(\sqrt{\frac{2E_b}{\eta}}\right)} \qquad (8.32)$$

which is identical to the BER for baseband polar signaling. Comparing Eq. (8.32) with Eq. (8.30) shows that the energy per bit in ASK must be twice that in PSK in order to have the same performance. In other words, ASK requires 3 dB more power than PSK. As a result, PSK is preferable to ASK in coherent detection.

8.4.3 Frequency-shift keying

For this case,

$$s(t) = \begin{cases} s_1(t) = A \cos \omega_1 t, & \text{for binary 1} \\ s_2(t) = A \cos \omega_2 t, & \text{for binary 0} \end{cases} \qquad (8.33)$$

where $0 \le t \le T$. For matched-filter reception, we apply Eq. (8.22). The difference energy signal is

$$\begin{aligned} E_d &= \int_0^T [A \cos \omega_1 t - A \cos \omega_2 t]^2 dt \\ &= A^2 \int_0^T \left[\cos^2 \omega_1 t + \cos^2 \omega_2 t - 2 \cos \omega_1 t \cos \omega_2 t\right] dt \\ &= \frac{A^2 T}{2} + \frac{A^2 T}{2} - A^2 \int_0^T \cos(\omega_1 - \omega_2)t \, dt \\ &= A^2 T - A^2 \frac{\sin(\omega_1 - \omega_2)T}{(\omega_1 - \omega_2)} \end{aligned} \qquad (8.34)$$

If we assume that $\omega_1 T \gg 1$, $\omega_2 T \gg 1$, and $(\omega_1 - \omega_2)T \gg 1$, then the second term on the right-hand side is negligible compared with A^2T. Alternatively, if we let $\omega_1 - \omega_2 = \pi n/T = \pi n R$, where n is an integer, $s_1(t)$ and $s_2(t)$ become orthogonal and the second term on the right-hand side goes to zero. Whichever way we look at it,

$$E_d = A^2 T$$

The energy per bit is

$$E_b = \frac{1}{2}\left(\frac{A^2 T}{2}\right) + \frac{1}{2}\left(\frac{A^2 T}{2}\right) = \frac{A^2 T}{2}$$

Thus, the BER is

$$P_e = Q\left(\sqrt{\frac{E_d}{2\eta}}\right) = Q\left(\sqrt{\frac{A^2 T}{2\eta}}\right)$$

or

$$\boxed{P_e = Q\left(\sqrt{\frac{E_b}{\eta}}\right)} \tag{8.35}$$

Note that the performance of FSK is identical to that of ASK signaling and is worse than PSK signaling by 3 dB.

EXAMPLE 8.4

In a coherent PSK system for which $P_e = 10^{-6}$, the PSD of the noise at the receiver input is $\eta/2 = 10^{-10}$ W/Hz. Find the required average energy per bit.

Solution

For PSK,

$$P_e = 10^{-7} = Q\left(\sqrt{2\gamma_b}\right) \quad \rightarrow \quad \sqrt{2\gamma_b} = 5.2$$

or

$$2\gamma_b = 2\frac{E_b}{\eta} = 5.2^2 = 27.04$$

Thus,

$$E_b = \frac{27.04}{2}\eta = 13.52 \times 2 \times 10^{-10} = 2.704 \text{ nJ}$$

PRACTICE PROBLEM 8.4

In a coherent ASK system, each ASK wave has a peak amplitude of $A = 5$ V and is transmitted over a channel where $\eta/2 = 10^{-7}$ W/Hz. If the bit rate is 2 Mbps, calculate the BER.

Answer: 3.861×10^{-5}.

8.5 NONCOHERENT DETECTION

Because of the cumbersome integral evaluation involved, the derivation of the probability of error or BER for noncoherent detection is more difficult than what we had in the previous section for coherent detection. As mentioned in earlier chapters, PSK cannot be detected noncoherently and so we will consider only ASK, FSK, and differential PSK (DPSK).

8.5.1 Amplitude-shift keying

The noncoherent detector of ASK (or OOK) signals is shown in Figure 8.10.
Again, the ASK signal is

$$s(t) = \begin{cases} s_1(t) = A \cos \omega_c t, & \text{for binary 1} \\ s_2(t) = 0, & \text{for binary 0} \end{cases} \tag{8.36}$$

where $0 \leq t \leq T$. The input to the envelope detector consists of the signal plus noise with a PSD of $\eta/2$. Hence,

$$r(t) = \begin{cases} r_1(t) = s_1(t) + n(t), & \text{for binary 1} \\ r_2(t) = s_2(t) + n(t), & \text{for binary 0} \end{cases} \tag{8.37}$$

The PDF of the envelope is Rayleigh distributed. Thus, the PDF of r when the signal is absent is

$$p(r/s_2) = \frac{r}{\sigma^2} e^{-r^2/2\sigma^2} \tag{8.38}$$

while the PDF of r when signal is present is Rician distributed so that

$$p(r/s_1) = \frac{r}{\sigma^2} e^{-(r^2+A^2)/2\sigma^2} I_o \left(\frac{rA}{\sigma^2} \right) \tag{8.39}$$

where σ^2 is the variance of the noise at the input of the envelope detector and I_0 is the modified Bessel function of the first kind of zeroth order, i.e.

$$I_0(x) = \frac{1}{2\pi} \int_0^{2\pi} e^{x \cos \theta} d\theta \tag{8.40}$$

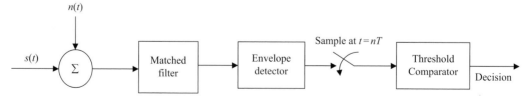

Figure 8.10. Noncoherent detector of ASK signals.

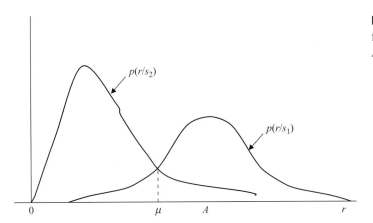

Figure 8.11. Conditional PDFs for noncoherent detection of ASK signals.

The optimum threshold $r = \mu$ occurs where the two PDFs in Eqs. (8.38) and (8.39) intercept as illustrated in Figure 8.11, i.e.

$$\frac{\mu}{\sigma^2} e^{-\left(\mu^2+A^2\right)/2\sigma^2} I_o\left(\frac{\mu A}{\sigma^2}\right) = \frac{\mu}{\sigma^2} e^{-\mu^2/2\sigma^2} \quad \rightarrow \quad \mu = \frac{A}{2}\sqrt{1 + \frac{8\sigma^2}{A^2}} \qquad (8.41)$$

Notice that the threshold is not fixed but varies with A and σ^2. The BER of noncoherent ASK receiver is

$$P_e = P(r > \mu/s_1 \text{ sent})P(s_1 \text{ sent}) + P(r < \mu/s_2 \text{ sent})P(s_2 \text{ sent}) \qquad (8.42)$$

Assuming 1s and 0s are equally probable and substituting Eqs. (8.38) and (8.39) into Eq. (8.42) leads to

$$P_e = \frac{1}{2}\int_0^\mu \frac{r}{\sigma^2} e^{-\left(r^2+A^2\right)/2\sigma^2} I_o\left(\frac{rA}{\sigma^2}\right) dr + \frac{1}{2}\int_\mu^\infty \frac{r}{\sigma^2} e^{-r^2/2\sigma^2} dr \qquad (8.43)$$

The integral involving the modified Bessel function cannot be evaluated in closed form. If we assume that $A/\sigma \gg 1$, then we can show that

$$P_e \simeq \frac{1}{2} e^{-A^2/8\sigma^2} \qquad (8.44)$$

Since the energy per bit $E_b = A^2 T/4$ and $\sigma^2 = \eta B = \eta/T$, then

$$\frac{A^2}{8\sigma^2} = \frac{4E_b/T}{8\eta/T} = \frac{E_b}{2\eta}$$

Thus, the BER for noncoherent detection of ASK is

$$\boxed{P_e \simeq \frac{1}{2} e^{-E_b/2\eta}, \quad \frac{E_b}{\eta} \gg 1} \qquad (8.45)$$

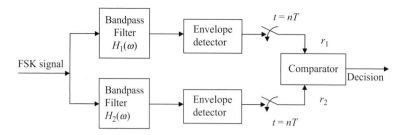

Figure 8.12. Noncoherent detection of FSK signals.

This indicates that the BER decreases with the SNR per bit $\gamma_b = E_b/\eta$. Comparing Eq. (8.45) with the formula for BER in Eq. (8.30) for a coherent detector (after Eq. (8.16) is applied) shows that the performance levels of both a noncoherent detector (envelope detector) and a coherent detector of ASK signals are similar for large γ_b.

8.5.2 Frequency-shift keying

The noncoherent detector of FSK signals is shown in Figure 8.12. Again, the FSK signal transmitted over $(0,T)$ is

$$s(t) = \begin{cases} s_1(t) = A\cos\omega_1 t, & \text{for binary 1} \\ s_2(t) = A\cos\omega_2 t, & \text{for binary 0} \end{cases} \qquad (8.46)$$

The bandpass filters $H_1(\omega)$ and $H_2(\omega)$ are assumed to be matched to the pulses. Due to the symmetry, the noise components of the output of the filters are identical Gaussian variables with variance σ^2 and the optimum threshold is $\mu = 0$. The output r_2 of the lower channel (the noise envelope) has Rayleigh PDF

$$p(r_2) = \frac{r_2}{\sigma^2} e^{-r_2^2/2\sigma^2} \qquad (8.47)$$

while the output r of the upper channel has Rician distribution with PDF

$$p(r_1) = \frac{r_1}{\sigma^2} e^{-(r_1^2+A^2)/2\sigma^2} I_o\left(\frac{r_1 A}{\sigma^2}\right) \qquad (8.48)$$

We keep in mind that the output of the envelope detector can only assume positive values. The probability of receiving s_1 when s_2 is sent is determined by the joint event $(r_1 > 0, r_2 > r_1)$, i.e.

$$P_{e0} = P(r_1 > 0, r_2 > r_1) \qquad (8.49)$$

Since r_1 and r_2 are independent,

$$
\begin{aligned}
P_{e0} &= \int_0^\infty \frac{r_1}{\sigma^2} e^{-(r_1^2+A^2)/2\sigma^2} I_o\left(\frac{r_1 A}{\sigma^2}\right) \int_{r_1}^\infty \frac{r_2}{\sigma^2} e^{-r_2^2/2\sigma^2} dr_1 dr_2 \\
&= \int_0^\infty \frac{r_1}{\sigma^2} e^{-(2r_1^2+A^2)/2\sigma^2} I_o\left(\frac{r_1 A}{\sigma^2}\right) dr_1
\end{aligned}
\qquad (8.50)
$$

If we change variables by letting $x = r_1\sqrt{2}$ and $y = A/\sqrt{2}$, then

$$P_{e0} = \frac{1}{2}e^{-A^2/4\sigma^2}\int_0^\infty \frac{x}{\sigma^2}e^{-(x^2+y^2)/2\sigma^2}I_o\left(\frac{xy}{\sigma^2}\right)dx \qquad (8.51)$$

Since the integrand is Rician PDF, the integral is unity. Hence,

$$P_{e0} = \frac{1}{2}e^{-A^2/4\sigma^2} \qquad (8.52)$$

Due to symmetry, we can similarly obtain

$$P_{e1} = \frac{1}{2}e^{-A^2/4\sigma^2} \qquad (8.53)$$

Assuming that the binary levels (1s and 0s) are equiprobable, the probability of error is given by

$$P_e = \frac{1}{2}(P_{e0} + P_{e1}) = P_{e0} = P_{e1}$$

or

$$P_e = \frac{1}{2}e^{-A^2/4\sigma^2} \qquad (8.54)$$

But for a matched filter,

$$E_b = A^2 T/2, \quad \sigma^2 = \eta B = \eta/T \quad \rightarrow \quad \frac{A^2}{4\sigma^2} = \frac{E_b}{2\eta}$$

Thus,

$$\boxed{P_e = \frac{1}{2}e^{-E_b/2\eta}} \qquad (8.55)$$

Notice that this is similar to the performance of a noncoherent ASK detector. Although noncoherent ASK and FSK have equivalent behavior, FSK is preferred over ASK for practical reasons because FSK has a fixed threshold whereas ASK has a variable threshold. Also, we observe that the performance levels of coherent and noncoherent FSK detectors are similar when $E_b/\eta \gg 1$. Although the analysis of the noncoherent FSK detector is cumbersome, it is easier to build than the coherent FSK detector.

8.5.3 Differential phase-shift keying

As mentioned earlier, it is impossible to demodulate PSK noncoherently (with an envelope detection). However, it is possible to demodulate PSK without the coherent local carrier by using the *differential coherent PSK* (DPSK) signaling technique, which consists of differentially encoded binary PSK. The DPSK demodulator is shown in Figure 8.13, where differential coding is provided by the delay and the multiplier.

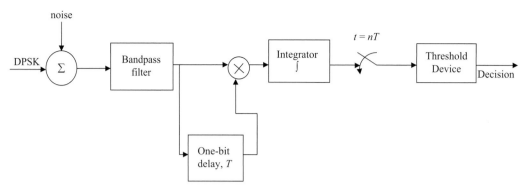

Figure 8.13. DFSK demodulator.

The derivation of the BER of DPSK receiver is lengthy and complicated. Only the result is presented here.

$$P_e = \frac{1}{2} e^{-E_b/\eta} \tag{8.56}$$

It should be noted that in practice DPSK is preferred over binary PSK in view of the fact that the DPSK receiver does not need carrier synchronization circuitry.

EXAMPLE 8.5

Most applications require a BER of 10^{-5} or less. Find the minimum SNR per bit γ_b in dB necessary to achieve that BER when ASK signaling is used.

Solution

For ASK signaling,

$$P_e \simeq \frac{1}{2} e^{-E_b/2\eta} \quad \rightarrow \quad -\frac{E_b}{2\eta} = \ln 2 P_e$$

or

$$\gamma_b = \frac{E_b}{\eta} = -2 \ln 2 P_e = -2 \ln\left(2 \times 10^{-5}\right) = 21.64$$

In decibels,

$$\gamma_b = 10 \log_{10}(21.64) = 13.35 \text{ dB}$$

PRACTICE PROBLEM 8.5

Rework Example 8.5 if DPSK signaling is used instead.

Answer: 10.34 (note the 3 dB difference).

8.6   COMPARISON OF DIGITAL MODULATION SYSTEMS

Like analog systems, the digital modulation methods covered in this chapter can be compared in many ways. The choice of digital modulation techniques depends primarily on error performance (P_e versus γ_b), bandwidth efficiency (in bps/Hz), and equipment complexity (cost). Figure 8.14 presents the equipment complexity of some representative modulation schemes. As we have noted in the previous sections, the bit error probability P_e is an important measure of performance. We now bring together for easy reference the various formulas we have derived. Table 8.1 compares P_e for the various signaling schemes covered in the previous sections. (The corresponding plots will be provided later using MATLAB.) All the results assume a matched-filtered (or optimum) receiver.

For baseband (or direct) transmission, polar signaling is better than unipolar signaling by a factor of 2 (or 3 dB) when $P_e \leq 10^{-4}$.

8.7   *M*-ARY COMMUNICATIONS

So far in this chapter, we have considered only binary signals, which are mostly used in today's communication systems. We now consider the *M*-ary signaling scheme, which may be used to send *M* possible signals, $s_1(t), s_2(t), \ldots, s_M(t)$, within the time interval of *T*. For practical purposes, $M = 2^n$, where *n* is an integer. The *M* signals are generated by changing the amplitude, phase, or frequency of a carrier thereby leading to *M*-ary ASK, *M*-ary PSK, and *M*-ary FSK modulation methods respectively.

We now determine the probability of error for *M*-ary signals transmitted over an AWGN channel.

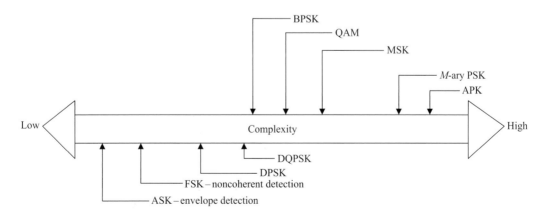

Figure 8.14. Relative complexity of digital modulation schemes. Adapted from J. D. Oetting, "A comparison of modulation techniques for digital radio," *IEEE Transactions on Communications*, vol. 27, Dec. 1979, p. 1757.

Table 8.1. **Comparison of digital signaling methods**

Type	Error probability P_e
Baseband signaling	
Unipolar	$Q(\sqrt{\gamma_b})$
Polar	$Q(\sqrt{2\gamma_b})$
Bipolar	$1.5Q(\sqrt{\gamma_b})$
Bandpass signaling	
Coherent ASK	$Q(\sqrt{\gamma_b})$
Coherent PSK	$Q(\sqrt{2\gamma_b})$
Coherent FSK	$Q(\sqrt{\gamma_b})$
Noncoherent ASK	$0.5e^{-\gamma_b/2}$
Noncoherent FSK	$0.5e^{-\gamma_b/2}$
Noncoherent DPSK	$0.5e^{-\gamma_b}$

Assuming that the *M*-ary symbols are equally likely,

$$P_e = \frac{1}{M}(P_{e1} + P_{e1} + \cdots + P_{eM}) \tag{8.57}$$

If we first consider the case of a quaternary ($M = 4$) polar signaling in the presence of an AWGN channel, the three thresholds for minimizing the error probability are at $r = -A$, 0, and $+A$. The two extreme symbols have to guard against only one symbol (like the binary case), whereas the remaining symbols must guard against neighbors on both sides. The error probabilities corresponding to the two extreme levels are

$$P_{e1} = P_{e4} = Q\left(\frac{A}{2\sigma}\right)$$

while

$$P_{e2} = P_{e3} = 2Q\left(\frac{A}{2\sigma}\right)$$

Thus,

$$P_e = \frac{1}{4}\left[2 \times Q\left(\frac{A}{2\sigma}\right) + 2 \times 2Q\left(\frac{A}{2\sigma}\right)\right] = 1.5Q\left(\frac{A}{2\sigma}\right) \tag{8.58}$$

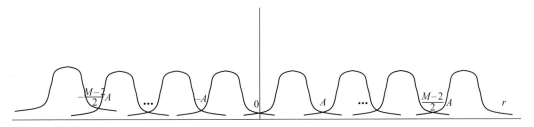

Figure 8.15. Conditional PDFs for *M*-ary signaling in the presence of AWGN.

This may now be generalized for the case of M symbols. There are $M - 1$ decision thresholds at

$$r = 0, \pm A, \pm 2A, \ldots, \pm \frac{M-2}{2}A \tag{8.59}$$

as illustrated in Figure 8.15. Again, the extreme levels have

$$P_{e1} = P_{eM} = Q\left(\frac{A}{2\sigma}\right)$$

while the $M - 2$ inner levels have

$$P_{e2} = \ldots = P_{eM-1} = 2Q\left(\frac{A}{2\sigma}\right)$$

Hence,

$$P_e = \frac{1}{M}\left[2 \times Q\left(\frac{A}{2\sigma}\right) + (M-2) \times 2Q\left(\frac{A}{2\sigma}\right)\right]$$
$$= \frac{2(M-1)}{M}Q\left(\frac{A}{2\sigma}\right) \tag{8.60}$$

For a matched-filter receiver,

$$\frac{A^2}{\sigma^2} = \frac{2E_p}{\eta} \tag{8.61}$$

where E_p is the energy of the pulse. Assuming that the pulses are $\pm p(t)$, $\pm 3p(t)$, $\pm 5p(t)$, $\ldots$, $\pm(M-1)p(t)$, the average pulse energy is

$$E_{Mp} = \frac{1}{M}\left[E_p + 9E_p + 25E_p + \cdots + (M-1)^2 E_p\right] \times 2$$
$$= \frac{2E_p}{M}\sum_{k=0}^{\frac{M-1}{2}}(2k+1) \tag{8.62}$$
$$= \frac{M^2-1}{3}E_p$$

Since an M-ary symbol carries $\log_2 M$ bits of information, the bit energy E_b is

$$E_b = \frac{E_{Mp}}{\log_2 M} = \frac{M^2 - 1}{3 \log_2 M} E_p \tag{8.63}$$

Substituting Eqs. (8.61) and (8.63) into Eq. (8.60) yields

$$\boxed{P_e = \frac{2(M-1)}{M} Q\left[\sqrt{\frac{6 \log_2 M}{M^2 - 1} \gamma_b}\right]} \tag{8.64}$$

where $\gamma_b = E_b/\eta$.

EXAMPLE 8.5

Calculate the error probability for an M-ary communication system with $M = 8$ and SNR per bit of 25.

Solution

Using Eq. (8.63),

$$P_e = \frac{2(M-1)}{M} Q\left[\sqrt{\frac{6 \log_2 M}{M^2 - 1} \gamma_b}\right]$$

$$= \frac{2 \times 7}{8} Q\left[\sqrt{\frac{6 \log_2 8}{63}(25)}\right]$$

$$= 1.75 Q(2.6726) = 6.6 \times 10^{-3}$$

PRACTICE PROBLEM 8.5

Consider an M-ary digital with $M = 16$ in the presence of white Gaussian noise with PSD $\eta/2 = 1.6 \times 10^{-8}$. Compute the error probability for $E_b = 0.5 \times 10^{-5}$.

Answer: 1.178×10^{-4}.

8.8 SPREAD-SPECTRUM SYSTEMS

A spread-spectrum (SS) system is one in which the transmitted signal is spread over a frequency range much wider than the minimum bandwidth required to send the signal. Using spread spectrum, a radio is supposed to distribute the signal across the entire spectrum. This way, no single user can dominate the band and collectively all users look like noise. The fact that such signals appear like noise in the band makes them difficult to find and jam, thereby increasing security against unauthorized listeners.

There are two types of spread-spectrum technology: frequency hopping and direct sequence. Our goal here is to determine the performance of the SS system in an AWGN environment.

8.8.1 Direct sequence

Direct sequence spread spectrum (DSSS) takes a signal at a given frequency and spreads it across a band of frequencies where the center frequency is the original signal. The spreading algorithm, which is the key to the relationship of the spread range of frequencies, changes with time in a pseudorandom sequence. The DSSS system takes an already modulated message and modulates it again using a pseudorandom sequence of ± 1s. The modulation for a DSSS system may be any coherent digital scheme, although BPSK, QPSK, and MSK are commonly used. For the purpose of illustration, we will examine the case of binary PSK (BPSK), shown in Figure 8.16.

To determine the probability of error for a DSSS system, let us assume that the modulation is BPSK. If $m(t)$ is the message data, the modulated carrier is

$$s_m(t) = Am(t)\cos\omega_c t \tag{8.65}$$

where $0 \leq t \leq T$ and $f_c = \omega_c/2\pi$ is the carrier frequency. The transmitted signal is

$$s(t) = a_o s_m(t) = a_o Am(t)\cos\omega_c t \tag{8.66}$$

where $a_o = \pm 1$ is the spreading signal. The received signal is corrupted by an additive inferring signal. Hence, the input to the receiver is

$$r(t) = a_o Am(t)\cos\omega_c t + n(t) + n_j(t) \tag{8.67}$$

where n_j is the jamming signal with PSD of $\eta_J = P_J/2f_c$ (in W/Hz) and $n(t)$ is white noise with PSD $\eta/2$. Since a_o only causes polarity reversal, which does not affect the power spectral density of the noise, the BER or probability of error is the same as the BER for the demodulator without spread-spectrum. The same conclusion holds for other modulation methods. Hence, for coherent matched-filter detection,

$$P_e = Q\left(\sqrt{\frac{2E_b}{\eta + \eta_J}}\right) \tag{8.68}$$

where $E_b = A^2 T/2$. If we assume that the jammer PSD η_J is much greater than the noise PSD η ($\eta_J \gg \eta$), then

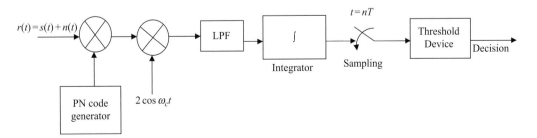

Figure 8.16. BPSK direct-sequence spread-spectrum system.

$$P_e = Q\left(\sqrt{\frac{2E_b}{\eta_J}}\right) \qquad (8.69)$$

Noting that $E_b = P_b T = P_b/R$, where P_b is the signal power per bit and $\eta_J = P_J/B$, where B is the signal bandwidth, Eq. (8.69) becomes

$$P_e = Q\left(\sqrt{\frac{2B/R}{P_J/P_b}}\right) \qquad (8.70)$$

where $B/R = BT$ is the processing gain and P_J/P_b is the jammer-to-signal power ratio (JSR).

8.8.2 Frequency hopping

Frequency-hopping spread spectrum (FHSS) uses a narrowband carrier that changes frequency in a pattern known to both transmitter and receiver. It is based on the use of a signal at a given frequency that is constant for a small amount of time and then moves to a new frequency. The sequence of different channels for the hopping pattern is determined in pseudorandom fashion. This means that a very long sequence code is used before it is repeated, over 65,000 hops, making it appear random. Thus it is very difficult to predict the next frequency at which such a system will stop and transmit/receive data as the system appears to be a noise source to an unauthorized listener. This makes the FHSS system very secure against interference and interception. FHSS is characterized by low cost, low power consumption, and less range than DSSS, although DSSS is more commercially important.

An FHSS receiver is shown in Figure 8.17. As shown in the figure, noncoherent detection is often used in order to avoid the difficulty of maintaining coherence with the frequency hopper. The probability of error for FSK noncoherent detection without jamming is

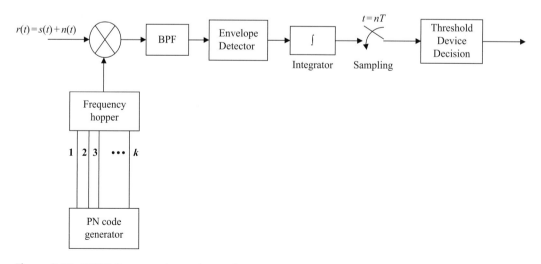

Figure 8.17. BPSK frequency-hopped spread-spectrum system.

$$P_e = \frac{1}{2}\exp\left(-\frac{E_b}{2\eta}\right) \tag{8.71}$$

If the jammer corrupts only a fraction β $(0 \leq \beta \leq 1)$ of the band and the PSD of the jammer over the entire band is η_J, then the PSD in the jammed frequency range is η_J/β. Hence,

$$P_e = \frac{1-\beta}{2}\exp\left(-\frac{E_b}{2\eta}\right) + \frac{\beta}{2}\exp\left(-\frac{E_b}{2(\eta + \eta_J/\beta)}\right) \tag{8.72}$$

Assuming that the jammer PSD dominates and that $E_b/\eta \gg 1$, Eq. (8.72) becomes

$$\boxed{P_e \simeq \frac{\beta}{2}\exp\left(-\frac{\beta E_b}{2\eta_J}\right)} \tag{8.73}$$

EXAMPLE 8.7

For a fixed processing gain of $B/R = 10$, we want to see the effect of the jammer-to-signal ratio (JSR) on the error probability of DS spread-spectrum receiver. Let JSR $= 1, 10, 100, 1000$.

Solution

$$P_e = Q\left(\sqrt{\frac{2B/R}{P_J/P_b}}\right) = Q\left(\sqrt{\frac{20}{P_J/P_b}}\right)$$

Substituting the values of JSR gives

$$P_e = \begin{cases} Q\left(\sqrt{\dfrac{20}{1}}\right) = 3.872 \times 10^{-6}, & \dfrac{P_J}{P_b} = 1 \\[2mm] Q\left(\sqrt{\dfrac{20}{10}}\right) = 0.0786, & \dfrac{P_J}{P_b} = 10 \\[2mm] Q\left(\sqrt{\dfrac{20}{100}}\right) = 0.3274, & \dfrac{P_J}{P_b} = 100 \\[2mm] Q\left(\sqrt{\dfrac{20}{1000}}\right) = 0.4438, & \dfrac{P_J}{P_b} = 1000 \end{cases}$$

showing an asymptotically exponential dependence.

PRACTICE PROBLEM 8.7

By differentiating Eq. (8.73) with respect to β, show that the worst-case β satisfies

$$\beta_{\text{opt}} = \frac{2\eta_J}{E_b}$$

Find the corresponding P_e.

Answer: Proof, $\frac{\eta_J}{eE_b}$.

8.9 COMPUTATION USING MATLAB

As a computing tool, MATLAB can be used to compute and plot the probability of error P_e versus SNR per bit γ_b. We will use MATLAB to compare the performance of the various modulation methods covered in this chapter. First, let us compare unipolar, polar, and bipolar baseband signaling. The MATLAB code is shown below. The program simply uses the appropriate formulas from Table 8.1. Since MATLAB does not have a command for the function $Q(x)$, we use Eq. (8.15) to get it. The **loglog** command is used to plot $\log(x)$ versus $\log(y)$ automatically. The performance levels of the three baseband signalings are plotted in Figure 8.18. We observe the polar signaling outperforms others.

```
% plot probability of error pe
x=0:0.1:40;
y1=sqrt(x);
pe1=0.5*erfc(y1/sqrt(2)); % unipolar case
y2=sqrt(2*x);
pe2=0.5*erfc(y2/sqrt(2)); % polar case
pe3=1.5*0.5*erfc(y1/sqrt(2)); % bipolar case
loglog(x,pe1,'r',x,pe2,'b',x,pe3,'g')
xlabel('E_b/eta')
ylabel('Probability of error, P_e')
```

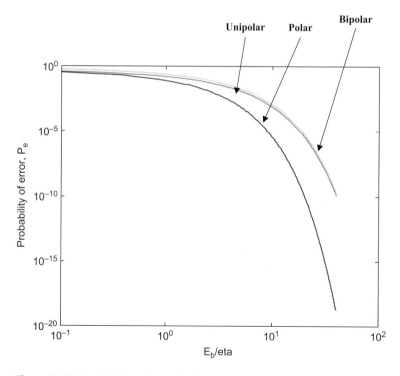

Figure 8.18. Probability of error for baseband signaling.

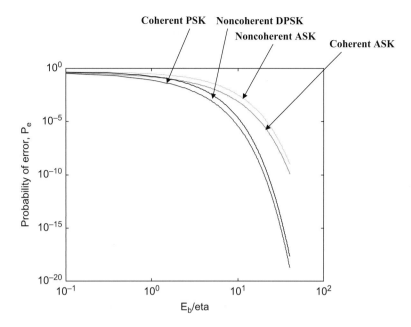

Figure 8.19. Probability of error for bandpass signaling.

We now compare bandpass signaling methods. The MATLAB code is presented below, while the plots of P_e for coherent ASK and PSK and noncoherent ASK and DPSK are shown in Figure 8.18.

```
% plots probability of error
x=0:0.1:40;
y1=sqrt(x);
pe1=0.5*erfc(y1/sqrt(2)); % coherent ASK
y2=sqrt(2*x);
pe2=0.5*erfc(y2/sqrt(2)); % coherent PSK
pe3=0.5*exp(-x/2); % noncoherent ASK
pe4=0.5*exp(-x); % noncoherent DPSK
loglog(x,pe1,'r',x,pe2,'b',x,pe3,'g',x,pe4,'k')
xlabel('E_b/eta')
ylabel('Probability of error, P_e')
```

It is evident from Figure 8.19 that there is no significant difference in the performance of coherent PSK and noncoherent DPSK. Also, the performance levels of coherent ASK and noncoherent ASK are close. Notice that coherent PSK is the best of them all.

Summary

1. If a signal $s(t)$ is corrupted in the presence of AWGN, the filter with impulse response matched to $s(t)$ maximizes the output SNR, i.e. $h(t) = s(t_o - t)$. The filter for optimum detection is matched to the difference between two signals, i.e. $h(t) = s_1(t) - s_2(t)$.

2. The performance of digital modulation systems is summarized in Table 8.1. The performance of the BPSK spread-spectrum system in an AWGN environment is the same as the performance of conventional BPSK coherence systems.

3. For M-ary communication, the error probability is

$$P_e = \frac{2(M-1)}{M} Q\left[\sqrt{\frac{6 \log_2 M}{M^2 - 1} \gamma_b}\right], \quad \gamma_b = \frac{E_b}{\eta}$$

4. Spread spectrum is a broad-band modulation scheme that uses pseudorandom code to spread the baseband sprectrum. The two major types of spread-spectrum system are direct sequence spread spectrum (DSSS) and frequency hop spread spectrum (HFSS).

5. For DSSS, the error probability is

$$P_e = Q\left(\sqrt{\frac{2B/R}{P_J/P_b}}\right)$$

and for FHSS,

$$P_e \simeq \frac{\beta}{2} \exp\left(-\frac{\beta E_b}{2\eta_J}\right), \quad E_b/\eta \gg 1 \tag{8.73}$$

6. MATLAB is a handy tool for plotting P_e versus γ_b and comparing the performance of various modulation schemes covered in this chapter.

Review questions

8.1 A filter matched to pulse $p(t) - q(t)$ is described by:
   (a) $h(t) = p(T - t) - q(T - t)$.   (b) $H(\omega) = P(\omega) - Q(\omega)$.
   (c) $h(t) = kp(t - T) - kq(t - T)$.   (d) $H(\omega) = P(\omega)e^{-j\omega T} - Q(\omega)e^{-j\omega T}$.

8.2 The output of a matched filter is:
   (a) $h(t)s(t)$.   (b) $s(t_o - t)h(t)$.   (c) $h(t)*s(t)$.   (d) $h(t)*s(t_o - t)$.

8.3 As far as the error probability is concerned, which of the modulation schemes performs best?
   (a) ASK.   (b) PSK.   (c) FSK.   (d) BPSK.

8.4 Which detector has a variable threshold?
(a) ASK coherent detector.   (b) PSK coherent detector.
(c) FSK coherent detector.   (d) ASK noncoherent detector.
(e) FSK coherent detector.

8.5 Which signaling scheme is practically impossible to demodulate noncoherently?
(a) ASK.   (b) FSK.   (c) PSK.   (d) DPSK.

8.6 The performance of a BPSK spread-spectrum system in the presence of AWGN is identical to that of a conventional BPSK coherence system.
(a) True.   (b) False.

8.7 How many binary digits are carried by an 8-ary communication system?
(a) 2.   (b) 3.   (c) 8.   (d) 16.   (e) 256.

8.8 A DS spread spectrum provides no performance improvement against additive white Gaussian noise.
(a) True.   (b) False.

8.9 The MATLAB instruction for plotting data on log scale along both x and y is:
(a) plot.   (b) semilog.   (c) log.   (d) loglog.

Answers: 8.1 a, 8.2 c, 8.3 b, 8.4 d, 8.5 c, 8.6 a, 8.7 b, 8.8 a, 8.9 d

Problems

Section 8.2 Matched-filter receiver
8.1 The input signal to a matched filter is

$$s(t) = \begin{cases} e^{-t}, & 0 < t < T \\ 0, & \text{otherwise} \end{cases}$$

Find the transfer function of the matched filter.

8.2 Determine the output of the matched filter over $(0,T)$ when the input is

$$s(t) = \begin{cases} t, & 0 < t < T \\ 0, & \text{otherwise} \end{cases}$$

8.3 Consider the signal shown in Figure 8.20.
(a) Determine the impulse response of a filter matched to the signal and sketch it.
(b) Plot the output of the matched filter.

8.4 Repeat the previous problem for the signal shown in Figure 8.21.

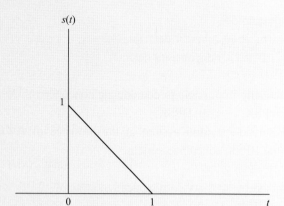

Figure 8.20. For Problem 8.3.

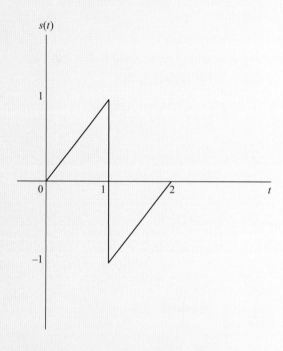

Figure 8.21. For Problem 8.4.

Section 8.3 Baseband binary systems

8.5 Orthogonal baseband signaling is one in which $s_1(t)$ and $s_2(t)$ are orthogonal over $(0, T)$, i.e.

$$\int_0^T s_1(t)s_2(t) = 0$$

Let

$$s(t) = \begin{cases} s_1(t) = A\cos\omega_c t, & \text{for binary 1} \\ s_2(t) = -A\cos\omega_c t, & \text{for binary 0} \end{cases}$$

and show that

$$P_e = Q\left(\sqrt{\frac{E_d}{\eta}}\right)$$

for an optimum (matched-filter) receiver.

8.6 A baseband communication system uses unipolar signaling with a matched-filter receiver. Obtain the probability of error for $E_b/\eta = 10$ dB.

8.7 A maximum BER of 10^{-6} is desired for a baseband system using polar signaling with a matched filter in the receiver. Find the minimum γ_b required.

8.8 Determine the required value of A that will achieve $P_e = 10^{-8}$ for a channel with $\eta/2 = 10^{-10}$ W/Hz and unipolar signaling with $R = 800$ kbps. Assume matched filtering.

8.9 A polar signal of amplitude ± 2 V is received in the presence of additive Gaussian noise of variance 0.1 V^2. Calculate the error probability.

Section 8.4 Coherent detection

8.10 An ASK system employs the following signals in the presence of AGWN.

$$s(t) = \begin{cases} s_1(t) = A\cos\omega_c t, & \text{for binary 1} \\ s_2(t) = \beta A\cos\omega_c t, & \text{for binary 0} \end{cases}$$

where $0 < \beta < 1$ and $0 < t < T$.

(a) Derive the probability of error assuming that 1 and 0 occur with equal probability.

(b) Determine the SNR required to achieve an error probability of 10^{-5} for $\beta = 0.4$.

8.11 A coherent PSK system transmits with a peak voltage of 5 V. If the bit intervals are 0.4 μs in duration and $\eta/2 = 10^{-6}$ W/Hz, calculate P_e.

Section 8.5 Noncoherent detection

8.12 A certain ASK system operates in the presence of AGWN with 10 dB SNR per bit. Find BER assuming: (a) coherent detection, (b) noncoherent detection.

8.13 For $\gamma_b = 15$ dB, calculate the error probability for: (a) noncoherent ASK; (b) noncoherent FSK; (c) noncoherent DPSK.

Section 8.7 *M*-ary communication

8.14 An *M*-ary digital system transmits 1500 symbols per second. Determine the equivalent transmission rate in bits per second where: (a) $M = 4$; (b) $M = 16$; (c) $M = 32$.

8.15 Consider a data sequence encoded and then modulated using coherent 16-ary PSK. If the received E_b/η is 12 dB, calculate the probability of error.

Section 8.8 Spread-spectrum systems

8.16 Consider the effect of increasing the processing gain on the probability of error for a DSSS system. Let $P_J/P_s = 10$, $R/B = 1$, 10, 100, and 1000 and find P_e.

8.17 Suppose we want to achieve an error rate performance of 10^{-8} or less with a DS spread-spectrum system. Calculate the jammer-to-signal ratio that will yield a processing gain of $B/R = 100$.

8.18 A DSSS system using BPSK has a processing gain of 500. What is the jammer-to-signal ratio if the desired error probability is 10^{-6}?

8.19 For a DSSS system, determine the JSR required to achieve $P_e = 10^{-5}$ when the processing gain is: (a) 10; (b) 100; (c) 1000.

8.20 A DS spread-spectrum system transmits at 1 kbps in the presence of a jammer. If the jammer power is 24 dB more than the desired signal, calculate the signal bandwidth required to achieve an error probability of 10^{-5}.

8.21 A DS spread-spectrum system uses PSK with a processing gain of 250. Determine the jamming margin against a continuous tone jammer when the desired $P_e = 10^{-6}$.

8.22 An FH spread-spectrum system transmits via FSK with noncoherent detection. Determine the probability of error for this system in an AWGN channel if the system operates at $E_b/\eta_J = 10$ and $\beta = 0.5$.

Section 8.9 Computation using MATLAB

8.23 Plot and compare the BER for coherent and noncoherent detections of ASK using MATLAB.

8.24 Use MATLAB to plot P_e versus γ_b for an M-ary communication system with $M = 4, 8, 16,$ and 32.

APPENDIX A
Mathematical formulas

This appendix – by no means exhaustive – serves as a handy reference. It does contain all the formulas needed to solve the problems in this book.

A.1   QUADRATIC FORMULAS

The roots of the quadratic equation $ax^2 + bx + c = 0$

$$x_1, x_2 = \frac{-b \pm \sqrt{b^2 - 4ac}}{2a}$$

A.2   TRIGONOMETRIC IDENTITIES

$$\sin(-x) = -\sin x$$

$$\cos(-x) = \cos x$$

$$\sec x = \frac{1}{\cos x}, \qquad \csc x = \frac{1}{\sin x}$$

$$\tan x = \frac{\sin x}{\cos x}, \qquad \cot x = \frac{1}{\tan x}$$

$$\sin(x \pm 90°) = \pm\cos x$$

$$\cos(x \pm 90°) = \mp\sin x$$

$$\sin(x \pm 180°) = -\sin x$$

$$\cos(x \pm 180°) = -\cos x$$

$$\cos^2 x + \sin^2 x = 1$$

$$\frac{a}{\sin A} = \frac{b}{\sin B} = \frac{c}{\sin C} \qquad \text{(law of sines)}$$

$$a^2 = b^2 + c^2 - 2bc \cos A \qquad \text{(law of cosines)}$$

$$\frac{\tan \frac{1}{2}(A - B)}{\tan \frac{1}{2}(A + B)} = \frac{a - b}{a + b} \quad \text{(law of tangents)}$$

$$\sin (x \pm y) = \sin x \cos y \pm \cos x \sin y$$

$$\cos (x \pm y) = \cos x \cos y \mp \sin x \sin y$$

$$\tan (x \pm y) = \frac{\tan x \pm \tan y}{1 \mp \tan x \tan y}$$

$$2 \sin x \sin y = \cos (x - y) - \cos (x + y)$$

$$2 \sin x \cos y = \sin (x + y) - \sin (x - y)$$

$$2 \cos x \cos y = \cos (x + y) - \cos (x - y)$$

$$\sin 2x = 2 \sin x \cos x$$

$$\cos 2x = \cos^2 x - \sin^2 x = 2 \cos^2 x - 1 = 1 - 2 \sin^2 x$$

$$\tan 2x = \frac{2 \tan x}{1 - \tan^2 x}$$

$$\sin^2 x = \frac{1}{2}(1 - \cos 2x)$$

$$\cos^2 x = \frac{1}{2}(1 + \cos 2x)$$

$$a \cos x + b \sin x = K \cos (x + \theta), \quad \text{where} \quad K = \sqrt{a^2 + b^2} \quad \text{and} \quad \theta = \tan^{-1}\left(\frac{-b}{a}\right)$$

$$e^{\pm jx} = \cos x \pm j \sin x \quad \text{(Euler's formula)}$$

$$\cos x = \frac{e^{jx} + e^{-jx}}{2}$$

$$\sin x = \frac{e^{jx} - e^{-jx}}{2j}$$

$$1 \text{ rad} = 57.296°$$

A.3 HYBERBOLIC FUNCTIONS

$$\sinh x = \frac{1}{2}(e^x - e^{-x})$$

$$\cosh x = \frac{1}{2}(e^x + e^{-x})$$

$$\tanh x = \frac{\sinh x}{\cosh x}$$

$$\coth x = \frac{1}{\tanh x}$$

$$\operatorname{csch} x = \frac{1}{\sinh x}$$

$$\operatorname{sech} x = \frac{1}{\cosh x}$$

$$\sinh(x \pm y) = \sinh x \cosh y \pm \cosh x \sinh y$$

$$\cosh(x \pm y) = \cosh x \cosh y \pm \sinh x \sinh y$$

$$\tan (x \pm y) = \frac{\tan x \pm \tan y}{1 \mp \tan x \tan y}$$

A.4   DERIVATIVES

If $U = U(x)$, $V = V(x)$, and a = constant,

$$\frac{d}{dx}(aU) = a\frac{dU}{dx}$$

$$\frac{d}{dx}(UV) = U\frac{dV}{dx} + V\frac{dU}{dx}$$

$$\frac{d}{dx}\left(\frac{U}{V}\right) = \frac{V\frac{dU}{dx} - U\frac{dV}{dx}}{V^2}$$

$$\frac{d}{dx}(aU^n) = naU^{n-1}$$

$$\frac{d}{dx}(a^U) = a^U \ln a \frac{dU}{dx}$$

$$\frac{d}{dx}(e^U) = e^U \frac{dU}{dx}$$

$$\frac{d}{dx}(\sin U) = \cos U \frac{dU}{dx}$$

$$\frac{d}{dx}(\cos U) = -\sin U \frac{dU}{dx}$$

A.5   INDEFINITE INTEGRALS

If $U = U(x)$, $V = V(x)$, and $a =$ constant,

$$\int a\, dx = ax + C$$

$$\int U\, dV = UV - \int V\, dU \quad \text{(integration by parts)}$$

$$\int U^n dU = \frac{U^{n+1}}{n+1} + C, \quad n \neq 1$$

$$\int \frac{dU}{U} = \ln U + C$$

$$\int a^U dU = \frac{a^U}{\ln a} + C, \quad a > 0, a \neq 1$$

$$\int e^{ax} dx = \frac{1}{a} e^{ax} + C$$

$$\int x e^{ax} dx = \frac{e^{ax}}{a^2}(ax - 1) + C$$

$$\int x^2 e^{ax} dx = \frac{e^{ax}}{a^3}\left(a^2 x^2 - 2ax + 2\right) + C$$

$$\int \ln x\, dx = x \ln x - x + C$$

$$\int \sin ax\, dx = -\frac{1}{a}\cos ax + C$$

$$\int \cos ax\, dx = \frac{1}{a}\sin ax + C$$

$$\int \sin^2 ax\, dx = \frac{x}{2} - \frac{\sin 2ax}{4a} + C$$

$$\int \cos^2 ax\, dx = \frac{x}{2} + \frac{\sin 2ax}{4a} + C$$

$$\int x \sin ax\, dx = \frac{1}{a^2}(\sin ax - ax \cos ax) + C$$

$$\int x \cos ax\, dx = \frac{1}{a^2}(\cos ax + ax \sin ax) + C$$

$$\int x^2 \sin ax \, dx = \frac{1}{a^3} \left(2ax \sin ax + 2 \cos ax - a^2 x^2 \cos ax \right) + C$$

$$\int x^2 \cos ax \, dx = \frac{1}{a^3} \left(2ax \cos ax - 2 \sin ax + a^2 x^2 \sin ax \right) + C$$

$$\int e^{ax} \sin bx \, dx = \frac{e^{ax}}{a^2 + b^2} \left(a \sin bx - b \cos bx \right) + C$$

$$\int e^{ax} \cos bx \, dx = \frac{e^{ax}}{a^2 + b^2} \left(a \cos bx + b \sin bx \right) + C$$

$$\int \sin ax \sin bx \, dx = \frac{\sin (a - b)x}{2(a - b)} - \frac{\sin (a + b)x}{2(a + b)} + C, \quad a^2 \neq b^2$$

$$\int \sin ax \cos bx \, dx = - \frac{\cos (a - b)x}{2(a - b)} - \frac{\cos (a + b)x}{2(a + b)} + C, \quad a^2 \neq b^2$$

$$\int \cos ax \cos bx \, dx = \frac{\sin (a - b)x}{2(a - b)} + \frac{\sin (a + b)x}{2(a + b)} + C, \quad a^2 \neq b^2$$

$$\int \frac{dx}{a^2 + x^2} = \frac{1}{a} \tan^{-1} \frac{x}{a} + C$$

$$\int \frac{x^2 \, dx}{a^2 + x^2} = x - a \tan^{-1} \frac{x}{a} + C$$

$$\int \frac{dx}{(a^2 + x^2)^2} = \frac{1}{2a^2} \left(\frac{x}{x^2 + a^2} + \frac{1}{a} \tan^{-1} \frac{x}{a} \right) + C$$

A.6 DEFINITE INTEGRALS

If m and n are integers,

$$\int_0^{2\pi} \sin x \, dx = 0$$

$$\int_0^{2\pi} \cos x \, dx = 0$$

$$\int_0^{\pi} \sin^2 x \, dx = \int_0^{\pi} \cos^2 x \, dx = \frac{\pi}{2}$$

$$\int_0^\pi \sin mx \sin nx \, dx = \int_0^\pi \cos mx \cos nx \, dx = 0, \quad m \neq n$$

$$\int_0^\pi \sin mx \cos nx \, dx = \begin{cases} 0, & m+n = \text{even} \\ \dfrac{2m}{m^2 - n^2}, & m+n = \text{odd} \end{cases}$$

$$\int_0^{2\pi} \sin mx \sin nx \, dx = \int_{-\pi}^\pi \sin mx \sin nx \, dx = \begin{cases} 0, & m \neq n \\ \pi, & m = n \end{cases}$$

$$\int_0^\infty \frac{\sin ax}{x} \, dx = \begin{cases} \dfrac{\pi}{2}, & a > 0 \\ 0, & a = 0 \\ -\dfrac{\pi}{2}, & a < 0 \end{cases}$$

$$\int_0^\infty \frac{\cos bx}{x^2 + a^2} \, dx = \frac{\pi}{2a} e^{-ab}, \quad a > 0, b > 0$$

$$\int_0^\infty \frac{x \sin bx}{x^2 + a^2} \, dx = \frac{\pi}{2} e^{-ab}, \quad a > 0, b > 0$$

$$\int_0^\infty \sin cx \, dx = \int_0^\infty \sin c^2 x \, dx = \frac{1}{2}$$

$$\int_{-\infty}^\infty e^{\pm j2\pi tx} \, dx = \delta(t)$$

$$\int_0^\infty e^{-a^2 x^2} \, dx = \frac{\sqrt{\pi}}{2a}, \quad a > 0$$

$$\int_0^\infty x^{2n} e^{-ax^2} \, dx = \frac{1 \cdot 3 \cdot 5 \cdots (2n-1)}{2^{n+1} a^n} \sqrt{\frac{\pi}{a}}$$

$$\int_0^\infty x^{2n+1} e^{-ax^2} \, dx = \frac{n!}{2a^{n+1}}, \quad a > 0$$

A.7 L'HÔPITAL'S RULE

If $f(0) = 0 = h(0)$, then

$$\lim_{x\to 0}\frac{f(x)}{h(x)} = \lim_{x\to 0}\frac{f'(x)}{h'(x)}$$

where the prime indicates differentiation.

A.8 TAYLOR AND MACLAURIN SERIES

$$f(x) = f(a) + \frac{(x-a)}{1!}f'(a) + \frac{(x-a)^2}{2!}f''(a) + \cdots$$

$$f(x) = f(0) + \frac{x}{1!}f'(0) + \frac{x^2}{2!}f''(0) + \cdots$$

where the prime indicates differentiation.

A.9 POWER SERIES

$$e^x = 1 + x + \frac{x^2}{2!} + \frac{x^3}{3!} + \cdots + \frac{x^n}{n!} + \cdots$$

$$\sin x = x - \frac{x^3}{3!} + \frac{x^5}{5!} - \frac{x^7}{7!} + \cdots$$

$$\cos x = 1 - \frac{x^2}{2!} + \frac{x^4}{4!} - \frac{x^6}{6!} + \frac{x^8}{8!} - \cdots$$

$$\tan x = x + \frac{x^3}{3} + \frac{2x^5}{15} + \frac{17x^7}{315} + \cdots$$

$$(1+x)^n = 1 + nx + \frac{n(n+1)}{2!}x^2 + \frac{n(n-1)(n-2)}{3!}x^3 + \cdots + \binom{n}{k}x^k + \cdots + x^n$$
$$\approx 1 + nx, \quad |x| \ll 1$$

$$\frac{1}{1-x} = 1 + x + x^2 + x^3 + \cdots, \quad |x| < 1$$

$$Q(x) = \frac{e^{-x^2/2}}{x\sqrt{2\pi}}\left(1 - \frac{1}{x^2} + \frac{1\cdot 3}{x^4} - \frac{1\cdot 3\cdot 5}{x^6} + \cdots\right)$$

$$J_n(x) = \frac{1}{n!}\left(\frac{x}{2}\right)^n - \frac{1}{(n+1)!}\left(\frac{x}{2}\right)^{n+2} + \frac{1}{2!(n+2)!}\left(\frac{x}{2}\right)^{n+4} - \cdots$$

$$J_n(x) \approx \sqrt{\frac{2}{\pi x}} \cos\left(x - \frac{\pi}{4} - \frac{n\pi}{2}\right), \quad x \gg 1$$

$$I_0(x) \approx \begin{cases} e^{x^2/4}, & x^2 \ll 1 \\ \dfrac{e}{\sqrt{2\pi x}}, & x \gg 1 \end{cases}$$

A.10 SUMS

$$\sum_{k=1}^{N} k = \frac{1}{2} N(N+1)$$

$$\sum_{k=1}^{N} k^2 = \frac{1}{6} N(N+1)(2N+1)$$

$$\sum_{k=1}^{N} k^3 = \frac{1}{4} N^2 (N+1)^2$$

$$\sum_{k=0}^{N} a^k = \frac{a^{N+1} - 1}{a - 1} \quad a \neq 1$$

$$\sum_{k=M}^{N} a^k = \frac{a^{N+1} - a^M}{a - 1} \quad a \neq 1$$

$$\sum_{k=0}^{N} \binom{N}{k} a^{N-k} b^k = (a+b)^N, \quad \text{where} \quad \binom{N}{k} = \frac{N!}{(N-k)!k!}$$

A.11 COMPLEX NUMBERS

$$e^{\pm j\pi/2} = \pm j$$

$$e^{\pm jn\pi} = \begin{cases} 1, & n \text{ even} \\ -1, & n \text{ odd} \end{cases}$$

$$e^{\pm j\theta} = \cos\theta \pm j\sin\theta$$

$$a + jb = re^{j\theta}, \quad r = \sqrt{a^2 + b^2}, \quad \theta = \tan^{-1}\left(\frac{b}{a}\right)$$

$$\left(re^{j\theta}\right)^k = r^k e^{jk\theta}$$

$$\left(r_1 e^{j\theta_1}\right)\left(r_2 e^{j\theta_2}\right) = r_1 r_2 e^{j(\theta_1 + \theta_2)}$$

APPENDIX B

MATLAB

MATLAB has become a powerful tool of technical professionals worldwide. The term MATLAB is an abbreviation for MATrix LABoratory, implying that MATLAB is a computational tool that employes matrices and vectors/arrays to carry out numerical analysis, signal processing, and scientific visualization tasks. Because MATLAB uses matrices as its fundamental building blocks, one can write mathematical expressions involving matrices just as easily as one would on paper. MATLAB is available for Macintosh, Unix, and Windows operating systems. A student version of MATLAB is available for PCs. A copy of MATLAB can be obtained from:

The Mathworks, Inc.
3 Apple Hill Drive
Natick, MA 01760–2098, USA
Phone:(508) 647–7000
Website: http://www.mathworks.com

A brief introduction on MATLAB is presented in this Appendix. What is presented is sufficient for solving problems in this book. Other information on MATLAB required in this book is provided on a chapter-by-chapter basis as needed. Additional information about MATLAB can be found in MATLAB books and from online help. The best way to learn MATLAB is to work with it after one has learned the basics.

B.1   MATLAB FUNDAMENTALS

The Command window is the primary area where you interact with MATLAB. A little later, we will learn how to use the text editor to create M-files, which allow one to execute sequences of commands. For now, we focus on how to work in the Command window. We will first learn how to use MATLAB as a calculator. We do so by using the algebraic operators in Table B.1.

To begin to use MATLAB, we use these operators. Type commands at the MATLAB prompt ">>" in the Command window (correct any mistakes by backspacing) and press the <Enter> key. For example,

Table B.1. **Basic operations**

Operation	MATLAB formula	
Addition	a + b	
Division (right)	a/b	(means $a \div b$)
Division (left)	a\b	(means $b \div a$)
Multiplication	a*b	
Power	a^b	
Subtraction	a − b	

```
» a=2; b=4; c=-6;
» dat = b^2 - 4*a*c
dat =
64
» e=sqrt(dat)/10
e =
0.8000
```

The first command assigns the values 2, 4, and −6 to the variables *a*, *b*, and *c* respectively. MATLAB does not respond because this line ends with a semicolon. The second command sets *dat* to $b^2 - 4ac$ and MATLAB returns the answer as 64. Finally, the third line sets *e* equal to the square root of *dat* and divides by 10. MATLAB prints the answer as 0.8. The function *sqrt* is used here; other mathematical functions listed in Table B.2 can be used. Table B.2 provides just a small sample of MATLAB functions. Others can be obtained from the online help. To get help, type

```
>> help
```

[a long list of topics come up]

and for a specific topic, type the command name. For example, to get help on *log to base 2*, type

```
>> help log2
```

[a help message on the log function follows]

Note that MATLAB is case sensitive so that sin(a) is not the same as sin(A).
  Try the following examples:

```
>> 3^(log10(25.6))
>> y=2* sin(pi/3)
>>exp(y+4-1)
```

In addition to operating on mathematical functions, MATLAB easily allows one to work with vectors and matrices. A vector (or array) is a special matrix with one row or one column. For example,

Table B.2. **Typical elementary mathematic functions**

Function	Remark
abs(x)	Absolute value or complex magnitude of x
acos, acosh(x)	Inverse cosine and inverse hyperbolic cosine of x in radians
acot, acoth(x)	Inverse cotangent and inverse hyperbolic cotangent of x in radians
angle(x)	Phase angle (in radians) of a complex number x
asin, asinh(x)	Inverse sine and inverse hyperbolic sine of x in radians
atan, atanh(x)	Inverse tangent and inverse hyperbolic tangent of x in radians
conj(x)	Complex conjugate of x
cos, cosh(x)	Cosine and hyperbolic cosine of x in radians
cot, coth(x)	Cotangent and hyperbolic cotangent of x in radians
exp(x)	Exponential of x
fix	Round toward zero
imag(x)	Imaginary part of a complex number x
log(x)	Natural logarithm of x
log2(x)	Logarithm of x to base 2
log10(x)	Common logarithms (base 10) of x
real(x)	Real part of a complex number x
sin, sinh(x)	Sine and hyperbolic sine of x in radians
sqrt(x)	Square root of x
tan, tanh(x)	Tangent and hyperbolic tangent of x in radians

```
>> a = [1 -3 6 10 -8 11 14 ];
```

is a row vector. Defining a matrix is similar to defining a vector. For example, a 3×3 matrix can be entered as

```
>> A = [1 2 3; 4 5 6; 7 8 9]
```

or as

```
>> A = [1 2 3
        4 5 6
        7 8 9]
```

Table B.3. **Matrix operations**

Operation	Remark
A'	Finds the transpose of matrix A
det(A)	Evaluates the determinant of matrix A
inv(A)	Calculates the inverse of matrix A
eig(A)	Determines the eigenvalues of matrix A
diag(A)	Finds the diagonal elements of matrix A
expm(A)	Exponential of matrix A

In addition to the arithmetic operations that can be performed on a matrix, the operations in Table B.3 can be implemented.

Using the operations in Table B.3, we can manipulate matrices as follows.

```
» B = A'
B =
1    4    7
2    5    8
3    6    9
» C = A + B
C =
2    6    10
6    10   14
10   14   18
» D = A^3 - B*C
D =
372  432  492
948  1131 1314
1524 1830 2136
» e= [1 2; 3 4]
e =
1 2
3 4
» f=det(e)
f =
-2
» g = inv(e)
g =
-2.0000    1.0000
 1.5000   -0.5000
» H = eig(g)
```

Table B.4. **Special matrices, variables, and constants**

Matrix/variable/constant	Remark
eye	Identity matrix
ones	An array of ones
zeros	An array of zeros
i or j	Imaginary unit or sqrt(-1)
pi	3.142
NaN	Not a number
inf	Infinity
eps	A very small number, $2.2e-16$
rand	Random element

```
H =
-2.6861
 0.1861
```

Note that not all matrices can be inverted. A matrix can be inverted if and only if its determinant is non-zero. Special matrices, variables, and constants are listed in Table B.4. For example, type

```
>> eye(3)
ans=
     1 0 0
     0 1 0
     0 0 1
```

to get a 3×3 identity matrix.

B.2   USING MATLAB TO PLOT

To plot using MATLAB is easy. For a two-dimensional plot, use the plot command with two arguments as

```
>> plot(xdata,ydata)
```

where xdata and ydata are vectors of the same length containing the data to be plotted.

For example, suppose we want to plot $y = 10*\sin(2*pi*x)$ from 0 to 5*pi, we will proceed with the following commands:

Table B.5 **Various color and line types**

y	yellow	.	point
m	magenta	o	circle
c	cyan	x	x-mark
r	red	+	plus
g	green	−	solid
b	blue	*	star
w	white	:	dotted
k	black	− .	dashdot
		− −	dashed

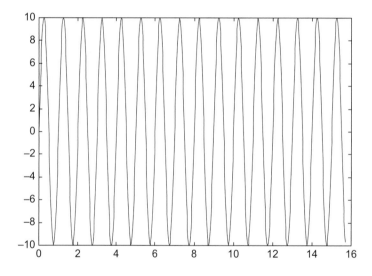

Figure B.1 MATLAB plot of y=10*sin(2*pi*x).

```
>> x = 0:pi/100:5*pi;     % x is a vector, 0 < x < 5*pi, increments of pi/100
>> y = 10*sin(2*pi*x);    % create a vector y
>> plot(x,y);             % create the plot
```

With this, MATLAB responds with the plot in Figure B.1

MATLAB will let you graph multiple plots together and distinguish with different colors.

This is obtained with the command plot(xdata, ydata, 'color'), where the color is indicated by using a character string from the options listed in Table B.5.

For example,

```
>> plot(x1, y1, 'r', x2,y2, 'b', x3,y3, '-');
```

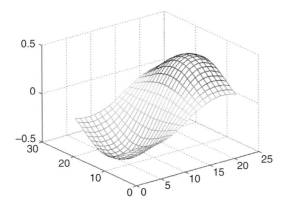

will graph data $(x1,y1)$ in red, data $(x2,y2)$ in blue, and data $(x3,y3)$ in a dashed line all on the same plot.

MATLAB also allows for logarithm scaling. Rather than the `plot` command, we use:

`loglog` $\log(y)$ versus $\log(x)$
`semilogx` y versus $\log(x)$
`semilogy` $\log(y)$ versus x

Three-dimensional plots are drawn using the functions `mesh` and `meshdom` (mesh domain). For example, draw the graph of $z = x*\exp(-x^2-y^2)$ over the domain $-1 < x, y < 1$, we type the following commands:

```
>> xx = -1:.1:1;
» yy = xx;
» [x,y] = meshgrid(xx,yy);
» z=x.*exp(-x.^2 -y.^2);
» mesh(z);
```

(The dot symbol used in `x.` and `y.` allows element-by-element multiplication.) The result is shown in Figure B.2.

Other plotting commands in MATLAB are listed in Table B.6. The `help` command can be used to find out how each of these is used.

B.3   PROGRAMMING WITH MATLAB

So far MATLAB has been used as a calculator; you can also use MATLAB to create your own program. The command-line editing in MATLAB can be inconvenient if one has several lines to execute. To avoid this problem, one creates a program which is a sequence of statements to be executed. If you are in the Command window, click `File/New/M-files` to open a new file in the MATLAB Editor/Debugger or simple text editor. Type the program and save the program in a file with an extension .m, say filename.m; it is for this reason it is called an M-file. Once the

Table B.6. **Other plotting commands**

Command	Comments
`bar(x,y)`	A bar graph
`contour(z)`	A contour plot
`errorbar (x,y,l,u)`	A plot with error bars
`hist(x)`	A histogram of the data
`plot3(x,y,z)`	A three-dimensional version of plot()
`polar(r, angle)`	A polar coordinate plot
`stairs(x,y)`	A stairstep plot
`stem(x)`	Plots the data sequence as stems
`subplot(m,n,p)`	Multiple (*m*-by-*n*) plots per window
`surf(x,y,x,c)`	A plot of three-dimensional colored surface

program is saved as an M-file, exit the Debugger window. You are now back in the Command window. Type the file without the extension .m to get results. For example, the plot that was made above can be improved by adding a title and labels and typed as an M-file called

```
example1.m
x = 0:pi/100:5*pi;           % x is a vector, 0 <= x <= 5*pi, increments of pi/100
y = 10*sin(2*pi*x);          % create a vector y
plot(x,y);                    % create the plot
xlabel('x (in radians)');     % label the x-axis
ylabel('10*sin(2*pi*x)');     % label the y-axis
title('A sine functions');    % title the plot
grid                          % add grid
```

Once it is saved as `example1.m` and we exit text editor, type

```
>> example1
```

in the Command window and hit <Enter> to obtain the result shown in Figure B.3.

To allow flow control in a program, certain relational and logical operators are necessary. They are shown in Table B.7. Perhaps the most commonly used flow control statements are `for` and `if`. The `for` statement is used to create a loop or a repetitive procedure and has the general form

```
for x = array

    [commands]

end
```

Table B.7. **Relational and logical operators**

Operator	Remark
<	less than
<=	less than or equal
>	greater than
>=	greater than or equal
= =	equal
~ =	not equal
&	and
|	or
~	not

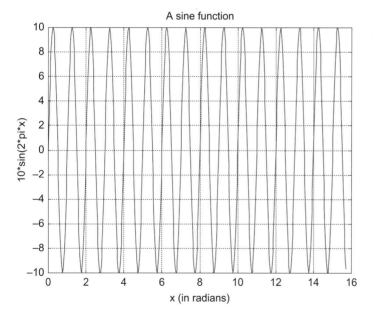

Figure B.3. MATLAB plot of $y = 10*\sin(2*pi*x)$ with title and labels.

The `if` statement is used when certain conditions need be met before an expression is executed. It has the general form

```
if expression
```

        [commands if expression is True]

```
else
```

[commands if expression is False]

```
end
```

For example, suppose we have an array $y(x)$ and we want to determine the minimum value of y and its corresponding index x. This can be done by creating an M-file as shown below.

```
% example2.m
% This program finds the minimum y value and its corresponding x index
x = [1 2 3 4 5 6 7 8 9 10]; %the nth term in y
y = [3 9 15 8 1 0 -2 4 12 5];
min1 = y(1);
for k=1:10
    min2=y(k);
    if(min2 < min1)
        min1 = min2;
        xo = x(k);
    else
        min1 = min1;
    end
end
diary
min1, xo
diary off
```

Note the use of `for` and `if` statements. When this program is saved as `example2.m`, we execute it in the Command window and obtain the minimum value of y as -2 and the corresponding value of x as 7, as expected.

```
» example2
min1 =
-2
xo =
7
```

If we are not interested in the corresponding index, we could do the same thing using the command

```
>> min(y)
```

The following tips are helpful in working effectively with MATLAB:

- Comment your M-file by adding lines beginning with a % character.
- To suppress output, end each command with a semi-colon (;), you may remove the semi-colon when debugging the file.
- Press up and down arrow keys to retrieve previously executed commands.
- If your expression does not fit on one line, use an ellipse (...) at the end of the line and continue on the next line. For example, MATLAB considers

$$y = \sin(x + \log10(2x + 3)) + \cos(x + \cdots$$
$$\log10(2x + 3));$$

as one line of expression
- Keep in mind that variable and function names are case-sensitive.

B.4 SOLVING EQUATIONS

Consider the general system of n simultaneous equations as

$$a_{11}x_1 + a_{12}x_2 + \cdots + a_{1n}x_n = b_1$$

$$a_{21}x_1 + a_{22}x_2 + \cdots + a_{2n}x_n = b_2$$

$$a_{n1}x_1 + a_{n2}x_2 + \cdots + a_{nn}x_n = b_n$$

$$\cdots \quad \cdots \quad \cdots$$

or in matrix form

$$AX = B$$

where

$$A = \begin{bmatrix} a_{11} & a_{12} & \cdots & a_{1n} \\ a_{21} & a_{22} & \cdots & a_{2n} \\ \cdots & \cdots & \cdots & \cdots \\ a_{n1} & a_{n2} & a_{n3} & a_{nn} \end{bmatrix}, \quad X = \begin{bmatrix} x_1 \\ x_2 \\ \cdots \\ x_n \end{bmatrix}, \quad B = \begin{bmatrix} b_1 \\ b_2 \\ \cdots \\ b_n \end{bmatrix}$$

A is a square matrix and is known as the coefficient matrix, while X and B are vectors. X is the solution vector we are seeking to get. There are two ways to solve for X in MATLAB. First, we can use the backslash operator (\) so that

$$X = A \backslash B$$

Second, we can solve for X as

$$X = A^{-1}B$$

which in MATLAB is the same as

$$X = \text{inv}(A)^* B$$

We can also solve equations using the command `solve`. For example, given the quadratic equation $x^2 + 2x - 3 = 0$, we obtain the solution using the following MATLAB command

```
>> [x ]=solve('x^2 + 2*x - 3 =0')
x =
   [-3 ]
   [1 ]
```

indicating that the solutions are $x = -3$ and $x = 1$. Of course, we can use the command `solve` for a case involving two or more variables. We will see that in the following example.

EXAMPLE B.1

Use MATLAB to solve the following simultaneous equations:

$$25x_1 - 5x_2 - 20x_3 = 50$$
$$-5x_1 + 10x_2 - 4x_3 = 0$$
$$-5x_1 - 4x_2 + 9x_3 = 0$$

Solution

We can use MATLAB to solve this in two ways:

Method 1

The given set of simultaneous equations could be written as

$$\begin{bmatrix} 25 & -5 & -20 \\ -5 & 10 & -4 \\ -5 & -4 & 9 \end{bmatrix} \begin{bmatrix} x_1 \\ x_2 \\ x_3 \end{bmatrix} = \begin{bmatrix} 50 \\ 0 \\ 0 \end{bmatrix} \quad \text{or} \quad AX = B$$

We obtain matrix A and vector B and enter them in MATLAB as follows.

```
» A = [25 -5 -20; -5 10 -4; -5 -4 9 ]
A =
 25 -5 -20
 -5 10 4
 -5 -4 9
» B = [50 0 0 ]'
B =
 50
 0
 0
» X = inv (A) *B
X =
 29.6000
 26.0000
 28.0000
» X=A\B
X =
 29.6000
 26.0000
 28.0000
```

Thus, $x_1 = 29.6$, $x_2 = 26$, and $x_3 = 28$.

Method 2

Since the equations are not many in this case, we can use the commond `solve` to obtain the solution of the simultaneous equations as follows:

```
[x1,x2,x3]=solve('25*x1 - 5*x2 - 20*x3=50', '-5*x1 + 10*x2 - 4*x3 =0', '-5*x1 - 4*x2 +
9*x3=0')
x1 =
  148/5
x2 =
  26
x3 =
  28
```

which is the same as before.

PRACTICE PROBLEM B.1

Solve the following simultaneous equations using MATLAB:

$$3x_1 - x_2 - 2x_3 = 1$$
$$-x_1 + 6x_2 - 3x_3 = 0$$
$$-2x_1 - 3x_2 + 6x_3 = 6$$

Answer: $x_1 = 3 = x_3.$ $x_2 = 2.$

B. 5   PROGRAMMING HINTS

A good program should be well documented, of reasonable size, and capable of performing some computation with reasonable accuracy within a reasonable amount of time. The following are some helpful hints that may make writing and running MATLAB programs easier.

- Use the minimum commands possible and avoid execution of extra commands. This is particularly true of loops.
- Use matrix operations directly as much as possible and avoid `for`, `do`, and/or `while` loops if possible.
- Make effective use of functions for executing a series of commands several times in a program.
- When unsure about a command, take advantage of the help capabilities of the software.
- It takes much less time running a program using files on the hard disk than on a memory stick.
- Start each file with comments to help you remember what it is all about later.
- When writing a long program, save frequently. If possible, avoid a long program; break it down into smaller subroutines.

B. 6   OTHER USEFUL MATLAB COMMANDS

Some common useful MATLAB commands which may be used in this book are provided in Table B.8.

Table B.8. **Other useful MATLAB commands**

Command	Explanation
diary	Save screen display output in text format
mean	Mean value of a vector
min(max)	Minimum (maximum) of a vector
grid	Add a grid mark to the graphic window
poly	Converts a collection of roots into a polynomial
roots	Finds the roots of a polynomial
sort	Sort the elements of a vector
sound	Play vector as sound
std	Standard deviation of a data collection
sum	Sum of elements of a vector

Complementary error function $Q(x)$

$$Q(x) = \frac{1}{\sqrt{2\pi}} \int_{x}^{\infty} e^{-\lambda^2/2} d\lambda = 0.5 - \text{erf}(x)$$

Table of $Q(x)$

x	$Q(x)$	x	$Q(x)$	x	$Q(x)$
0.0	0.5000	1.55	0.0606	3.05	0.00114
0.05	0.4801	1.60	0.0548	3.10	0.00097
0.10	0.4602	1.65	0.0495	3.15	0.00082
0.15	0.4404	1.70	0.0446	3.20	0.00069
0.20	0.4207	1.75	0.0401	3.25	0.00058
0.25	0.4013	1.80	0.0359	3.30	0.00048
0.30	0.3821	1.85	0.0322	3.35	0.00040
0.35	0.3632	1.90	0.0287	3.40	0.00034
0.40	0.3446	1.95	0.0256	3.45	0.00028
0.45	0.3264	2.00	0.0228	3.50	0.00023
0.50	0.3085	2.05	0.0202	3.55	0.00019
0.55	0.2912	2.10	0.0179	3.60	0.00016
0.60	0.2743	2.15	0.0158	3.65	0.00013
0.65	0.2578	2.20	0.0139	3.70	0.00011
0.70	0.2420	2.25	0.0122	3.75	0.00009
0.75	0.2266	2.30	0.0107	3.80	0.00007
0.80	0.2169	2.35	0.0094	3.85	0.00006

(cont.)

x	Q(x)	x	Q(x)	x	Q(x)
0.85	0.1977	2.40	0.0082	3.90	0.00005
0.90	0.1841	2.45	0.0071	3.95	0.00004
0.95	0.1711	2.50	0.0062	4.00	0.00003
1.00	0.1587	2.55	0.0054	4.25	10^{-5}
1.05	0.1469	2.60	0.0047	4.75	10^{-6}
1.10	0.1357	2.65	0.0040	5.20	10^{-7}
1.15	0.1251	2.70	0.0035	5.60	10^{-8}
1.20	0.1151	2.75	0.0030		
1.25	0.1056	2.80	0.0026		
1.30	0.0968	2.85	0.0022		
1.35	0.0885	2.90	0.0019		
1.40	0.0808	2.95	0.0016		
1.45	0.0735	3.00	0.00135		
1.50	0.0668				

APPENDIX D

Answers to odd-numbered problems

Chapter 1

1.1 See text.

1.3 (a) -14.43 dB. (b) 16.23 dB. (c) 27.06 dB. (d) 54.77 dB.

1.5 (a) 0.5 mW. (b) 0.063 W. (c) 3162.3 W. (d) 3162.3 W.

1.7 (a) 90 dBrn. 88.5 dBrn. 30 dBrn. (b) Proof.

1.9 (a) 11.26 Mbps. (b) 4.7 Mbps.

1.11 3.6118 kHz.

Chapter 2

2.1 (a) An analog signal is a continuous-time signal in which the variation with time is analogous to some physical phenomenon. (b) A digital signal is a discrete-time signal that can have a finite number of values (usually binary). (c) A continuous-time signal takes a value at every instant of time. (d) A discrete-time signal is defined only at particular instants of time.

2.3 See Figure D.1.

2.5 See Figure D.2.

2.7 It is not possible to generate a power signal in a lab because such a signal would have infinite duration and infinite energy. Signals generated in the lab have finite energy and are energy signals.

2.9 (a) A system is linear when its output is linearly related to its input. It is nonlinear otherwise. (b) A nonlinear system is one in which the output is not linearly related to its input. (c) A continuous-time system has input and output signals that are continuous-time. (d) A discrete-time system has input and output signals that are discrete-time.

2.11 (a) Linear. (b) Nonlinear. (c) Linear.

2.13 $a_0 = 3.75,\ a_n = \dfrac{-5}{n\pi}\sin\dfrac{n\pi}{2},\quad b_n = \dfrac{5}{n\pi}\left(3 - 2\cos n\pi + \cos\dfrac{n\pi}{2}\right).$

2.15 $f(t) = \displaystyle\sum_{n=1}^{\infty}\left[\dfrac{4}{n^2\pi^2}(\cos n\pi - 1)\cos\dfrac{n\pi}{2}t - \dfrac{2}{n\pi}(\cos n\pi - 1)\sin\dfrac{n\pi}{2}t\right].$

2.17 See Figure D.3.

2.19 $a_1 = 40, a_2 = -10, a_3 = 10, \omega_1 = 5\pi, \omega_2 = 7\pi, \omega_3 = 3\pi, \theta_1 = 0, \theta_2 = -60^2, \theta_3 = -120°.$

2.21 (a) $f(t) = \dfrac{2}{\pi}\displaystyle\sum_{n=1}^{\infty}\dfrac{(-1)^{n+1}}{n}\sin 2nt.$ (b) 0.2839 W.

(a)

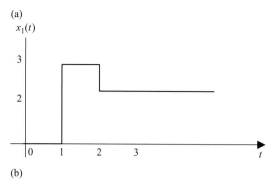

(b)

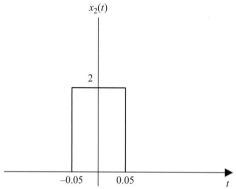

(c)

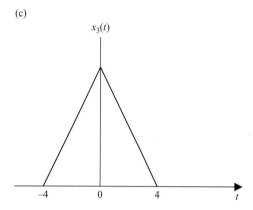

2.23 $f(t) = \sum_{n=-\infty}^{\infty} \frac{1}{jn\pi} \left(e^{-j2n\pi\tau/T} - 1 \right) e^{jn\pi t/T}$.

2.25 (a) $C_n' = C_n e^{-j2n\omega_0}$. (b) $C_n' = 2jn\omega_0 C_n$. (c) $C_n' = -C_n \left(n^2 \omega_0^2 + jn\omega_0 \right)$.

2.27 (a) $X(\omega) = 8\text{sinc}\,\omega$. (b) $Y(\omega) = \frac{1}{j\omega}\left(2 - e^{-j\omega} - e^{-j2\omega} \right)$. (c) $Z(\omega) = \text{sinc}^2(\omega/2)$.

2.29 $H(\omega) = \frac{A}{\omega^2 \tau} \left[1 - j\omega\tau - e^{-j\omega\tau} \right]$.

2.31 $\frac{2\tau}{\pi} \frac{\cos(\omega\tau/2)}{1 - \left(\frac{\omega\tau}{\pi} \right)^2}$.

(a)

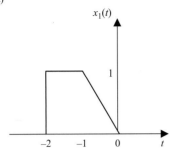

Figure D.2. For Problem 2.5.

(b)

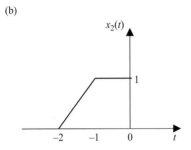

(c)

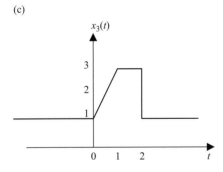

(d)

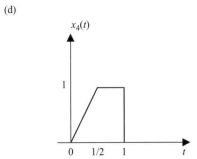

(e)

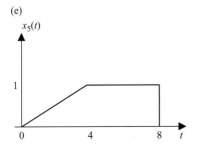

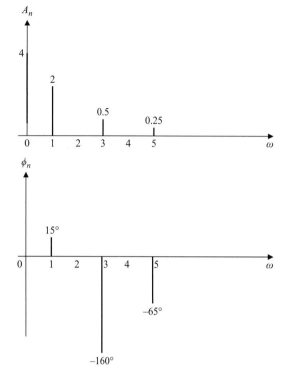

2.33 (a) $f_1(t) = \frac{1}{2}[\delta(t + \pi/4) - \delta(t - \pi/4)]$. (b) $f_2(t) = e^{-4(t-2)}u(t - 2)$. (c) $f_3(t) = \frac{1}{2}e^{-|t|}$.

2.35 $f(t) = \dfrac{A\tau}{\pi}\sin(\tau t)[1 + \cos(\omega_o t)]$.

2.37 (a) $\dfrac{20}{2 - j\omega}$. (b) $\dfrac{20e^{j\omega}}{(1 + j\omega)^2}$. (c) $-\dfrac{20}{(1 + j\omega)^2}$. (d) $\left[\dfrac{10}{1 + j(\omega + \pi)} + \dfrac{10}{1 + j(\omega - \pi)}\right]$.

2.39 (a) π. (b) $\dfrac{\pi}{16}$.

2.41 $G(\omega) = 20\cos(2\omega)\text{sinc }\omega$.

2.43 $B = 5441.4$ rad/s.

2.45 $H(s) = \dfrac{1}{s^3 + 2s^2 + 2s + 1}$.

2.47 See Figure D.4.

2.49 See Figure D.5.

2.51 See Figure D.6.

2.53 Poles $=$
    -0.3090 + 0.9510i
    -0.3090 - 0.9510i
    -1.0000 + 0.0000i
    -0.8090 + 0.5879i
    -0.8090 - 0.5879i

(a)

(b)

(c)

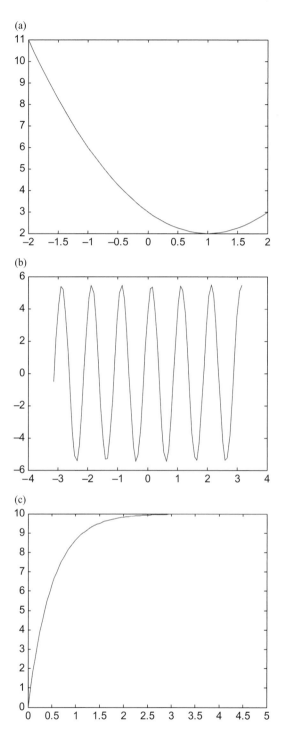

Figure D.4. For Problem 2.47.

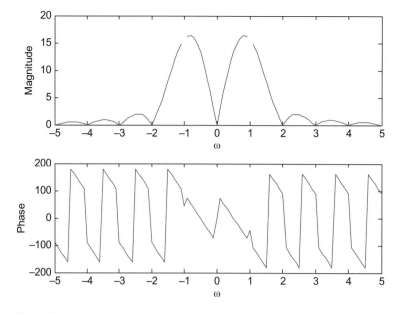

Figure D.5. For Problem 2.49.

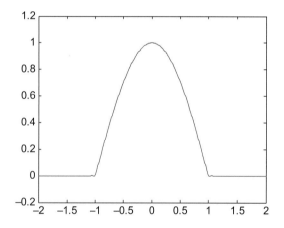

Figure D.6. For Problem 2.51.

Chapter 3

3.1 (a) $L_{min} = 10^3$ m or 1 km. (b) $L_{min} = 10$ m.

3.3 $\mu = 0.4$; Sketch is similar to Figure 3.2(a) but $A_{max} = 14$ and $A_{min} = 6$. $\mu = 1.0$; Sketch is similar to Figure 3.2(b) but $A_{max} = 20$ and $A_{min} = 0$. $\mu = 2.0$; Sketch is similar to Figure 3.2(b) but $A_{max} = 30$ and $A_{min} = -10$.

3.5 (a) $\eta = 0.077$ or 7.7%. (b) $\eta = 0.25$ or 25%.

3.7 $\mu = 0.6$ or 60%; $\eta = 0.107$ or 10.7%.

3.9 $\eta = 0.111$ or 11.1%.

3.11 $y(t) = [(3 + \alpha) + m(t)]2\beta R A_c \cos \omega_c t$.

3.13 (a) $R = 1 \text{ k}\Omega$. (b) (i) $R_{\max} = 1.013 \text{ k}\Omega$. (ii) $R_{\max} = 523 \text{ k}\Omega$.

3.15 Envelope detector bandwidth is 100 kHz. The bandwidth of the rectifier detector low-pass filter is 10 kHz. The different bandwidth requirements show that the two methods are not identical.

3.17 (a) $\phi(t) = \dfrac{3A_c}{2}[\cos(\omega_c - \omega_m)t + \cos(\omega_c - \omega_m)t] + \dfrac{A_c}{2}[\cos(\omega_c - \omega_m)t + \cos(\omega_c - \omega_m)t].$

$\Phi(\omega) = \dfrac{3\pi A_c}{8}\{[\delta(\omega - \omega_c + \omega_m)t + \delta(\omega + \omega_c - \omega_m)t] + [\delta(\omega - \omega_c - \omega_m)t$

$\qquad + \delta(\omega + \omega_c + \omega_m)t]\} + \dfrac{\pi A_c}{8}\{[\delta(\omega - \omega_c + 2\omega_m)t + \delta(\omega + \omega_c - 2\omega_m)t]$

$\qquad + [\delta(\omega - \omega_c - 2\omega_m)t + \delta(\omega + \omega_c + 2\omega_m)t]\}$

(b) $\phi(t) = \dfrac{3A_c}{8}[\cos(\omega_c - \omega_m)t + \cos(\omega_c + \omega_m)t] + \dfrac{1}{8}[\cos(\omega_c - 3\omega_m)t + \cos(\omega_c - 3\omega_m)t].$

$\Phi(\omega) = \dfrac{3\pi A_c}{8}\{[\delta(\omega - \omega_c + \omega_m)t + \delta(\omega + \omega_c - \omega_m)t] + [\delta(\omega - \omega_c - \omega_m)t$

$\qquad + \delta(\omega + \omega_c + \omega_m)t]\}.$

$\qquad + \dfrac{\pi A_c}{8}\{[\delta(\omega - \omega_c + 3\omega_m)t + \delta(\omega + \omega_c - 3\omega_m)t] + [\delta(\omega - \omega_c - 3\omega_m)t$

$\qquad + \delta(\omega + \omega_c + 3\omega_m)t]\}.$

The spectra are similar to the spectrum for Problem 3.10, but they do not contain carrier impulses and the amplitudes are different.

3.19 (a) $\Phi_{DSB}(\omega) = \dfrac{\tau}{2}\left[\text{sinc}\left(\dfrac{(\omega - \omega_c)\tau}{2}\right) + \text{sinc}\left(\dfrac{(\omega + \omega_c)\tau}{2}\right)\right]$

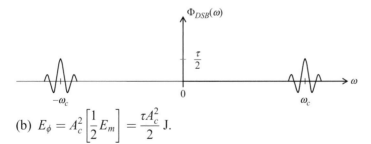

(b) $E_\phi = A_c^2\left[\dfrac{1}{2}E_m\right] = \dfrac{\tau A_c^2}{2}$ J.

3.21 (a) $\Phi_{DSB}(\omega) = 2\text{sinc}(\omega - \omega_c)\cos[5(\omega - \omega_c)] + 2\text{sinc}(\omega + \omega_c)\cos[5(\omega + \omega_c)].$

(b) $\Phi_{DSB}(\omega) = \dfrac{2\pi}{W}\Pi\left(\dfrac{\omega - \omega_c}{W}\right) + \dfrac{2\pi}{W}\Pi\left(\dfrac{\omega + \omega_c}{W}\right).$

3.23 (a) $\phi_{DSB}(t) = \dfrac{jA\omega_0}{\pi}\text{sinc}\left(\dfrac{\omega_0 t}{2}\right)\sin\left(\dfrac{\omega_0 t}{2}\right)\cos\omega_c t.$ (b) $\omega_0 A^2.$

3.25 $y(t) = \dfrac{4}{\pi} m(t) \cos \omega_c t.$

3.27 (a) $y_d(t) = \frac{1}{2} m(t)$. (b) $y_d(t) = \left(\frac{1}{2} e^{j\alpha}\right) m(t)$. (c) $y_d(t) = \left(\frac{1}{2} e^{j\alpha} \cos \alpha\right) m(t)$.

The demodulated signal is an undistorted replica of the message signal, save for a scaling in amplitude, for the three local oscillator carriers.

3.29 (a) $m(t) \cos 2\pi \Delta f t$. (b) $Y_d(f) = \frac{1}{2} [M (f - \Delta f) + M (f + \Delta f)]$.

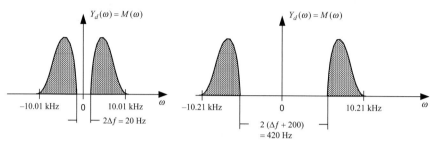

3.31 (a) $x(t) = 3m(t) \cos \left(0.8\pi \times 10^5 t\right) + m(t) \cos \left(3.2\pi \times 10^5 t\right) + 2m(t) \cos \left(4.8\pi \times 10^5 t\right).$

(b) $3m(t) \cos \left(0.8\pi \times 10^5 t\right)$; $f_L = 30$ kHz; $f_H = 50$ kHz.

3.33 (a) $\phi_{QAM}(t) = A_1 \left[\sin \left(2.08\pi \times 10^5 t\right) - \sin \left(1.92\pi \times 10^5 t\right) \right]$

$$- A_2 \left[\cos \left(2.12\pi \times 10^5 t\right) + \cos \left(1.88\pi \times 10^5 t\right) \right]$$

(b) $y_1(t) = 2A_1 \sin \left(8\pi \times 10^3 t\right) \sin \alpha + 2A_2 \cos \left(12\pi \times 10^3 t\right) \sin \alpha.$

3.35 (a) $y_1(t) = \dfrac{1}{2} [m_1(t) + m_2(t)]$; $y_2(t) = \dfrac{1}{2} [m_2(t) - m_1(t)].$

(b) $y_1(t) = \dfrac{m_1(t)}{2} [\cos \alpha + j \sin \alpha] + \dfrac{m_2(t)}{2} [j \cos \alpha - \sin \alpha].$

$y_2(t) = \dfrac{m_1(t)}{2} [\sin \alpha - j \cos \alpha] + \dfrac{m_2(t)}{2} [\cos \alpha + j \sin \alpha].$

3.37 (a) $\phi_{USB}(t) = -A_c A_m \cos (11\omega_m t)$. (b) $\phi_{USB}(t) = A_c A_2 \sin (13\omega_m t) - A_c A_1 \cos (11\omega_m t).$

3.39 $\phi_{USB}(t) = 8 \sin (11\omega_m t) + 4 \cos (12\omega_m t)$. $\phi_{LSB}(t) = 8 \sin (9\omega_m t) + 4 \cos (8\omega_m t).$

3.41 (a) $y(t) = m(t) \cos (\Delta \omega t + \alpha) - m_h(t) \sin (\Delta \omega t + \alpha).$

(b) (i) $y(t) = m(t) \cos \alpha - m_h(t) \sin \alpha.$ (ii) $y(t) = m(t) \cos \Delta \omega t - m_h(t) \sin \Delta \omega t.$

3.43 (a) $m(t) = 2 \cos \omega_m t.$ (b) $A_c \geq 31.623.$

3.45 $\phi_{VSB}(t) = 0.6 \cos (\omega_c - \omega_m)t + 1.4 \cos (\omega_c + \omega_m)t.$ Or

$\phi_{VSB}(t) = 2 \cos \omega_m t \cos \omega_c t - 0.8 \sin \omega_m t \sin \omega_c t.$

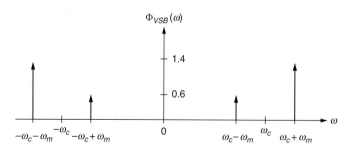

3.47 $\phi_{VSB}(t) = 0.7\cos(\omega_c - \omega_m)t + 0.3\cos(\omega_c + \omega_m)t.$ Or
$\phi_{VSB}(t) = \cos\omega_m t \cos\omega_c t + 0.4\sin\omega_m t \sin\omega_c t.$

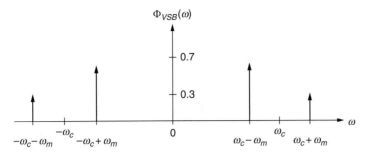

Chapter 4

4.1 (a) $f_{i,\max} = 5.25\,\mathrm{MHz}, f_{i,\min} = 5.05\,\mathrm{MHz}$; sketch is similar to Figure 4.2(d).

   (b) $f_{i,\max} = 5.2\,\mathrm{MHz}, f_{i,\min} = 4.8\,\mathrm{MHz}$; sketch is similar to Figure 4.2(c).

4.3 $f_{i,\max} = 8.04\,\mathrm{MHz}, f_{i,\min} = 7.96\,\mathrm{MHz}.$

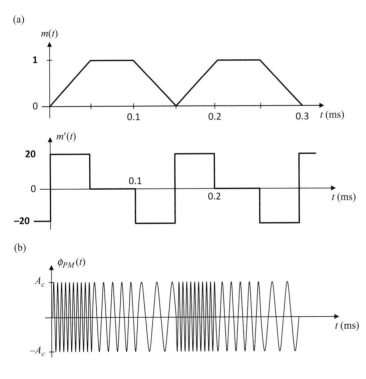

4.5 (a) $k_f \le 4.5\pi \times 10^3$.   (b) $\Phi_{FM}(\omega) = 5\pi[\delta(\omega + \omega_c) + \delta(\omega - \omega_c)] + 0.25\pi[\delta(\omega + \omega_c + \omega_m) + \delta(\omega - \omega_c - \omega_m)] - 0.25\pi[\delta(\omega + \omega_c - \omega_m) + \delta(\omega - \omega_c + \omega_m)].$
The spectrum is similar to that of Figure 4.6, except for the differences in sideband frequencies and amplitudes.

4.7 (a) $N = 3$. (b) $B_{FM} = 30$ kHz.

4.9 (a) (i) $P_o = 0.01134$ W. (ii) $P_o = 1.134$ W. (b) (i) $P_o = 0.02507$ W. (ii) $P_o = 2.507$ W.

4.11 (a) (i) $B_{FM} = 140$ kHz. (ii) $B_{PM} = 240$ kHz. (b) (i) $B_{FM} = 220$ kHz. (ii) $B_{PM} = 400$ kHz.

4.13 (a) $\Delta f = 100$ kHz. (b) $k_p = 2.5\pi$. (c) $B_{FM} = 260$ kHz. (d) $B_{FM} = 320$ kHz.

4.15 $C = 0.3166$ nF; $R = 2.513$ Ω.

4.17 (a) $\Delta f_1 = 100$ Hz; $\Delta f_2 = 3$ kHz; $f_{c2} = 6$ MHz; $f'_{c2} = 4$ MHz.
(b) $n_1 = 30 = 2 \times 3 \times 5$; $n_2 = 25 = 5^2$; $f_{LO} = 10$ MHz or 2 MHz.

4.19 (a) $n_1 = n_2 = 80 = 2^4 \times 5$; $f_{LO} = 5.25$ MHz or 2.75 MHz. (b) $n_1 = n_2 = 81 = 3^4$; $f_{LO} = 5.285$ MHz or 2.815 MHz. (c) $\Delta f_1 = 12.193$ Hz.

4.21 (a) $C_{vo} = 12.771$ pF. (b) $C_{v,max} = 12.903$ pF; $C_{v,min} = 12.639$ pF.

4.23 (a) Employing the scheme of Figure 4.11, $n = 3$; $f_{LO} = 210$ MHz or $f_{LO} = 30$ MHz. (b) Employing a scheme similar to Figure 4.11 but with multiplication preceded by frequency translation, $n = 3$; $f_{LO} = 70$ MHz or $f_{LO} = 10$ MHz.

4.25 (a) 2.3148 ns $\leq \tau_{max} \leq$ 2.8409 ns. (b) 0.0783%.

4.27 $C_1 = 61.468$ pF; $R_1 = 12.496$ kΩ; $C_2 = 65.269$ pF; $R_2 = 12.192$ kΩ.

4.29 (a) $G = 485.527$. (b) (i) $\psi_{eq} = 22.844°$ when $\Delta f = 30$ Hz; $\psi_{eq} = 31.174°$ when $\Delta f = 40$ Hz. (c) (i) $\psi_{eq} = 11.193°$ when $\Delta f = 30$ Hz; $\psi_{eq} = 15.0°$ when $\Delta f = 40$ Hz.

4.31 (a) $\omega_o = 200$ rad/s; $\xi = 0.5$. (b) $\omega_o = 282.84$ rad/s; $\xi = 0.3536$.
(c) $\omega_o = 282.84$ rad/s; $\xi = 0.7071$.

4.33 (a) Bypass the pulse-shaping circuit and set the VCO frequency to $f_v = f_c$.
(b) Include the pulse-shaping circuit and set the VCO frequency to $f_v = 2f_c/3$.

4.35 $R_1 = 26.302$ kΩ; $C = 2.854$ nF.

4.37 $f_1 = 2.368$ kHz; $f_2 = 28.894$ kHz; $H_p(j0) = 0.08197$; i.e. $H_p(j\infty) = 1.00015$.

4.39 (a) 109.4 MHz $\leq f'_c \leq$ 129.4 MHz. (b) $C_{min} = 1.0836$ nF; $C_{max} = 1.635$ nF; $R_{max} = 4.605$ MΩ; $R_{min} = 3.057$ MΩ.

Chapter 5

5.1 5.6 kHz.

5.3 1.

5.5 (a) Proof. (b) 50 dB.

5.7 0.009549.

5.9 See Figure D.7.

5.11 See Figure D.8.

5.13 (a) 0.5181%. (b) 2.155%. (c) 1.234%.

5.15 (a) 0.647 μs. (b) 1.544 Mbps. (c) 772 kHz.

5.17 8000 samples/s.

5.19 (a) Six lines. (b) 80.36 %.

Chapter 6

6.1 0.5177.

6.3 (a) 0.1. (b) 0.5. (c) 0.3. (d) 0.6.

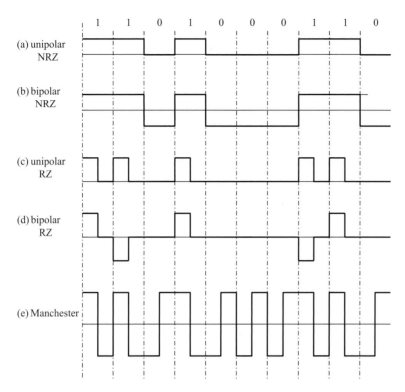

Figure D.7. For Problem 5.9

Guard bands

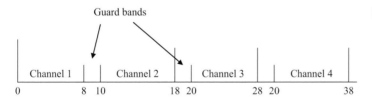

Figure D.8. For Problem 5.11.

6.5 (a) 0.19. (b) 0.41.

6.7 (a) 0.4. (b) 0.08. (c) 0.2.

6.9 $\dfrac{1}{4}$, $\dfrac{1}{19}$.

6.11 (a) $\dfrac{2}{15}$, (b) $\dfrac{1}{15}(x^2 - 1)$, (c) 0.35.

6.13 $F_X(x) = \begin{cases} \sqrt{x}, & 0 < x < 1 \\ 0, & \text{otherwise} \end{cases}$, 0.1507.

6.15 $F_X(x) = \dfrac{1}{2} + \dfrac{1}{\pi}\tan^{-1}x.$

6.17 −7432.3.

6.19 0.2858.

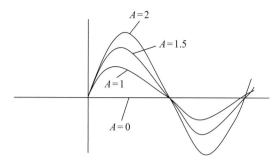

Figure D.9. For Problem 6.33(a).

6.21 2.5, 0.75.

6.23 (a) 0.2835, 0.6065. (b) 0.6694.

6.25 140.288.

6.27 0.1855.

6.29 (a) 9.837. (b) 0.1129.

6.31 $E(X) = \dfrac{a+b+c}{3}$, $\quad \mathrm{Var}(X) = \dfrac{(a^2 + b^2 + c^2) - ab - ac - bc}{18}$.

6.33 (a) See Figure D.9. (b) $\sin 4t$, $\quad \dfrac{4}{3}\sin^2 4t$.

6.35 (a) $2.5 + 4.5e^{-|\tau|} - 2.021e^{-\sqrt{3}\tau}$. (b) 4.9793.

6.37 (a) $R_X(\tau) = \dfrac{AB}{2\pi}\,\mathrm{sinc}^2(\tau B/2)$

   (b) $R_Y(\tau) = \dfrac{1}{\pi}[A\omega_2\,\mathrm{sinc}\,\omega_2\tau - A\omega_1\,\mathrm{sinc}\,\omega_1\tau + B\omega_1\,\mathrm{sinc}\,\omega_1\tau]$.

6.39 (a) 5. (b) $\dfrac{2 + j\omega}{(2 + j\omega)^2 + \omega_0^2}$. (c) $2e^{-\omega^2/4}$. (d) $2\Pi\left(\dfrac{\omega}{4\pi}\right)$.

6.41 (a) $S_Z(\omega) = \dfrac{\omega^2}{\omega^2 + 4} + \dfrac{4}{\omega^2 + 4} - 4\pi\delta(\omega)m_X m_Y$. (b) $S_{XY} = 2\pi m_X m_Y \delta(\omega)$.

   (c) $S_{YZ}(\omega) = 2\pi m_X m_Y \delta(\omega) - \dfrac{4}{\omega^2 + 4}$.

6.43 $R_X(\tau) = 1.047e^{-2|\tau|} + 0.481e^{-5|\tau|}$.

6.45 3.75.

6.47 $S_Y(\omega) = \dfrac{S_X(\omega)}{R^2 + \omega^2 L^2}$.

6.49 See Figure D.10.

6.51 See Figure D.11.

6.53 See Figure D.12.

Chapter 7

7.1 The difference between them is that thermal noise is present in metallic resistors, while shot noise is not.

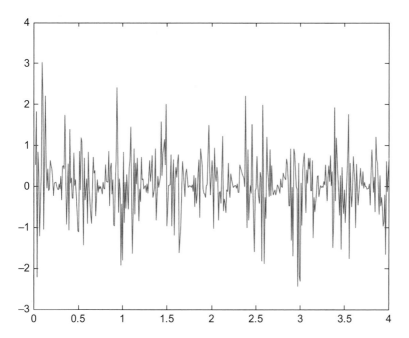

Figure D.10. For Problem 6.49.

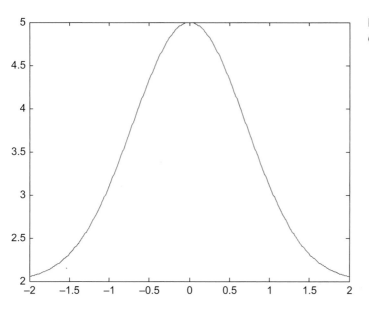

Figure D.11. For Problem 6.51.

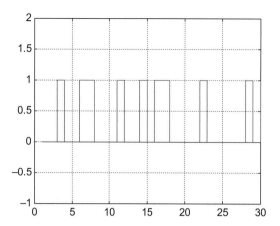

Figure D.12. For Problem 6.53.

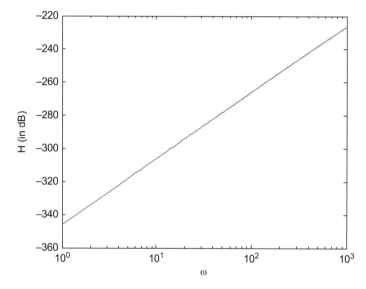

Figure D.13. For Problem 7.15.

7.3 9.968 μV.

7.5 97.98 nA.

7.7 628.3 μW.

7.9 (a) 0.1111γ. (b) 4.1414 dB.

7.11 Preemphasis is a filter used in FM broadcasting to improve SNR, while deemphasis is a filter used for FM reception to restore the preemphasized signal. They are not used in AM because the effect of noise is uniform across the spectrum of the modulating signal.

7.15 See Figure D.13.

Chapter 8

8.1 $H(\omega) = \frac{1}{1-j\omega}\left(1 - e^{-(1-j\omega)T}\right)e^{-j\omega t_o}$.

8.3 See Figure D.14.

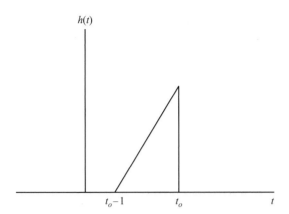

Figure D.14. For Problem 8.3.

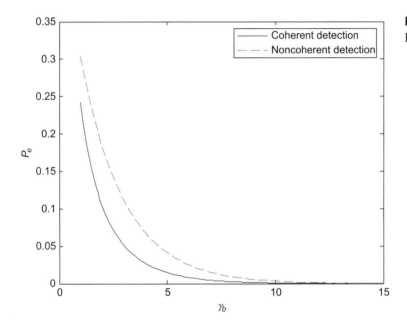

Figure D.15. For Problem 8.23.

8.5 Proof.

8.7 11.28.

8.9 1.27×10^{-10}.

8.11 0.0127.

8.13 (a) 6.795×10^{-8}. (b) 6.795×10^{-8}. (c) 8.876×10^{-15}.

8.15 0.2081.

8.17 6.3776.

8.19 (a) 1.1073. (b) 11.073. (c) 110.73.

8.21 22.16.

8.23 See Figure D.15.

8.25 See Figure D.16.

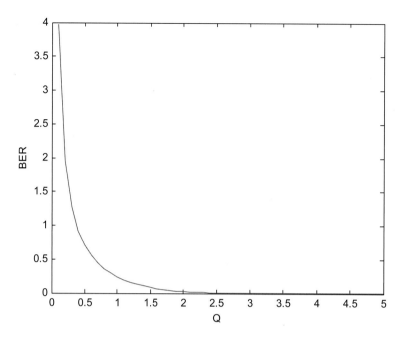

Figure D.16. For Problem 8.25.

INDEX